FUNDAMENTALS OF HEARING

FUNDAMENTALS OF HEARING

An Introduction

FOURTH EDITION

William A. Yost

Parmly Hearing Institute
Loyola University Chicago
Chicago, Illinois

Academic Press
San Diego New York Boston
London Sydney Tokyo Toronto

Copyright ©2000, Elsevier Science (USA).

Academic Press
An imprint of Elsevier Science
525 B Street, Suite 1900, San Diego, California 92101-4495, USA
http://www.academicpress.com

Academic Press
84 Theobald's Road, London WC1X 8RR, UK
http://www.academicpress.com

Library of Congress Catalog Card Number: 00-103469

International Standard Book Number: 0-12-775695-7

PRINTED IN THE UNITED STATES OF AMERICA
03 04 05 06 07   9 8 7 6 5 4 3

To my friends, colleagues, and teachers

Gilbert Johns
Don Robinson
Dave Green
Don Teas

Contents

PART III
AUDITORY PERCEPTION OF SIMPLE SOUNDS

PART IV
COMPLEX SOUNDS PROCESSING, THE CNS, AND AUDITORY DISORDERS

Preface

Almost 30 years ago as part of my job at the University of Florida, I was asked to teach a course, Fundamentals of Hearing. Much to my dismay, there was no single textbook that reviewed the topics I had been asked to cover. Thus, Don Nielsen agreed to collaborate with me to write a textbook that covered the fundamentals of hearing from sound to perception with a clear description of basic auditory anatomy and physiology. Given the introductory level of the course, the textbook could not require a high level of science or math background. This fourth edition of *Fundamentals of Hearing: An Introduction* has the same basic aims to cover sound, auditory anatomy and physiology, and auditory perception for students with a limited science background. This edition, like the third, is written by me, but includes many of Don Nielsen's original contributions.

In the early 1970s it was not easy to write an introductory textbook on hearing given the wealth of knowledge that existed. The task is even more daunting today. There is no problem deciding what should be included in such a book, but the challenge is deciding what to omit. The guiding concept for putting the book together has always been to concentrate on the *fundamentals* of hearing. What are the basic facts and theories that form the foundation for understanding auditory processing?

My starting point is that the major role of the auditory system, like all sensory systems, is to determine objects in the environment. Objects can vibrate and, as a result, the sound produced by this vibration is a potential physical cue for a sensory system to determine something about the originating sound source. From this perspective, the fundamentals of hearing concern three major components: sound, the structure and function of the auditory system, and the basic perceptions that result when a sound is present. These three topics make up the first three parts of *Fundamentals of Hearing: An Introduction*: *The Auditory Stimulus*, *Peripheral Auditory Anatomy and Physiology*, and *Auditory Perception of Simple Sounds*. These sections are dominated by a discussion of the function and structure of the normal peripheral auditory system and the basic code of sound it provides. In terms of the structures of the auditory system and considering all of our auditory perceptions, hearing is much more that just the auditory periphery.

Thus, the fourth part of the fourth edition, *Complex Sounds Processing, The Central Auditory Nervous System, and Auditory Disorders*, reviews these topics. This part reflects one of the major changes in this edition. Previous editions included a chapter on noise and hearing loss. This chapter is now part of the final chapter in the book, *The Abnormal Auditory System*, in which the effects on hearing of noise along with age, ototoxic drugs, disease, and heredity are briefly covered. This chapter also succinctly reviews topics such as inner and outer haircell loss, haircell regeneration, and neural plasticity. The other changes primarily reflect updating the book to capture the recent enormous explosion of knowledge in the field.

The major limitations in writing a comprehensive introduction to hearing are space and the scientific background required to understand these fundamentals. The book is still written with the expectation that any undergraduate with good skills in algebra and high school science can benefit from studying this book. As in past editions, six appendices are intended to provide additional background information to assist the reader with crucial concepts.

Over the years, many colleagues have commented that they keep a copy of *Fundamentals of Hearing: An Introduction* on their bookshelves because, in addition to being a useful textbook, it contains many features that make it a useful reference. The supplemental sections to each chapter, the definition of symbols, the references and index of names, and the glossary and subject index (based on American National Standard Institute Standard S32.0, 1978) are maintained in this edition to assist both the student and the professional. It is impossible to provide a comprehensive primary-reference bibliography for the fundamentals of hearing in the space allowed. Thus, the bibliography contains many other secondary sources and references to the literature

cited in the textbook, with an emphasis on the early or original works on a topic.

In addition to the textbook, a workbook, *Fundamentals of Hearing: An Instructor's Workbook*, and a CD-ROM are available. The workbook contains problem sets with answers for each chapter of the textbook along with a few other aids for teaching the fundamentals of hearing. The CD-ROM is for PC users and contains files of all the figures in the textbook (in *gif* or *jpg* formats); files containing all of the data used in the textbook (in ASCII text format, readable by most spreadsheets and word processors); three programs executable in Windows 95/98/NT to generate simple tones, complex sounds, neural tuning curves, and to run a simple psychophysical discrimination experiment; all figures in the textbook readable by Microsoft's PowerPoint software (PowerPoint slide, *pps*, files); and a toolbox of MathWorks Matlab files (*m* files). PowerPoint is a copyrighted program of Microsoft, Inc.; and Matlab is a copyrighted program of MathWorks, Inc. Each is required to execute the files provided on the CD-ROM.

The most difficult decision in writing this edition concerned coverage of the recent explosion in cellular and molecular biology, especially at the genetic level. In the end, I decided not to devote as much of this edition to these topics as some may think I should have. I felt that covering these topics would require a level of background in college biology that I had not previously assumed for students. I also felt that, while knowing about the cellular and molecular basis of audition is a crucial part of understanding hearing, I am not convinced at this time that this knowledge is necessary and sufficient to understand the fundamentals of sound, auditory anatomy and physiology, and perception as defined in this book. However, I am sure that, if there is a fifth edition of *Fundamentals of Hearing: An Intro-*

duction, it will have to cover these topics in far greater detail.

An effort like putting out a book cannot be done without the help of many people. The staff, students, and faculty of the Parmly Hearing Institute, Loyola University Chicago have been wonderful in helping me with this project. I am especially grateful to Marilyn Larson, Jim Collier, Noah Jurcin, Ned Avejic, Linda Emmerich, Rick Beauchamp-Nobbs, and Sara Polis for all the time they spent working with me on the book and workbook. I learned so much from my colleagues Dick Fay, Sheryl Coombs, Stan Sheft, Bill Shofner, Toby Dye, John New, J. D. Trout, Andy Lotto, Jim Brennan, Rich Bowen, and Anne Sutter. My graduate students and postdocs have also helped shape my understanding of hearing: Mark Stellmack, Sandy Guzman, Dan-Mapes Riordan, Bill Whitmer, Chris Brown, Greg Sandel, Chris Braun, and Roel Delhayne. Very special thanks are due to my lovely wife, Lee, and our two daughters, Kelley and Alyson, and their families.

William A. Yost

CHAPTER 1

The World We Hear, An Introduction

HEARING

Sound constantly surrounds us and informs us about many objects in our world. Determining the sources of sounds is one of our most important biological traits. Any animal's ability to locate food, avoid predators, find a mate, and communicate depends on being able to determine the sources of sounds. In order to determine sound sources we need to know that an object exists, what it is, where it is, if it is moving, and so on.

When we listen, sounds from various sources are combined into one complex sound field and do not reach us as individual sounds. We do not have separate "pipelines" to our senses for each object in our environment, as the ancient Greeks suggested. We receive one sound input that is made up of sounds from all sources in our environment. Figure 1.1 depicts this situation for several musical instruments. Whether we listen at a concert (where all the instruments exist as sound sources) or to the radio (where only the radio loudspeaker is the source of the sound), we can determine the various instruments in the band. In order to determine sound sources, the nervous system first processes the complex sound field by translating (transducing) the various physical aspects of sound into

a *neural code* for the physical characteristics of the complex sound field. This neural code is then further processed to provide neural subsets that aid us in determining the sound sources. Finally, this auditory neural information is combined with that from other sensory systems and from that provided by experience to elicit appropriate behaviors in response to the presence of sound sources. The entire sequence of coding, processing, integration, and responding to sound defines *hearing* as depicted in Figure 1.2.

NORMAL HUMAN HEARING AND THE PLAN OF THE BOOK

This book is primarily concerned with the neural coding of sound and the direct consequences of that coding for the normal auditory system. A great deal is known about neural coding of sound, but far less is known about processing, integration, and response output (see Figure 1.2). To comprehend this coding, we need to understand something about the basic physical attributes of sound—which we will learn are *frequency, intensity,* and *time.* It will also be useful to learn a little

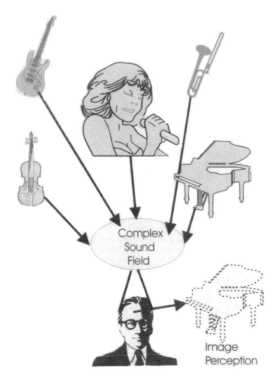

FIGURE 1.1 A schematic diagram indicating a number of objects (musical instruments) that could produce sound. The sounds from all these sources are combined into one complex sound field that is received by the listener. The auditory nervous system of the listener first provides a neural code of the basic physical attributes of the complex sound field, and then this neural code is further processed to aid the listener in determining the various sources. The listener perceives an auditory image of each sound source (e.g., the piano).

about physical systems that process sound before we turn to a study of biological processors. Thus, physical concepts of filtering and the nonlinear response to sound stimulation will be briefly discussed. These topics constitute the first section (The Auditory Stimulus) of the book, which includes four chapters. The next section (Peripheral Auditory Anatomy and Physiology) contains five chapters describing the structure (*anatomy*) and function

(*physiology*) of the parts of the nervous system that provide the neural code for frequency, intensity, and time. The four chapters of the third section (Perception of Simple Sounds) cover humans' perception of the basic attributes of sound: frequency, intensity, and time. In addition, this section will cover the ability of humans to locate the source of sounds. The three chapters of the last section (Complex Sound Processing, Central Nervous System, and Abnormal Auditory System) provide a brief introduction to sound source determination, the basics of the central auditory nervous system, and abnormalities of auditory processing. Other aspects of complex auditory perception, such as speech perception, will also be briefly covered in this last section.

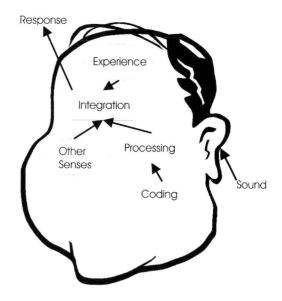

FIGURE 1.2 The stages of processing that lead to hearing. The physical attributes of sound (frequency, intensity, and time) are first coded by the peripheral auditory nervous system. This neural code is then processed by higher neural centers to help the listener determine the sources of sound. This neural information is integrated with other sensory information and that based on experience, and all of this neural processing leads to behavioral responses.

As indicated earlier, the book primarily describes the processes and perceptions of the normal auditory system. Chapter 16 does provide an overview of some of the ways in which the auditory system becomes damaged. While many students and readers of this book have interests in non-normal auditory systems, the premise of this book is that an understanding of normal auditory processing is crucial to fully appreciate the abnormal system and the various ways of dealing with hearing impairment.

The general area of medicine devoted to the study of auditory problems is call *otology*, which is part of a larger area called *otorhino-laryngology or otolaryngology*. "Oto" refers to the ear, "rhino" to the nose, and "laryngology" to the throat; thus, this area of medicine is often referred to as the ear, nose, and throat (ENT) specialty. In addition to persons with an MD degree, those trained in the field of *clinical audiology* with either an MA, AuD (Doctor of Audiology), or PhD degree work with patients having auditory problems. The otologist undertakes special training after receiving a medical degree; the audiologist receives training as part of the work toward an MA, AuD, or PhD and by meeting certification requirements. The otologist and audiologist often work together in aiding someone with an auditory problem. The audiologist provides the diagnostic testing and rehabilitation, while the otologist may provide additional diagnostics and any necessary medical treatment.

Hearing scientists come from just about every field of science: biology, psychology, engineering, physics, and chemistry, as well as otology and audiology. However, the majority of hearing scientists are probably best labeled as neuroscientists, scientists who are interested in those parts of the nervous system that serve the sense of hearing.

Most tests for hearing impairment provided by otologists and audiologists as well as cures and rehabilitation procedures deal with aspects of the neural coding of the basic attributes of sound. For example, a hearing aid is usually fitted based on a number of hearing tests, the most important of which is the *audiogram*, which determines an individual's sensitivity to sounds of different frequencies. The hearing aid primarily amplifies sound such that listeners are better able to detect sounds with frequencies to which they are insensitive. A firm background in the fundamentals of hearing (i.e., understanding the neural code for the basic attributes of sound) is required to understand why an audiogram measures hearing sensitivity, how and why a hearing aid might help a person hear better, and new advancements in hearing aid design.

The neural code is primarily provided by the early or peripheral stages of the auditory nervous system. Thus, the book will emphasize the anatomy and physiology of the peripheral auditory system. Again, most is known about the peripheral stages of the neural basis of hearing. You will be introduced to the central auditory system (those neural stages that come after the peripheral ones) and complex sound processing, but we will not cover these topics in as much detail as the peripheral system and the perception of simple sounds.

The ability to communicate is perhaps the most important attribute of hearing for humans. A *communication system* is often described as consisting of a *sender*, *a message*, and *a receiver*; for humans this represents *speech*, *language*, and *hearing*. Each of the three parts are essential for full communication. This book focuses on a basic understanding of the hearing part of this system. With such knowledge one should be better able to understand the role hearing plays in communication, especially speech communication.

Although the book's aim is to describe the basics of normal human hearing, much of what is known about auditory processing comes from the study of hearing in other animals. A significant amount of the basic knowledge that we have about hearing, hearing impairment, and relief from hearing impairment is a result of our ability to study a wide variety of animals. Understanding how all animals cope with their environments through the use of their sensory systems provides valuable information about and insights into a wealth of crucial scientific and societal questions.

HISTORY

Documentation of an interest in the senses probably goes back as far as recorded time. The Greeks were fascinated with the senses, and philosopher/scientists such as Pythagoras made important contributions that are still useful today. Among other discoveries, Pythagoras worked out the relationship between the length of strings and the pitch of musical sounds. However, most of what we currently know about hearing was developed in the middle of the 19th century, although some of the gross anatomy of the inner ear was studied as early as the 17th century. The mid to late 1800s saw the full development of biology and the beginnings of the modern study of behavioral sciences. During this time, refinements of the microscope provided a significant tool for biology, and the philosophical debate concerning mind/body dualism was waning, making the empirical study of behavior possible. In the early part of 20th century, development of the oscilloscope for measuring electrical activity and vacuum tube devices that generated sounds produced quantum increases in our knowledge of hearing.

During the late 1800s and early 1900s, physicists such as Fechner, Rayleigh, and

Helmholtz and biologists such as Corti, Cajal, Deiter, and Muller performed important experiments and suggested crucial theories that form the basis for much of what we know about hearing today. A great deal of the early work on hearing, as with the other senses, began with an interest in object or source determination. As a part of this interest, scientists and philosophers reasoned that objects were perceived because we processed the attributes of the object. Thus, a great deal of the study of hearing hinged on a study of the attributes of sound: frequency, intensity, and time. Helmholtz crystallized this approach when he proposed a theory for auditory processing of frequency as the key ingredient of our perception of pitch. As we will learn, pitch perception is a crucial element of hearing. Without our ability to perceive pitch, music would be little more than drumbeats and speech would probably sound something like Morse Code. The fascination with frequency coding as the key element in pitch perception permeated the field of hearing for nearly a century. Classic hearing textbooks often described *"Theories of Hearing."* These were really more theories of frequency processing than theories of all of hearing, but since pitch perception is such a crucial aspect of hearing, a theory of frequency processing could almost be considered a full theory of hearing. From the 1940s to the 1960s, interests in these theories focused on the *place theory* (the place in the nervous system with the most neural activity codes for the perception of pitch), of von Bekesy versus the *temporal theory* ("volley theory" or a theory of the temporal properties of the neural response determines pitch) of Wever. Although his volley theory was really a combination of place and temporal theories, Wever (1949) is often cited as a proponent of the temporal theory. Von Bekesy won the Nobel Prize for Medicine and Physiology in 1961 for his work on the place theory of frequency coding, in particular for his work on how the mechanics of the inner ear provides a code for frequency that can be used by the nervous system. We will learn a great deal more about this work later in this book. The debate concerning theories of hearing has dwindled to arguments about details. Most hearing scientists now recognize the important role that both the place and temporal theories play in understanding hearing.

During the 1930s and 40s, a great deal of research in hearing was conducted in order to develop and improve the telephone, and thus some of the key studies were performed at Bell Laboratories. During World War II, interests in sonar detection spurred additional knowledge of hearing, especially at the Psychoacoustics Lab and later at the Eaton Peabody Labs at Harvard University. During the 20th century, especially during its later decades, significant contributions to the hearing sciences were provided by researchers from Europe and Japan. With the development of the digital computer and its use in the biological and behavioral sciences during the 1960s, a large number of new experiments and theories were performed and developed. Today the arsenal of tools available, including those emerging from cell and molecular biology, to hearing scientists is awesome and growing at a rapid rate. Thus, it is not surprising that our knowledge of hearing is also growing rapidly.

HEARING AND SCIENCE

The empirical method of science is the basis for most of what we know about hearing. To fully appreciate the facts, data, and theories pertaining to hearing, it is important that one have some acquaintance with how those facts, data, and theories are acquired. There are two general ways of conducting science: the *empirical method* and the *observational method*. Empirical methods involve conducting experiments in which scientists directly manipulate the objects they want to understand. Observational methods are used when the scientist cannot directly manipulate these objects, such as in the study of astronomy.

In an experiment a scientist determines the functional relationship between two variables: the *independent variable* (IV) and the *dependent variable* (DV). The scientist seeks the following:

$$DV = f(IV), \qquad (1.1)$$

that is, to determine the functional relationship [$f( )$] between the DV and the IV; or how does the dependent variable change when the independent variable changes?

The independent variable is something the experimenter manipulates, while the dependent variable is something the experimenter measures that might vary as a consequence of changing the independent variable. Thus, a scientist who wants to know how the steepness of an incline affects the speed at which a ball rolls down the incline will vary the steepness (slope) of the incline (IV) and measure the speed (DV) of the rolling ball for each setting of the incline's slope. The scientist could establish the following relationship:

$$S = f(SL), \qquad (1.2)$$

where S is the ball's speed, say in meters per second, SL is the slope as a ratio of the height of the incline to the horizontal width of the incline, and $f( )$ is the relationship that was found (e.g., if for each unit increase in the slope the ball rolled twice as fast, the relationship could be written as: $S = 2SL$).

In most experiments, variables in addition to the independent variable may also lead to a change in the dependent variable. These other variables are especially important if they co-vary with the independent variable. Suppose, for instance, that in the ball-rolling experiment the experimenter increased the slope (IV) by keeping the width of the incline fixed and raising its height. In this case, both the slope and height of the incline change together. Thus, the experimental outcome (equation (1.2)) could be written as either $S = f(SL)$ or $S = f(H)$, where H is the incline height. If the scientist is really interested in the effect of slope on ball speed, then the fact that height covaried with slope means that the results are confounded, and thus height becomes a *confounding variable* (CV) for this experiment.

The experimenter needs to *control* for confounding variables. In the present example this could be done in a number of ways. The experimenter could, for example, do two experiments: (1) keep the slope the same but change the height of the incline (that is, change the width in proportion to height), and (2) keep height the same but change the slope (by changing the width). The outcome from these two experiments would indicate how the two variables affect the ball's speed. A wide variety of experimental methods exist that allow scientists to control for confounding variables in order to determine the relationship between the independent variable and the dependent variable without the relationship being affected by confounding variables.

A key aspect of science is communicating the outcome of an experiment as clearly as possible. This means displaying the functional relationship (e.g., equation (1.1)) as clearly as possible. This is usually done in one of four ways: (1) show the raw data, (2) table the data, (3) draw a figure of the data, or (4) write an equation that describes the functional relationship.

Table 1.1 and Figure 1.3 show the results of the ball rolling experiment if equation (1.2) represented the hypothetical experimental outcome. Notice that the DV is shown as the entries in the cells of the table and along the vertical axis (the *y*-axis or *ordinate*) of the graph, whereas the IVs are represented by the rows and columns of the table and along the horizontal axis (the *x*-axis or *abscissa*) of the graph. If there is more than one independent variable, then one IV can be represented along the horizontal axis and the other can be represented by having different curves on the figure.

TABLE 1.1 Hypothetical Relationship between the Slope of an Incline and the Speed (in meters/sec) of a Ball Rolling Down the Incline

Slope	Speed (m/sec)
1	2
2	4
3	6
4	8
5	10

The data from experiments are used to form hypotheses, models, theories, and laws in order to account for a number of relationships between independent and dependent variables and to *predict* how untested relationships between independent and dependent variables might turn out. Whether a particular combination of relationships is called a hypotheses, model, theory, or law often depends on the scientist making the proposal. Usually the term "law" is reserved for statements that apply to a wide range of conditions and are unlikely to change very quickly. There are few laws of hearing, but many hypotheses, models, and theories.

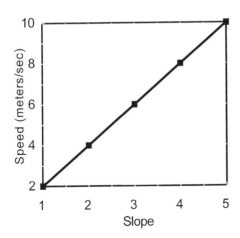

FIGURE 1.3 A graph showing the hypothetical relationship between the speed of the rolling ball (the DV) and the slope of the incline (the IV). The curve is based on the relationship shown in equation (1.2) and Table 1.1.

SUMMARY

This book concentrates on the neural coding of the physical attributes of sound—which are frequency, intensity, and time—and on the perceptual consequences of that coding. An emphasis is placed on normal human hearing so that future scientists and practitioners such as audiologists and otologists will understand the basic concepts of hearing. The study of hearing has a rich history, in which a great deal of the work over the past 100 years dealt with various "theories of hearing." Knowledge of hearing is based on the empirical scientific method, in which scientists seek to find functional relationships between independent and dependent variables that are not confounded.

SUPPLEMENT

The book by Bregman (*Auditory Scene Analysis*, 1990) and articles by Yost (1992a,b, 1993) cover in more detail the idea that sound source determination is a crucial aspect of hearing. The Acoustical Society of America has reprinted many of the classic textbooks in acoustics, including some in hearing: Stevens and Davis (reprinted in 1983) and von Bekesy (1989). These are excellent books that serious students of hearing should read. Gulick, Gescheider, and Frisina (1989) provide a brief historical review of hearing, especially of the various "theories of hearing." A complete history of perception can be found in Boring's (1942) classic book on the history of experimental psychology. Some of Helmholtz's works have been translated by Warren and Warren (1968) and provide fascinating reading for anyone interested in the history of physics as it relates to sensory processing. The book by Shaughnessy and Zechmeister (2000) provides an excellent discussion of the scientific method used in the behavioral sciences.

The alert student would have noticed that one more control experiment is required in order to determine the exact relationship between incline slope and the speed of the rolling ball, since the two control experiments that were mentioned still leave the length of the incline as a confounding variable.

The preferred way to display a functional relationship between the DV and IV is with an equation. Since it is not always possible to write such an equation, plotting the data is the second best preferred method for displaying data.

PART I

THE AUDITORY STIMULUS

CHAPTER 2

Sinusoids, The Basic Sound

*I*n the most general sense, when we say we hear, we usually mean we are sensitive to the sounds in our environment. Sound may be defined in either psychological or physical terms. Let us first consider the physical definition. *Vibration* of an object makes sound possible. Any object with the properties of *inertia* and *elasticity* may be set into vibration and hence produce a sound. Thus, any vibrating object has the potential to produce sound. If we "hear" the vibration, the sound is "audible." Vibration is the movement of an object from one point in space to another point and usually back again to the first point. The fact that a force must be exerted on a body to make it move defines *inertia*, whereas the ability of the object to return to a starting or initial state after it is deformed or moved by this force defines *elasticity*. In practice, almost every object has inertia and elasticity, and so almost every object can be set into vibration. Few restrictions are placed on the type of vibration required to produce a sound. As long as an object moves from one point to another and back again, it has the potential to produce a sound. The object may move regularly back and forth or randomly; it may complete a vibratory cycle in seconds or in fractions of a second, or it may move only once or millions of times. All of these vibrations are capable of producing sound.

If our knowledge of vibration and sound production were limited to these general descriptions of vibration, the area of *acoustics* (the study of sound) would be very complicated. We would have no easy way of describing different vibrations and sounds. Fortunately, a method of categorizing different vibrations has been established. Joseph Fourier, a Frenchman who lived during the time of Napoleon I, derived important theorems for the flow of heat; his analysis can be applied to the types of vibrations we have described. Fourier derived a theorem that specified that *any* vibration can be resolved into a sum of a particular type of vibration, called a *sinusoidal vibration*. This sum of sinusoidal vibrations, called a *Fourier series*, can describe almost any arbitrary vibration. The derivation of these sinusoids is called *Fourier analysis*. Sinusoidal vibrations are the basic building blocks of all vibrations and, hence, of all possible sounds.

SINUSOIDS

A sinusoid, also called a sine wave, describes a particular relationship between *displacement* and *time*, that is, a particular vibration. (Appendix A contains additional discussion of sinusoids.) Figure 2.1 is a diagram of a sinusoid. Displacement simply means the dis-

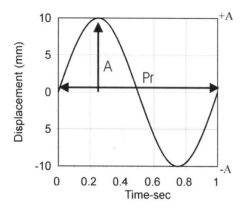

FIGURE 2.1 A sinusoidal relation between displacement and time. From the starting position at 0, the peak amplitude is +10 (+*A*); the period (*Pr*) is 1 second; the frequency is 1 Hz.

tance an object moves, and a sine wave describes the continuous, regular back-and-forth displacement of a vibrating object. Notice that from the starting position (displacement *D* = 0) the motion in this example goes upward to a maximum positive distance (*D* = + *A*, +*A* = +10), then back to the starting position, then downward to a maximum negative distance (*D* = –*A*, –*A* = –10), and finally back to the starting position over the time period of 1 second. The sinusoidal vibration is symmetric in displacement about the starting distance, and the vibratory pattern repeats itself perfectly. Theoretically, this back-and-forth vibration can go on forever, but because it repeats itself we need to diagram only one complete transition of the motion. A *cycle* is one complete transition of a sinusoidal function.

Three properties (parameters) characterize a sinusoid: *frequency*, *starting phase*, and *amplitude*. Once we have specified the values of these three parameters, we have uniquely described a single sinusoid. Unless all three parameters are specified, there can be more than one sine wave. Because any vibration consists of a sum of one or more sinusoidal vibrations, we can completely describe any vibration by specifying the amplitude, frequency, and starting phase of the various sinusoids that constitute the vibration. Any vibration consisting of more than one sinusoid is called a *complex vibration*, whereas a single sine wave is called a *simple vibration*. Vibrations are also called *waves* or *waveforms*; therefore, waves or waveforms can be either simple or complex.

Generally, amplitude is a measure of displacement; frequency is a measure of how often per unit of time an object moves back and forth (oscillates); and starting phase measures the relative position of the object at the instant in time it begins to vibrate. The sinusoidal relationship between displacement and time can be written in equation form as

$$D(t) = A \sin (2\pi ft + \theta), \quad (2.1)$$

where $D(t)$ is instantaneous amplitude, A is a measure of maximum amplitude, f is a measure of frequency, t is a measure of time, θ is a measure of starting phase, and sin is the sine function (see Appendix A).

Sinusoidal motion is often referred to as *simple harmonic motion* (additional discussion of harmonic motion will be covered at the end of this chapter and in Chapter 3). It can be described as a swinging pendulum, like that shown in Figure 2.2, where D is the amplitude of the sinusoidal motion, the rate of swing is the frequency, and the starting point of the pendulum relative to point P_0 is the starting phase of the simple harmonic motion (e.g., starting the pendulum at P_1 rather than at P_0 would represent a change in starting phase).

Equation (2.1) is an algebraic description of the waveform drawn in Figure 2.1. The parameters of the sine wave that we have been discussing refer to a physical description of

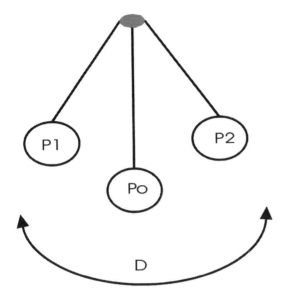

FIGURE 2.2 A pendulum swinging from P_0 to P_1 to P_0 to P_2 plots out a sinusoidal function. D is the amplitude of the sinusoid, the rate of swing is proportional to frequency, and a starting point relative to P_0 corresponds to starting phase.

the vibration. A large area of psychophysics is concerned with human perception of these parameters. On a subjective (psychological) basis, we generally label changes in the amplitude of a sine wave as *loudness* and changes in frequency as *pitch*. We do not have a single subjective term for starting phase because we are sensitive to changes in starting phase only under particular conditions. For instance, changes in the difference of starting phase at the two ears result in changes in the perceived location of the stimulus in space. Thus, for those conditions in which two ears are stimulated with a difference in starting phase, changes in the starting phase correspond to changes in the *locus*, or *location*, of the stimulus.

The relations of the physical descriptions to subjective descriptions of the parameters of a

sine wave are somewhat complex. For instance, although changes in intensity are almost always perceived as loudness changes, variations in frequency, which are independent of intensity changes, can also result in perceived loudness changes. (Other interactions will be discussed in Chapter 13.) Because of these complex interactions between the physical and subjective dimensions of sine waves, we must be careful when describing the acoustical stimulus. In referring to the physical stimulus, we should use only the terms *frequency, amplitude,* and *starting phase.* In referring to how a person perceives the stimulus, we can use the subjective terms *pitch, loudness,* and, in some cases, *perceived location.*

FREQUENCY

The frequency of a sinusoid is the *number of cycles* it completes per *second*. The symbol Hz, standing for *hertz* (hertz is in cycles per second), is used to denote this number. If a sine wave completes 100 complete cycles in 1 second, then the sinusoid is said to have a frequency of 100 Hz. As mentioned previously, a complete cycle occurs when the sine wave begins and ends at the same point of displacement after having taken on all possible values of $D(t)$. Thus, frequency in hertz is the same as the number of cycles per second a sinusoid vibrates. The sinusoid in Figure 2.1 has completed one cycle in 1 second, and so its frequency is 1 Hz. The sinusoid in Figure 2.3 has completed five cycles in 1 second, and so its frequency is 5 Hz.

The amount of time a sinusoid takes to complete one cycle is called its *period (Pr)*. Thus, the period of the sinusoid in Figure 2.1 is 1 second, and the period of the sinusoid in

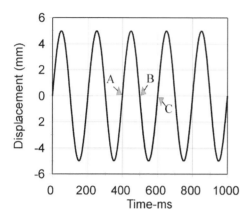

FIGURE 2.3 A sinusoid with a frequency of 5 Hz ($Pr = 100/200$ ms); the starting phase is $0°$, the peak amplitude is $+5$. One period is the time between points A and C and not between points A and B or B and C.

Figure 2.3 is 1/5 second. Since the periods of the sinusoids we will study are short, we usually express fractions of a second (abbreviated sec) in terms of milliseconds (thousandths of a second, abbreviated msec or ms; 1 sec = 1000 msec or 1 msec = 0.001 sec). The period of the sine wave in Figure 2.3 is 200 msec or 0.2 sec). Notice that, although we defined the period in terms of one complete cycle, the period of a sinusoid may be found by determining the time between any two identical points on the waveform. In Figure 2.3 the period may be determined, for example, by finding the time between points A and C on the waveform. The time between points A and B is equal to only half a period. Notice that point B is not exactly like point A (the sine wave is going up at points A and C and down at point B).

Frequency and period are reciprocally related, and therefore

$$f = 1/Pr \qquad (2.2)$$

when Pr is expressed in units of seconds and f in units of hertz, and

$$Pr = 1/f.$$

For periods measured in milliseconds the relations are

$$f = 1000/Pr \quad \text{and} \quad Pr = 1000/f. \qquad (2.3)$$

Either frequency or period may be used to describe oscillation of sinusoidal vibration, although it is usually easier to determine the period and compute the frequency, using equation (2.2) or (2.3). These two terms (frequency and period) are also used to describe the repetition occurring in many non-sinusoidal vibrations (i.e., complex vibrations). The only requirement of these non-sinusoidal vibrations is that they be periodic; whatever pattern of vibration exists, it must be repeated. The complex vibration described in Figure 2.4 is periodic, but it is not a sinusoid. Its period

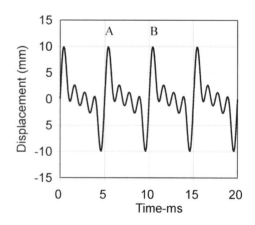

FIGURE 2.4 A complex periodic vibration that is not a sinusoid. The frequency of repetition is 200 Hz. The period is shown by the time indicated between points A and B (5 ms). Its peak amplitude is $+10$.

is 5 msec because the time between two identical points (such as A and B) is 5 msec. Its frequency is 200 Hz because 200 Hz = 1/0.005 sec (or 200 Hz = 1000/5 msec), or because the vibration went through one cycle (one pattern of vibration between points A and B) in 0.005 s. For complex vibrations with a repetition rate, the frequency of vibration is called *repetition frequency*, and for sinusoids, *sinusoidal frequency*.

STARTING PHASE

The starting phase of a sinusoid corresponds to the point in the displacement cycle at which the object begins to vibrate. Starting phase is usually defined in terms of *degrees of angle*. A sinusoid is said to start at zero phase, or to start "in phase," if when the time is equal to zero ($t = 0$) the displacement D is equal to zero. The sine wave in Figure 2.1 starts at zero phase. The sinusoid in Figure 2.5 starts one-half of a period later than the zero-starting-phase (Figure 2.1) condition. In Figure 2.6 the starting phase is one-quarter of a period later.

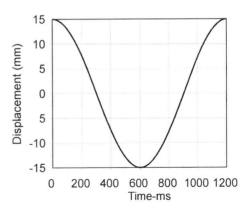

FIGURE 2.6 A sinusoid that begins at 90° starting phase. Compare this sinusoid to those in Figures 2.1 and 2.5.

The description of starting phase is the *phase angle*. This term stems from the fact that a sinusoid that has completed one cycle of vibration has completed a circle (see Appendix A). Because a circle has 360° (or 2π radians) a sine wave has gone through 360° (2π radians) when it has completed one cycle, 180° (π radians) when it has completed one-half of a cycle, 90° ($\pi/2$ radians) in one-quarter of a cycle, and so on. Thus, degrees of circular angle can define the starting phase of a sinusoid. The sinusoid in Figure 2.5 has a starting phase angle of 180°, those in Figures 2.1 and 2.4 start at a phase angle of 0°, and the one in Figure 2.6 starts at a phase angle of 90°. If the wave begins at time equal to zero when the displacement is positive, the starting phase must be between 0 and 180°, whereas if it begins at a negative displacement the starting phase lies between 180 and 360°. Notice that, although the sinusoids in Figures 2.1 and 2.3 have different frequencies, their starting phase in terms of phase angle is the same (0°).

The starting-phase angle is determined by noting the change in the starting displacement

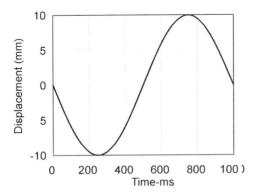

FIGURE 2.5 A sinusoid that begins at a 180° starting phase. Compare this sinusoid to the one in Figure 2.1, which begins at 0° starting phase.

of a sinusoid *relative* to the 0°-starting-phase condition shown in Figures 2.1 and 2.3. Starting phase is a relative term; all starting phases are stated in degrees relative to the 0°-starting-phase condition. If, for instance, a sinusoid starts at maximum displacement at a time equal to zero (see Figure 2.6), then it has begun a quarter of a period later than the 0°-starting-phase condition. Thus, the starting phase is one-quarter of 360°, or 90°. Figure 2.7 shows a sinusoid divided into fractions of a period. The time or horizontal axis is divided into eight equal divisions of phase. This type of figure can be used to determine starting phase. If the sinusoid starts at one of these eight positions when time equals zero, then the starting phase can be determined by using the appropriate value on the horizontal axis.

Sometimes two sinusoids are said to be "out of phase" with each other. If the two are out of phase and have the same frequency, then they must have different starting phases. In Figure 2.8, sinusoid A is 90° out of phase with sinusoid B because sinusoid A starts at 0° phase and sinusoid B at 90° phase. Because both sinusoids have the same frequency, how-

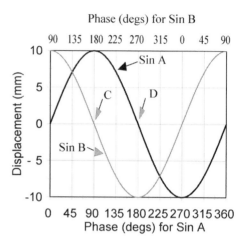

FIGURE 2.8 Two sinusoids of the same frequency. Sinusoid A starts at 0° starting phase and sinusoid B at 90° starting phase. This same 90° phase difference exists at all time points, including points C and D.

ever, the phase difference between them could be determined at any two points along the time axis. At both points C and D, for instance, the difference between the two waveforms is also one-quarter of a period, or 90° (360°/4). Sinusoid B is said to *lead* sinusoid A in phase because sinusoid B has reached its maximal point of displacement before sinusoid A. Conversely, sinusoid A *lags* sinusoid B in phase.

If the two sinusoids are of different frequencies, as in Figure 2.9 (where the frequencies differ by a factor of two), then the relationship between the difference in phase is not so simple. The two sine waves in Figure 2.9 have the same starting phase. At points D and E, sine wave A leads sine wave B, but at points F and G on sine wave B leads sine wave A.

In order to avoid confusing the phase difference computed at the start and that figured at some other points in time, we speak of *starting phases* and *instantaneous phases*. Starting phase is figured at time *t* = 0 (when the sine

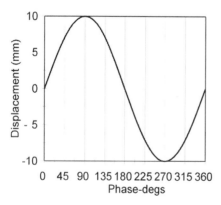

FIGURE 2.7 A sinusoid is divided into eight divisions of phase, expressed relative to the 0°-starting phase condition.

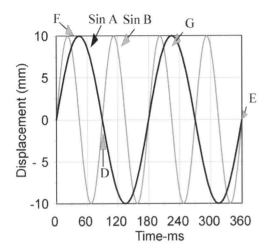

FIGURE 2.9 Two sinusoids (sin A, sin B) that differ in frequency by a factor of 2. The phase differences computed at different points (D, E, F) are not all the same, although the starting phase for both sinusoids is 0°. The phase difference is 180° at point D, 0° at point E, and 90° at points F and G.

wave begins); instantaneous phase refers to the phase difference figured at any time point. Therefore, if two sinusoids have the same frequency, the starting-phase difference always equals the instantaneous-phase difference. If the two sinusoids have different frequencies, then the starting phase will generally not be equal to the instantaneous phase.

If a sine wave starts at a phase of 90°, it is called a *cosine (cos) wave*. The wave in Figure 2.7 is a cosine wave. In terms of equation (2.1),

$$\cos x = \sin (90° - x),$$

where x is equal to $2\pi ft$.

AMPLITUDE

Amplitude refers to the amount of vibratory displacement; how far an object moves describes its amplitude. The displacement of a sine wave varies with time. This time-varying displacement is called the *instantaneous amplitude* of the sine wave. $D(t)$ in equation (2.1) describes the instantaneous amplitude of a sinusoid. Definitions of amplitude that do not vary with time include *peak amplitude, peak-to-peak amplitude*, and *root-mean-square amplitude*.

Peak amplitude refers to the maximum positive displacement the waveform achieves in one period. For a sine wave, such as that in Figure 2.1, the peak amplitude is A. Peak-to-peak amplitude is the total distance from the maximum positive peak to the maximum negative displacement peak the waveform achieves in one period. For the sine wave in Figure 2.1 this is $2A$. The amplitude of non-sinusoidal (Figure 2.4) waveforms can also be described in terms of peak or peak-to-peak amplitude.

Peak and peak-to-peak amplitude appear to be adequate descriptions of the amplitude of sinusoids. However, for non-sinusoidal waveforms we might not wish to describe the amplitude with these values. For instance, the complex waveform in Figure 2.4 might have much larger peak and peak-to-peak amplitudes than the sinusoid in Figure 2.1. Yet over most of the period of the complex waveform in Figure 2.4 very little displacement occurs. Peak and peak-to-peak amplitudes are insufficient to summarize the amplitude of the waveform over an entire period. However, we might wish to compare the average amplitude of the waveform in Figure 2.4 over one period with the identical average amplitude of the sine wave in Figure 2.2. If we simply take the average of the instantaneous amplitudes over one period of a *sine wave*, we would always have zero average amplitude because half of the sine wave has positive amplitude and the other half negative amplitude. If, however, we square each instantaneous amplitude, all of the negative numbers become positive. We could then compute the average of these squared instantaneous amplitudes and then take the square root of this average. This result would not be zero and would become larger as the peak (or peak-to-peak) amplitude increased. Such an averaging operation is called computing the *root-mean-square* (rms) amplitude of a waveform. In equation form the rms amplitude, A_{rms}, may be expressed as

$$A_{rms} = \sqrt{[(\sum_{t=0}^{t=T} E_1^2)/N]}$$

or

$$A_{rms} = \sqrt{[1/T \int_0^T D^2(t)dt]}, \qquad (2.4)$$

where $\int$ is the integral sign, T is the period, N is the number of instantaneous amplitudes that are added, and D_t is an instantaneous amplitude. The summing operation (Σ) must be performed over one complete waveform period (T).

In the case of a sine wave, the rms amplitude equals approximately 0.707 times the peak amplitude, or $0.707A$. When we use the term "amplitude," we will be using primarily the rms values, because this technique enables us to compare simple and complex waveforms.

VIBRATIONS

The pendulum example in Figure 2.2 describes one type of vibration referred to as simple harmonic motion. The mass and spring example shown in Figure 2.10 may be used to describe, in a general way, all vibrations including the swinging pendulum. Recall that, if an object has inertia and elasticity, it can vibrate, and this vibration has the potential to produce sound. In Figure 2.10 the mass

would require a force to move it, and so the mass represents the property of inertia. The spring has a restoring force and so represents the property of elasticity (the opposite of elasticity is *stiffness*). If some force moves the mass out from its resting state (called *state of equilibrium*, as shown in Figure 2.10A) to the position shown in Figure 2.10B and then the force is removed, the spring would pull the mass back through the point of equilibrium to a position such as that shown in Figure 2.10C. The mass would continue to oscillate back and forth, that is, the mass would vibrate with some amplitude and frequency. If there was no *resistance*, such as friction, to the motion of the mass, it would continue to oscillate back and forth between the two extremes forever. A

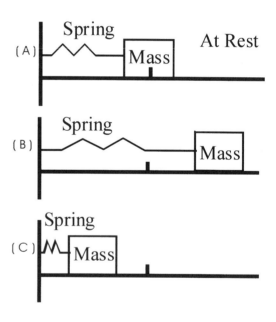

FIGURE 2.10 A diagram of the mass and inertia (spring) components necessary for sound production. When the mass is set into motion, it will move back and forth (left and right as you view from panels 2.10A to 2.10B to 2.10C) in a sinusoidal motion, assuming there is no friction. This is the description of a free-vibration system.

sinusoidal function describes this motion of the mass. Thus, the example shown in Figure 2.10 is a type of model for vibratory systems. In this case, because no force is applied to the system after it is set in motion, the vibration is called a *free vibration*.

Obviously, the real world contains sources of resistance to the motion of the mass. Friction is one source of resistance that prevents the mass from continuing to vibrate forever. The mass would slowly but surely lose its amplitude of vibration. Figure 2.11 describes how the amplitude of vibration would die out due to resistive forces. In this case, the sinusoidal vibration is said to be *damped*. In real systems, the rate of damping is such that the ratio of each successive peak remains constant. The second peak in the sinusoidal function might be one-half the first peak, and the third peak might be one-half the second, and so on. The ratio (in the example above, 0.5) describes

how quickly the vibration is damped. With vibrations in nature the properties of mass (inertia), elasticity (spring), and resistance all contribute to the final description of the vibration and, hence, to any sound we might experience.

SUMMARY

Any vibration is capable of producing an audible sound. Any vibration is equal to a sum of sinusoidal vibrations. A sinusoid (simple wave) has three physical parameters: amplitude (a measure of displacement in terms of peak, peak-to-peak, or rms amplitude), frequency (a measure of the number of vibrations per unit time), and starting phase (a measure of where the wave begins). Fourier analysis shows that a complex wave consists of many sinusoids. The properties of mass (inertia), elasticity, and resistance all contribute to a complete description of a vibration.

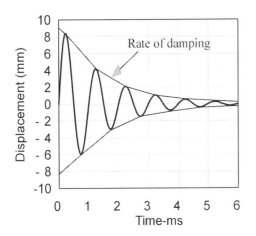

FIGURE 2.11 A plot of the amplitude of sinusoidal vibration, given there is a frictional force acting against the free vibration. Notice that the amplitude of successive peaks decreases at a constant ratio of 0.5. Friction causes the vibration to die out (damp) over time.

SUPPLEMENT

Because sound is a well-studied area of physics, most introductory physics textbooks contain a section on sound and acoustics. Rosen and Howell (1991, Chapters 2 and 3), Moore (1997, Chapter 2), and Pickles (1988, Chapters 1) provide reviews of sound as it pertains to hearing that go beyond the material presented in this book. For the advanced reader, the book by Hartmann (1998) provides a much more thorough discussion of sound.

A variety of topics in this chapter deserve additional discussion so that the interested student can more fully appreciate the material and be better prepared for the material in the next chapters. The sinusoid is not the only function or waveform that can describe sound or free vibrations. However, the sinusoidal function has a variety of properties that make it ideal for analyzing vibrations, including those that relate to Fourier's theorem (see Appendix C) and harmonic motion.

In using the circular analogy (see Appendix A) to describe sinusoids, frequency can be expressed as the rate at which a point moves along a circle's circumference. Thus, w becomes the angular velocity (angular distance divided by time) of the moving point:

$$w = 360°/t = 2\pi f,$$

where $360°$ is 2π, f is frequency; and t is time (recall that $f = 1/t$). For this reason, the sinusoidal term is sometimes expressed as sin (wt).

The concepts of velocity, acceleration, and force have already been mentioned. The definitions of these terms are as follows:

$$v = \text{velocity} = d/t,$$

where d is distance and t is time.

$$a = \text{acceleration} = v/t,$$

where v is velocity and t is time; and

$$F = \text{force} = ma,$$

where m is the mass of the object and a is acceleration.

It follows that

$$a = d/t/t \quad \text{and} \quad F = mv/t.$$

These concepts help in derivation of a quantitative description of the free vibration shown in Figures 2.10 and 2.11. A force is needed to move the object, and the spring applies a restoring force. From the equations above:

$$F = ma \quad \text{and} \quad F = -sx,$$

where the force $F = ma$ is the moving force; the force $F = -sx$ is the restoring force; s represents stiffness; and a and x represent acceleration terms.

The free vibration relationship can be written as $ma + sx = 0$. This equation has as one of its solutions

$$a(t) = A \sin [\sqrt{(s/m)}t + \theta],$$

where A, s, m, t, and θ are defined as previously stated in this chapter.

Hence, a sinusoidal motion can be used to describe the motion or vibration associated with a free vibration. From this concept the frequency of vibration equals $\sqrt{(s/m)}$, which in turn means that for a vibrating object the frequency of vibration is proportional to the square root of the stiffness of the restoring force and inversely proportional to the square root of the mass of the object being vibrated. Thus, increasing the mass of an object four times will reduce the frequency by a factor of one-half, and making the stiffness four times as great will increase the frequency by a factor of two.

In this chapter, the root-mean-square amplitude of a sinusoid was stated as equal to 0.707 times the peak amplitude. This can be seen by applying the definition of rms to a sinusoidal function and using integration:

$$A_{rms} = \sqrt{1/T \int_0^T [A \sin wt]^2 \, dt}$$

(see equation (2.4)),

$$= \sqrt{[A^2/T \int_0^T \sin^2 wt \, dt]},$$

where A_{rms} is the rms amplitude, A is peak amplitude, $w = 2\pi f$, and T is one period of vibration.

Applying integral calculus for a sin-squared function yields

$$A_{rms} = \sqrt{[A^2/T(t/2 - 1/4(\sin 2wt))]} \,|_0^T$$

that is, the integral of $\sin^2(wt)$ is equal to $t/2 - 1/4 (\sin^2 wt)$.

We must evaluate the integral at $t = 0$ and $t = T$. At $t = 0$, the entire right-hand side of the equation is zero. At $t = T$,

$$A_{rms} = \sqrt{[A^2/2 - A^2/4(\sin wt)]}.$$

Under this condition, $\sin wt = \sin(2\pi fT)$, and because $f = 1/T$, $\sin (2\pi ft) = \sin (2\pi T/T) = \sin 2\pi = 0$. (See Appendix A for sin tables and notice that $\sin 2\pi$ is equal to zero.)

Therefore,

$$A_{rms} = \sqrt{(A^2/2 - 0)} = \sqrt{(A^2/2)} = A/\sqrt{2} = 0.707\,A,$$

since

$$1/\sqrt{2} = 0.707.$$

For the purposes of this book, amplitude and displacement will be considered synonymous. Chapter 3 discusses other measures of the magnitude of sound and introduces some of the concepts that must be considered when sound magnitude is described. For the present, displacement and amplitude may be used interchangeably, but they must not be viewed as synonymous with other terms such as level and intensity, which will be covered in Chapter 3.

CHAPTER 3

Sound Transmission

SOUND PROPAGATION

In everyday life we do not hear an object vibrate directly; rather, the vibrating object causes a wave motion in air, which then causes our eardrum to vibrate, starting the process of hearing. Sound can travel through any elastic medium that has inertia. In other words, sound can travel through air, water, steel, and so on, but not through a vacuum.

Because air is the medium for sound transmission in most everyday situations, we will consider the transmission of sound through air. Air consists of molecules in constant random motion. When an object vibrates in air, the molecules tend to move in the direction the object moves rather than with their normal random motion. The air molecules next to the object move first and then pass this movement on to adjacent molecules. The molecules themselves do not move all the way from the object to the receiver; they only pass along a wave of motion. In this train-like progression (much like a falling row of dominoes), the motion of the air molecules is *propagated* (transferred) through the air toward the ear. When the air molecules next to the ear are moved by this motion, the eardrum (tympanic membrane) is vibrated, and this eventually results in our experiencing audible sound.

Let us examine more closely the changes that take place in air when an object is vibrated. We will use the mass and spring analogy shown in Figure 2.10 to describe the air molecule's mass and elasticity. When a vibrating object moves in one direction, air molecules are pushed in the same direction (assuming no frictional forces interfere). The molecules next to the vibrating object are compressed together as the object moves outward away from its resting state, creating an area with a greater *density* (mass per unit of volume) of air molecules. Certain laws of chemistry/physics, the *gas laws*, state that, as the density of air molecules increases, the pressure increases. Thus, as the object moves outward, the density of air molecules next to the vibrating object increases and creates high pressure in this area: an area of *condensation* had been generated. As the vibrating object moves in the opposite direction (back toward its resting state), the air molecules obey another property of the gas laws: gases, like air, will evenly fill the space they occupy. Thus, the air molecules fill the space vacated by the vibrating object moving in the opposite direction. As the vibrating object moves back past its resting state, an even larger vacated area is generated for the air molecules to fill. Now the density of air molecules has decreased, and, thus, the pressure is lower. An area of *rarefaction* has been generated.

The mere presence of molecules in air creates a pressure, or *static air pressure*, that is proportional to the density of the molecules. The changes in pressure due to the vibrating object are changes in the existing static air pressure. As the object vibrates, the static pressure increases at any one location, then decreases, and then increases again, and so on, generating a changing pressure. This changing air pressure moves away from the vibrating object. In other words, areas of condensation are alternating *over space* with areas of rarefaction. That is, molecules next to those initially moved are moved back and forth through stages of condensation and rarefaction. This pattern continues across space as the wavelike motion is *propagated* away from the vibrating source.

Recall that air molecules are in constant random motion, so the changes elicited by a vibrating object change the tendency of the motion of these molecules into the pattern of condensations and rarefactions described above. Imagine that we photographed the air molecules when the object was vibrating to freeze this pattern at a moment in time. Figure 3.1 represents the cartoon of what we might see at the *instant in time* the picture was taken. The molecules appear to cluster at some points in space (condensation), and they are farther apart at other places (rarefaction). Although such a picture would not reveal the direction of motion of the air molecules, the molecules tend to move in the direction of the arrows above the source in Figure 3.1. That is, as the object pushes out, the air molecules tend to move away from the object; as the object moves back toward its starting point, the molecules tend to move back in the other direction. Thus, the motion at a condensation tends to be away from the source, and the motion at a rarefaction tends to be toward the

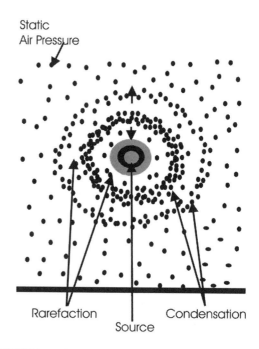

FIGURE 3.1 Diagram of what one might see if air molecules were photographed as a sound source vibrated. Rarefactions and condensations are shown, as well as the direction (grey arrows above the source) in which the molecules were moving at the instant the picture was taken. The wave moves out in circular manner (actually as a sphere in the three-dimensional real world). As the wave moves out from the source, the density of molecules at rarefactions and condensations varies from low to high. The area around the border of the figure represents the static air motion before the propagated wave reaches this area.

source. The wave propagates out from a source in a circular manner (actually in three dimensions the wave propagates in a spherical manner).

Viewing waves on water provides an approximation (but only an approximation) to what happens in air when an object vibrates. If we push water (for instance, by dropping a stone in a pool), waves are formed. The water waves are similar to the air waves associated with sound propagation. The crests of the wave are the condensations, and the valleys are the rarefactions. If you have ever watched a wave "move" through the water, you have observed that the water itself does not move across the surface; rather, the wave motion is passed (propagated) along the water surface. A similar, *but not identical*, type of motion is associated with air molecules that are moved by a vibrating object.

The distance between each successive condensation (or rarefaction) is the *wavelength* (λ) of sound. Actually, the distance between any successive identical points in the propagated wave is the wavelength (i.e., like the distance between adjacent wave peaks in the water analogy). Wavelength is usually expressed in meters. The more frequently an object oscillates (higher frequency), the closer together the rarefactions and condensations become, and the shorter the wavelength. The speed with which the wave motion is propagated through the medium (i.e., the *speed of sound*) also affects the wavelength. If the speed of sound is fast, then as the object vibrates the first wave moves from the object quickly and the second wave lags far behind. This causes the distance between successive rarefactions, and hence the wavelength, to be long. If the speed of sound is slow, the first wave travels only a short distance before the second wave passes an identical spot, so the wavelength is short. Therefore, both the frequency and speed of sound affect wavelength. The equation relating wavelength (λ) in meters to speed of sound (*c* in meters per second) and to frequency (*f* in Hz) is

$$\lambda = c/f. \tag{3.1}$$

Wavelength is directly proportional to the speed of sound and is inversely proportional to the frequency of vibration.

The speed of sound in air is approximately 345 meters per second, although it can vary as a function of the temperature, density, and humidity of air. The speed of sound is higher in a hot, humid area than in a cold, dry place. As the density of the air molecules increases, the speed of sound decreases.

PRESSURE AND INTENSITY

The amplitude of vibration affects the pressure generated at rarefactions and condensations. The greater the amplitude, the greater the instantaneous pressure at a condensation and the lower the pressure at a rarefaction. The relationship between the vibrating object and changes in pressure is further defined in Figure 3.2. Figure 3.2a describes the sinusoidal displacement of air molecules, or particles, caused by the vibrating object, while Figure 3.2b describes changes in air particle velocity, and Figure 3.2c describes changes in pressure. As the air particles move away from their point of equilibrium, the velocity of particle motion increases. That is, as a air particle begins to move out, it is moving a great distance in a short period of time, and, hence, its velocity (distance/time) is fast. As the particle reaches its maximum distance, it must decelerate before it reverses direction; when this occurs, the velocity reaches zero. Thus, when displacement is smallest, velocity is greatest, and when displacement is greatest, velocity is smallest; this explains the 90° phase relation between Figures 3.2a and 3.2b. Because particle pressure is proportional to particle velocity, changes in pressure are also proportional to changes in particle velocity, as shown in Figure 3.2c. The *instantaneous* pressure is changing above and below the existing static air pressure and in relation to particle displacement, as shown in Figure 3.2.

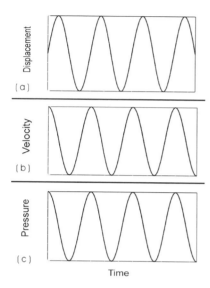

FIGURE 3.2 Three relationships of sinusoidal vibration: (a) An object's (such as an air particle's) displacement as a function of time. (b) Velocity of the particle as a function of time (note 90° phase shift). (c) Instantaneous pressure as a function of time, because pressure is proportional to particle velocity.

The instantaneous pressure $p(t)$ that a vibrating object exerts on an area is directly proportional to the vibrating object's velocity and inversely proportional to the area on which the vibrating object is pushing:

$$p(t) = mv/tAr, \qquad (3.2)$$

where m is mass, v is velocity, t is time, and Ar is area. This equation describes the instantaneous pressure one object can generate. We can describe the change in the air molecules' instantaneous pressure caused by a vibrating object if we let m equal the mass of air molecules. Thus, the relationship shown in Figure 3.2c represents pressure changes for air molecules (pressure is proportional to velocity). If we compute the rms value of these instantaneous pressures in the same way that we computed the rms value of displacement in Chapter 2, rms pressure (p) describes the pressure that the vibrating object exerts in some known area.

Another way to define pressure is in terms of force:

$$p = F/Ar \quad \text{since} \quad F = mv/t, \qquad (3.3)$$

where p is pressure, F is force, m is mass, t is time, and Ar is area. Pressure is thus equal to *force* per unit area. That a vibrating object exerts a force means that the force could move an object through some distance. This is a definition of *work*. The rms pressure has established the necessary conditions for work to be performed. *Energy* (E) is the ability to do work. *Power* (P) is the rate at which work is done. Therefore,

$$P = E/T \text{ and } E = P T, \qquad (3.4)$$

where T is the time in seconds over which the work is done.

Sound intensity (I) is a measure of power for situations involving sound. There is an exact relationship between sound intensity (I) and rms pressure (p):

$$I = p^2/(\rho_0 c),$$

and since

$$p^2 = (F/Ar)^2$$

(see equation (3.3)),

$$I = (F/Ar)^2/(\rho_0 c). \qquad (3.5)$$

where ρ_0 is the density of the medium (such as air), and c is the speed of sound in the medium.

Because sound intensity (I) and power (P) are the same type of term and power is related to energy (E) by equation (3.4), pressure, energy, and power are related as follows:

$$I \equiv p_2/(\rho_0 c) \; P \equiv E/T, \qquad (3.6)$$

where $\equiv$ means "the same as."

Each of these measures will have a different numerical value, but the relationships do not affect the basic description of the sinusoidal vibration of the object (see Table 3.1). An object that vibrates with a frequency of 200 Hz might have a pressure of 100 micropascal or 10^{-12} watts/cm^2 of sound power, but its frequency remains 200 Hz. In this book, *intensity* is used to describe vibration when the measurement is in units of sound intensity (energy or power), and *amplitude* will refer to measures of pressure and displacement.

DECIBELS

Our description of the intensity and amplitude of a sound wave is still incomplete. If the ear were not as sensitive as it is, the foregoing definitions of pressure, energy, and power would be sufficient. Imagine that we measured the smallest amount of sound intensity required to just detect a sound; this amount could be called 1 unit of intensity. Now imagine measuring the greatest intensity that can

be presented before the ear is destroyed. The range of intensities from the smallest required to detect a sound to the largest amount tolerated before ear damage occurs is called the *dynamic range of the auditory system*. If one unit of intensity represents the smallest intensity, then 100,000,000,000,000 (10^{14}) units represent the largest tolerated intensity. The auditory system thus has a dynamic range of 10^{14} intensity units. This range is so large that it is impossible to work with in practical situations (imagine trying to plot a function on graph paper from 1 to 10^{14} in 10-unit steps). Reducing the dynamic range to more manageable numbers is advisable, therefore. One way to accomplish this is to change the type of scale used to describe the dynamic range from an interval scale, such as that used above, to a ratio scale. Logarithms can be used to describe the dynamic range of the auditory system when sound intensity is considered on a ratio scale (see Appendix B for a discussion of ratio scales and logarithms). The starting point is the logarithm of the ratio of two intensities, I_1 and I_2. This logarithm is labeled a *bel*; a bel is $\log I_1/I_2$. Because a bel is too large an interval to work with conveniently, the *decibel* (dB or one-tenth of a bel) is used. The ratio of two intensities expressed in decibels is $10 \log I_1/I_2$. If I_1 is equal to 10^{14} and I_2 is equal to 1 (which covers the dynamic range of hearing), then the dynamic range in decibels is as follows:

TABLE 3.1 Units Used to Measure Quantities Related to Sound Magnitude

Quantity	Units of measurement
Force (F)	1 newton = 10^5 dynes: dyne = (gram $\times$ cm)/sec^2
Potential difference	volt (V, a pressure-like measure)
Pressure (P)	$P = F/Ar$; 1 newton/cm^2 = 10^5 dynes/cm^2 = 10^4 pascals (Pa) 1 micropascal (μPa) = 10^{-6} pascals.
Energy (E)	force through distance; 1 joule = 10^7 ergs; ergs = 10^7 dyne-cm = 1 newton-meter
Power (P)	$P = E/T$, T = time; 1 watt = 10^7 ergs/sec = 1 joule/sec
Sound intensity (I)	watts/cm^2

$$10 \log (10^{14}/1) = 10 \log (10^{14})$$
$$= 10 \times 14 = 140 \text{ dB}.$$

The dynamic range of the auditory system can now be described in terms of decibels, yielding a convenient range of numbers (140 dB). Notice that the original interval scale of intensities ranging from 1 to 10^{14} has been reduced to a *new* interval scale of decibels ranging from 1 to 140 by transforming intensity units to decibels by use of logarithms.

Because P is the same type of measure as I and $P = E/T$, the decibel formulae for intensity, power or energy are

$$\text{decibel (dB)} = 10 \log (I_1/I_2)$$
$$= 10 \log (P_1/P_2)$$
$$= 10 \log (E_1/T)/(E_2/T)$$
$$= 10 \log (E_1/E_2). \quad (3.7)$$

Thus, the decibel can be the ratio of two intensities, powers, or energies.

Because $P = p^2/k$, where $k = \rho_0 c$ as described in equation (3.5), then the formula for decibels in terms of pressure (p) is (see Appendix B)

$$\text{decibel (dB)} = 10 \log P_1/P_2$$
$$= 10 \log (p_1^2 1/k)/(p_2^2/k)$$
$$= 10 \log (p_1^2/p_2^2)$$
$$= 10 \log (p_1/p_2)^2$$
$$= 20 \log (p_1/p_2). \quad (3.8)$$

The decibel, therefore, is *10 times the log of the ratio of two intensities, two powers, or two energies and 20 times the log of the ratio of two pressures*. Once amplitude or intensity has been specified in terms of decibels, then the decibel has the same meaning regardless of whether it represents intensity, power, energy, or pressure. To convert from sound intensity, power, energy, or pressure to decibels, or vice versa, we must keep in mind the units involved.

It is important to remember that the decibel is the *ratio of two quantities*. Because the decibel

expresses how many units one intensity or amplitude is above or below another intensity or amplitude, saying that a sound had an intensity of 60 dB is meaningless because this statement does not indicate whether the intensity is 60 dB above the most intense sound you can tolerate, 60 dB above the lowest sound you can hear, 60 dB above what your friend can hear, or anything else. Always the sound has a *relative* intensity of 60 dB, but each of these sounds has a different *absolute* intensity. A decibel is a relative, not an absolute measurement.

Two conventions are commonly used to define decibels in relative terms. Experiments conducted in the 1930s determined that a pressure of 20 micropascals, or 20 μPa, was the smallest amount required for the average young adult to detect the presence of a 1000- to 4000-Hz sinusoid. When decibels are expressed relative to 20 μPa, they are expressed in terms of *sound pressure level* (SPL). Hence, if intensity is expressed as 60 dB SPL, then the level is 60 dB above 20 μPa. Equivalently, dB SPL means decibels above the threshold for normal detection of a 1000- to 4000-Hz sinusoid. Another way to describe the decibel is in terms of *sensation level* (SL). Sensation level refers to the least intense sound a particular subject can detect in a particular experimental situation (for example, at a particular frequency). Thus, 60 dB SL means a sound was 60 dB more intense than that required for detection in another experimental situation. SPL is based on the reference of 20 μPa, whereas SL is based on the reference of the lowest level a particular subject can detect in a specific experimental context. Although many other conventions are used to define reference levels for the decibel, SPL and SL will be used primarily throughout this book. For this book we will use *sound pressure level* or *pressure level* when the magnitude of sound is expressed in decibels of amplitude (pressure or displacement) and *intensity level* when the magnitude is expressed in decibels of intensity (sound intensity, power, or energy). Thus, *level* will refer to a decibel measure.

INTERFERENCE

Now that we have studied the general properties of the propagation of a sound wave through air, let us consider some consequences of sound propagation in the real world. One of the more obvious properties of sound propagation is that, the farther away we are from a sound source (the vibrating object), the softer sound becomes. In terms of the situation displayed in Figure 3.1, the sound source is considered a point, and sound radiates in a spherical fashion from the source. Notice from equation (3.5) that sound intensity (*I*) is proportional to pressure squared (power) divided by area. As we move away from the center of the sphere (away from the sound source), the surface area of the sphere becomes larger. Because the sound source is assumed to be producing a constant power and the area is increasing, sound intensity (*I*) must be decreasing as one moves away from its source. The area of the surface of a sphere is $4\pi r^2$, where *r* is the radius of the sphere or, in our case, the distance from the sound source to the point of measurement.

From this relationship and the definition of sound intensity (*I*), we obtain the following relationship:

$$I \propto P/(4\pi r^2), \qquad (3.9)$$

where ∝ means proportional to, *P* is the power of the sound at its source, and *r* is the distance from the source to the point of measurement. This in turn means that sound intensity (*I*) is

inversely proportional to distance from the sound source squared (r^2). Thus, if distance is doubled, sound intensity is decreased by a factor of four. This is referred to as the *inverse square law*.

Recall from equations (3.5) and (3.6) that sound intensity (*I*) is proportional to pressure (*p*) squared. Therefore,

$$p = K\sqrt{I} = K\sqrt{(P/(4\pi r^2))} = k/r, \qquad (3.10)$$

where *k* and *K* are constants based on the density of the sound media and speed of sound. Equation (3.10) reveals that *sound pressure* is inversely proportional to distance (*r*). Thus, if the distance from a sound source is doubled, the *sound intensity level* decreases by a factor of four, or 6 dB (10 log 4), and the *sound pressure level* decreases by a factor of two or, also, 6 dB (20 log 2). Thus, when expressed in decibels, level decreases by 6 dB for each doubling of the distance from the source to the point of measurement.

These relationships hold only in situations in which the sound encounters no obstacles. Sound waves most often do encounter obstacles. In general, objects impede the propagation of a sound wave. This impedance to sound transmission occurs whenever there is a change in the medium through which sound must travel (e.g., from air to a brick wall). The term *impedance* has an exact definition, and its properties are important for an understanding of sound propagation. To discuss impedance, let us return to the mass and spring analogy used in Chapter 2 and again at the beginning of this chapter. In general, there are two forms of impedance: *reactance* and *resistance*. The symbol *Z* will be used for impedance, the symbol *X* for reactance, and the symbol *R* for resistance. Both forms of impedance involve properties of a vibrating object that oppose its vibratory motion. Those properties that impede motion because of the properties of the mass and spring are the reactance components of impedance. There are thus two types of reactance: *mass reactance* (Xm) and *spring* (or *stiffness*) *reactance* (Xs). The amount of reactance (i.e., the amount the object's motion is impeded) depends on the frequency of vibration for both mass and spring reactance. Resistive forces to vibratory motion do not depend on frequency. Friction is one type of resistance. The following equation describes the relationship among impedances (Z), reactance (both Xm and Xs), and resistance (R):

$$Z = \sqrt{[R^2 + (Xm - Xs)^2]}. \qquad (3.11)$$

In general, the impedance of any medium is called the *characteristic impedance* (Zc) of the medium. Characteristic impedance is

$$Zc = \rho_0 c, \qquad (3.12)$$

where ρ_0 is the density of the medium, and c is the speed of the sound in the medium. Notice that Zc is the same as the denominator of the definition of sound intensity given in equation (3.5). Therefore, we can redefine sound intensity (I) as

$$I = p^2/Zc, \qquad (3.13)$$

where p is rms pressure. Thus, sound intensity can also be expressed as sound pressure squared divided by the characteristic impedance of the medium through which the sound is transmitted.

Sound intensity varies according to the medium through which it is transmitted because of the characteristic impedance of the medium. If a sound wave encounters a change in medium and, thus, a change in impedance, a portion of the sound wave will be *reflected* from the surface with the greater impedance, that is, the sound wave bounces back from the surface. The same type of sound wave propagation as described earlier takes places for the reflected wave as it moves away from the reflective surface. The amount of sound intensity reflected depends on the difference between the characteristic impedances of the two media. The greater this difference, the greater the sound intensity of the reflected wave. That portion of the sound wave *not* reflected away from the medium is either *transmitted* to the new medium (and the sound continues to propagate through the new medium) or *absorbed* in the new medium.

The reflected sound wave may encounter the original sound wave as it propagates away from the barrier, as shown in Figure 3.3. In this case, two types of interactions can occur between the original and reflected sound waves. Two points of condensation can occur together, resulting in an area of greater condensation, called *constructive interference*. Points of condensation and rarefaction can overlap, resulting in a reduction in pressure, called *destructive interference* (Figure 3.3). That is, the net result can be the summing of two waveforms (*reinforcement*, as in constructive interference) or the subtraction of two waveforms (*cancellation*, as in destructive interference). Thus, a person sitting in front of a wall might hear a sound equal in intensity to the oncoming wave (no reflections), a more intense sound (in the case of reinforcement), or a less intense sound (cancellation).

Figure 3.3 illustrates the sound coming from one point. If sound is coming from several points or from a large area, many reinforcements and cancellations are possible. Aside from the wall, anything more dense (having a different characteristic impedance) than air will reflect sound. Determining the intensity at any one place in a room, therefore, becomes very complicated.

Notice that in Figure 3.3 the frequency is the same for all waves. If two waves of differ-

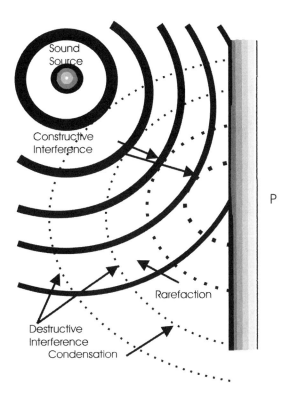

FIGURE 3.3 A sound wave reflects off a wall at point P. As the sound wave returns from a reflection, it can add to the oncoming wave (*constructive interference*, in which condensation peaks meet or rarefaction valleys meet) or cancel the oncoming wave (*destructive interference*, in which a condensation peak meets a rarefaction valley).

ent frequencies were reflected off a wall, complex interaction patterns might result. As a consequence of these interactions, a sound of one frequency might be easier to hear than another because its intensity is greater (reinforcement) at the point of measurement. The intensity of a sound may vary not only as a function of where in a room it is measured, but also as a function of frequency.

In summary, we must be careful in assuming that the waveform arriving at some point in a room (e.g., a person's ear) is identical to the waveform that leaves the source (a loud-

speaker). The two should be nearly the same *only if* reflections off walls, floors, and other objects are discounted and the distance from the source is used in the calculation.

A *sound shadow* is an additional influence an object can exert on a sound wave. The sound shadow can be viewed in the context of waves on water (Figure 3.4). Imagine throwing a pebble into water and watching waves radiate from the spot of impact. As described previously, this is analogous (but not identical) to sound waves radiating from a sound source. A very large object in water causes waves to bounce off the object, as described

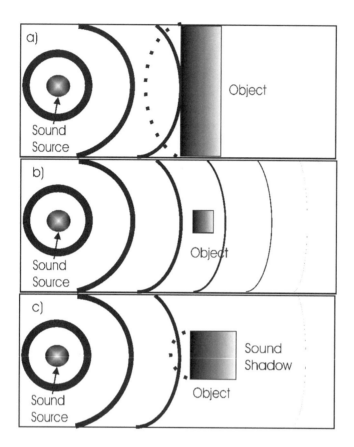

FIGURE 3.4 Simplified diagrams of sound waves passing objects. (**a**) Objects larger than the wavelength: most of the wave is reflected. (**b**) Objects much smaller than the wavelength: most of the wave passes the object. (**c**) Objects close in size to the wavelength: a sound shadow of reduced sound intensity is produced beyond the object.

previously and in Figure 3.4a, but a wave passes over a very small object with little change (Figure 3.4b). A medium-sized object produces some reflections from the object, but at some distance beyond the object the waves appear to have been unaffected. In fact, if we look carefully at this last situation, we notice that just past the medium-sized object is an area with no waves at all or reduced wave magnitude. This area is referred to as the

sound shadow (Figure 3.4c). What is meant by "small," "medium-sized," and "large" objects? These sizes are expressed in relation to the wavelength (recall that wavelength is measured in units of distance) of sound; objects much larger than the wavelength of sound and with a different characteristic impedance reflect sound, objects that are much smaller than the wavelength of sound do not affect wave motion, while objects whose sizes are approximately equal to the wavelength of sound may produce sound shadows. The term *diffraction* is used to describe the property by which sound "passes around or is scattered by" (is diffracted by) small or medium-sized objects. The area of reduced wave motion produced by a sound shadow will be at least as large as the wavelength of the oncoming sound. Thus, if a sound with a 2-meter wavelength (approximately 175 Hz) passes a 1-meter object, then an approximately 2-meter conical area of little or no sound intensity will be produced on the other side of the object. The fact that the human head can produce a sound shadow has important implications for our ability to use our two ears to locate sound. These implications will be encountered when binaural (two-ear) hearing is discussed (see Chapter 12).

SOUND FIELDS

Any environment that contains sound is called a *sound field*. A sound field without any reflections is called a *free field*. A truly free-field environment is almost impossible to obtain. An *anechoic* room is an environment in which every attempt is made to reduce reflections. This is generally accomplished by using materials and shapes that absorb rather than reflect sound. Sometimes a room with reflections is desired. Such environments are called *reverberation* (or *echoic*) rooms. One type of reverberation room or field is a *diffuse* field. In a diffuse field, an attempt is made to arrange the reflecting surfaces so that sound intensity will be as uniform as possible throughout the room. The reflecting surfaces are arranged so that constructive and destructive interference leads to constant sound intensity throughout the sound field. In any room, whether or not it is a reverberation room, measuring reflections can be useful. If the room has surfaces that

reflect a large proportion of the sound, then both the original sound and the reflected sound can be heard (e.g., an echo might occur). This is usually not desirable. In a room there will be many reflections, many at some distance from the source. Thus, the reverberate sound in the room can last a long time (often many seconds). However, after a sound is terminated, the reverberation will die away over time. The primary value used in measuring this aspect of reverberation is *reverberation time*. Reverberation time is the time it takes for the reverberant sound pressure in a room to reach some proportion (usually defined as one-thousandth or 60 dB = 20 log 1000 = 20 × 3) of its original pressure. Both the amount of absorption of sound in the room and the volume of the room help determine the reverberation time. High absorption and small volume lead to short reverberation times (that is, the pressure of the echo or reverberation has quickly been reduced). That is,

$$RT \propto Vol/Ab, \qquad (3.14)$$

where *RT* is reverberation time, *Vol* is room volume, *Ab* is total room absorption, and ∝ means "proportional to."

As we have already discussed, sound waves and their reflections can interact in a variety of ways (as indicated in Figure 3.3). In enclosed spaces, these interactions produce several effects that are of interest to acousticians, musicians, architectural designers, as well as hearing scientists. One of these interactions is the *standing wave*. The concept of a standing wave is perhaps easiest to explain in terms of a vibrating string, shown in Figure 3.5. Imagine that a string is attached to a wall and that the string is vibrated by flicking it once. A wave motion travels down the string and is reflected off the wall (Figure 3.5a). Notice that as the wave comes back from the reflecting surface (the wall), it is inverted rela-

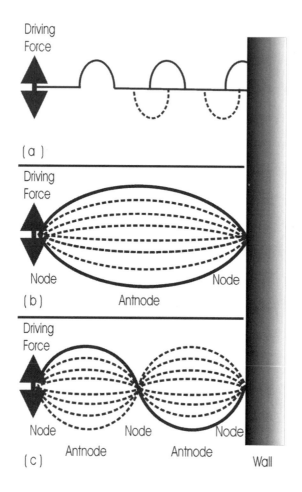

FIGURE 3.5 A diagram of the vibratory pattern of a string when it is attached at one end (e.g., to a wall) and vibrated at the other end. (**a**) With a single movement of the string, a wave moves to the wall and is inverted as it reflects from the wall. (**b**) When the string is vibrated, a standing-wave motion may exist with a node at either end and an antinode in the middle. (**c**) The frequency of vibration is twice that in Figure 3.5b, yielding three nodes and two antinodes.

tive to the original oncoming wave. Now imagine that you continuously "vibrate" the string. At the correct frequency, cancellations and reinforcements will cause the entire string to move up and down without an apparent wave traveling to one end and back. In this case (Figure 3.5b), locations of minimum displacement (*nodes*) alternate with locations of maximum displacement (*antinodes*). If you increase the frequency of vibration of the string, you can generate a wave pattern with two antinodes, as shown in Figure 3.5c. At even higher frequencies wave patterns with more than two antinodes can be produced. Such wave patterns with fixed locations of nodes and antinodes are called standing waves. Remember that the existence of a standing wave does not mean an absence of wave motion but that a wave motion does not travel left to right (*transversely*) along the string. The string moves in an up–down direction maximally at the antinode and minimally at the node.

A standing wave with one antinode (Figure 3.5b) is called the *fundamental mode* of vibration. The fundamental mode occurs when the vibrating frequency has a wavelength (λ) equal to twice the length of the string. Hence,

$$f_0 = c/(2L), \tag{3.15}$$

where f_0 is the fundamental frequency of the mode, c is the speed of sound in the medium, and L is the length of the string. Recall that $\lambda = c/f$ from equation (3.1), so the length of string (L) for one mode of a standing wave to occur is $\lambda_0/2$, where λ_0 is the wavelength of f_0, and $\lambda_0 = 2L$.

Thus, f_0 becomes the fundamental frequency in producing the fundamental mode of vibration. Vibrating frequencies at integer multiples of the fundamental vibrating frequency ($2f_0$, $3f_0$, $4f_0$, etc.) produce standing waves with two, three, four, and so forth, times the number of antinodes (Figure 3.5c). However, as the vibrating frequency increases above its fundamental frequency, the amplitude of the standing wave decreases. Standing waves are generated by a fixed pattern of constructive and destructive interference. These patterns of interference result from interactions of the inverted reflected wave and the original ongoing wave when a continuous frequency of vibration is set up.

Standing waves are important properties of stringed instruments. Standing waves can also exist in enclosed air spaces (such as an organ pipe, or a tube such as the outer ear canal) when air is forced into the space (such

as occurs when sound is generated and travels down the outer ear canal). The same types of interference described above can generate a standing wave of air movement within the pipe or tube. If the tube is closed at both ends or open at both ends, the standing-wave fundamental mode will occur when the wavelength of the fundamental frequency is twice the length of the tube, the same as for the string example of Figure 3.5 and equation (3.15). If the tube is closed at one end, the wavelength of the fundamental frequency is four times the length of the tube ($f_0 = c/4L$, $L = \lambda_0/4$, $\lambda_0 = 4L$). In this case, the higher modes of vibration only occur at odd integer multiples of the fundamental frequency (i.e., at $3f_0$, $5f_0$, $7f_0$, and so on).

Standing waves can exist in any environment in which a unique pattern of interference can be set up. Rooms can generate standing waves under certain conditions of vibration. These standing waves produce fixed locations of large pressure changes (antinodes) and of minimal pressure changes (nodes). In general, such standing waves are to be avoided in designing a room for acoustic purposes. Likewise, standing waves may exist in auditory structures such as the ear canal. For instance, the outer ear canal is like a tube closed at one end. As such, standing waves may be generated within the outer ear canal when sound is present, and, if so, the frequencies of these standing waves will influence what we hear.

SUMMARY

Sound is propagated through air from its vibrating source to a receiving object. The vibrating object causes a movement of air molecules that corresponds to the vibration

of the object. The wavelength of vibration is a measure of the distance between successive points (e.g., rarefactions or condensations) in the sound wave. Sound intensity is measured in units of energy or power derived from pressure. The decibel is 10 times the logarithm of the ratio of two intensities (powers or energies) and 20 times the logarithm of the ratio of two pressures. Sound waves can encounter various forms of interference. There can be cancellations and additions, standing waves can be formed, objects can create sound shadows in the sound wave field, and objects might introduce impedance to the transmission of sound. Impedance has reactance and resistive components. The intensity of sound decreases by a factor of the distance squared as a listener moves away from the sound source. Various sound fields can be constructed to control reflections. The frequency of a standing wave is related to the length of the space being vibrated as well as to other properties of the space. Reverberation time indicates the time it takes for reverberation to lose its intensity a specified amount.

SUPPLEMENT

In addition to the readings suggested in Chapter 2, textbooks on architectural acoustics may provide some informative discussion of topics touched on in this chapter. An excellent textbook is the classic by L. L. Beranek (1954; repr. 1988) titled *Acoustics*, reproduced by the Acoustical Society of America. The textbook by Rossing (1990) on *The Science of Sound* provides excellent coverage of the physics of sound at an elementary level.

Although we have used the analogy of waves in water to describe sound waves in air,

the reader should not assume that wave propagation is the same in both media. Caution should be used before drawing any strong parallels between waves in water (or other media) and sound waves in air.

Molecules in any gas, including air, are in constant motion, which causes the molecules to collide with one another. The gas laws describe the properties of this molecular motion. Two aspects of this motion, sometimes referred to as *Brownian motion*, relate to discussions in this chapter. The movement and collision of molecules produce a pressure, called *static pressure*. The changes in pressure associated with introducing a sound source cause an additional change in this static air pressure. Temperature and humidity can also change static air pressure. The changes in air pressure due to temperature and humidity occur at a much slower rate than the air pressure changes usually associated with sound production. Therefore, it is usually easy to separate sound pressure changes from these other air pressure changes. When air molecules collide with each other, energy is given off. Some of this energy results in potential sound waves. The sound energy of this Brownian motion noise (sometimes called *thermal noise*) is very small, probably 20 or 30 db below the faintest sound a human can hear. Thus, except for a few very sensitive humans and perhaps a few animals with greater auditory sensitivity than humans, this thermal noise is not influencing an animal's ability to detect weak sounds (see D. M. Green's *An Introduction to Hearing*, 1976, for additional information).

Frequency is a variable that is like velocity or speed, since it means cycles per second. However, frequency of vibration should not be confused with speed of sound. Frequency refers to the rate at which a vibrating object goes through one cycle of oscillation, while the speed of sound refers to the rate at which a sound wave is propagated through air (or any other elastic medium). When applying these terms to the motion of air molecules, frequency refers to the rate at which the air molecules at one point in space change from a particular pattern of condensation or instantaneous pressure back to this same pattern or instantaneous pressure (i.e., frequency refers to changes that takes place over time at one location in space). Speed of sound refers to the rate at which the pattern of condensation is passed from one location in space to the next location in space (i.e., speed of sound refers to the wave traveling across space over time).

Chapter 2 introduced the terms *velocity* and *acceleration*, and Chapter 3 has used these terms repeatedly. Velocity is a measure of change. Calculus is used to analyze such relationships. In calculus, velocity and acceleration are defined as follows: $v = dx/dt$, where dx is the change in distance and dt the change in time. Acceleration becomes $a = d^2x/d^2t$. Velocity is the first derivative of distance, and acceleration is the second derivative of distance. This type of calculus nomenclature is used to describe free and damped vibrations. By using the concepts discussed in the Supplement to Chapter 2, we can describe free vibration as follows:

$$m(d^2x/d^2t) = -sx,$$

so,

$$m((d^2x)/(d^2t)) + sx = 0.$$

By solving this differential equation for x (since s and m are constant), we arrive at the equation

$$A \sin (\sqrt{s/m}\,t + \theta),$$

which was described in the Supplement to Chapter 2. Adding the frictional forces that lead to a damped vibration, the differential equation becomes

$$m((d^2x)/(d^2t)) + sx + r_f(dx/dt) = 0,$$

where r_f is the influence of friction. The solution to this differential equation yields a function with the shape shown for the damped sinusoidal vibration in Figure 2.10 of Chapter 2.

A differential equation called the *wave equation* is used to describe wave propagation in any media:

$$\delta^2(p/\delta^2t) = c^2[(\delta^2p)/\delta^2x)],$$

where p is sound pressure, c is the speed of sound, x is distance, and t is time. With two independent variables, space (x) and time (t), the differential equation involves partial derivatives (δ). Solutions to the wave equation describe the type of wave possible in any medium. In air, solutions to the wave equation describe the propagation of sound.

Table 3.1 describes the units of measurement for most of the terms defined in this chapter. In electricity, impedance, resistance, and reactance are usually measured in ohms, although from equation (3.13) *acoustic impedance* is defined with the following units: (grams/cm³)(cm/sec). For definitions of power, energy, impedance, resistance, and reactance, there is a strong parallel between acoustic conditions and electrical conditions. In general, the definitions from one condition can be used to interpret the terms in the other condition. This must be done with some caution, however, and a careful study of physics is required to make exact comparisons. The term *voltage*, used in electrical circuits, is a pressure-like term (see Table 3.1).

As we stated in Chapter 2 and in this chapter, we will use the term *intensity* for measures of sound intensity (power or energy), *amplitude* for pressure or displacement, and *level* when intensity or amplitude has been converted to decibels. However, it is common practice in the hearing sciences to use the terms *intensity* and *level* more loosely and often interchangeably.

When reinforcement and cancellation were discussed, we described these actions as addition and subtraction. Adding and subtracting sound waveforms is treated as vector addition (see Appendix A), with the amplitude of the sound wave being the vector's magnitude and the wave's starting phase representing the vector phase. It should also be emphasized that, when addition and subtraction are used to determine the amount of reinforcement or cancellation, the calculation is done in units of intensity or amplitude and *not* in decibels (adding decibels is multiplication of intensity, and subtracting decibles is division as explained in Appendix B).

CHAPTER 4

Complex Stimuli

COMPLEX STIMULI

Although Fourier's theorem states that sinusoids are the basic vibrations of sound, they are not the types of vibrations experienced in everyday life. Most acoustic stimuli are complex, consisting of the sum of many sinusoids. As with sinusoidal stimuli, there are various ways to describe a complex waveform. The waveforms drawn in Chapter 2 are defined in the *time domain*. The time-domain description relates the instantaneous amplitude or pressure of a waveform to time. When a complex waveform is described in terms of the individual sinusoids that are added, the waveform is being described in the *frequency domain*. In the frequency-domain, the amplitude, frequency, and starting phase of each sinusoid in the complex waveform must be described.

Each sinusoid that constitutes a complex waveform is characterized by its frequency. Thus, if we say that a complex sound consists of four sinusoids, we mean that four sinusoids with four *different* frequencies exist. The plot of the amplitude of each sinusoid as a function of its frequency is called the *amplitude spectrum* (or power spectrum depending on the measure of sound magnitude used), whereas

the plot of the starting phase of each sinusoid is called the *phase spectrum*. When the phase and amplitude spectra of a complex waveform are described, the waveform has been completely defined.

Figure 4.1 demonstrates how a complex wave is portrayed in both the time and frequency domains. Figure 4.1a is a time-domain plot of a complex wave. This complex wave consists of the sum of three sinusoids with frequencies 100, 200, and 300 Hz. Figures 4.1b and 4.1c show the amplitude and phase spectra of the complex sound. Figure 4.2 demonstrates how sinusoids are added to produce a complex sound. In Figure 4.2, the instantaneous amplitudes of the three sinusoids with frequencies of 100, 200, and 300 Hz have been added at each successive point in time (e.g., at points a, b, and c). The summed waveform at the bottom of Figure 4.2 is a perfect match to that in Figure 4.1a. Thus, this graphical application of Fourier's theorem demonstrates that the complex time-domain waveform can be created by adding the instantaneous amplitudes of all of the sinusoids present in the spectrum of the complex sound.

Thus, when we wish to ascertain the variation of a complex wave in time, we emphasize the waveform in the time domain. If the

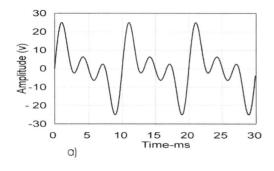

a)

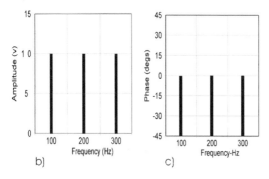

b)   c)

FIGURE 4.1 A complex waveform consisting of the sum of three sinusoids (with frequencies of 100, 200, and 300 Hz) in the time domain (**a**) and frequency domain (amplitude spectrum, **b**; phase spectrum, **c**).

between certain limits are present. For many line spectra, only integer multiples (2 times, 3 times, 4 times, and so forth) of the lowest frequency exist. For instance, in Figure 4.1 the lowest frequency was 100 Hz, and its higher-integer multiples existed at 200 and 300 Hz. In this situation, the lowest frequency is called the *fundamental frequency*, and each higher-integer multiple of the fundamental frequency is called a *harmonic*. Thus, 100 Hz is the fundamental frequency, 200 Hz is the *second harmonic*, and 300 Hz is the *third harmonic*. Many complex sounds consist of a fundamental frequency and higher harmonics.

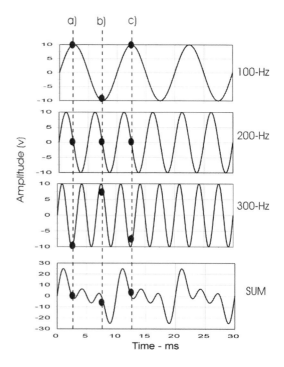

FIGURE 4.2 The time-domain representations of each sinusoid described in Figure 4.1a,b are shown at the top. The *SUM* was constructed by adding together at each successive point in time (e.g., a, b, c) the amplitudes of the three sinusoids. The result (SUM) is exactly the same as that in Figure 4.1a.

frequency content of the complex wave is important, we emphasize the amplitude and phase spectra. As shown in Figure 4.1, it is not difficult to reconstruct the time-domain description of a wave from the amplitude and phase spectra, but deriving the frequency-domain spectra from the time-domain waveform requires the use of Fourier analysis. Although the technique of Fourier analysis is beyond the scope of this book, it is briefly described in Appendix C (see also Chapter 5).

A *line spectrum* describes a complex wave consisting of a discrete number of sinusoids, whereas a *continuous spectrum* describes a wave in which all frequencies (a continuum)

Figure 4.3 demonstrates another interesting ramification of Fourier analysis: a sinusoid turned on and off is not a simple stimulus but a complex sound consisting of the sum of many sinusoids (it has a continuous spectrum). The shorter the sinusoid is in duration, the more noticeable these other sinusoids become. Perhaps you have heard the click that sometimes occurs when a sound is abruptly turned on and off. The perception of a click represents the ear's sensitivity to frequencies other than that of the sound that is switched on and off. A sinusoid sounds more like a tone and less like a click the longer it remains on. This is because the longer a sinusoid remains

on, the smaller the amplitudes of the other frequency components become relative to the amplitude of the frequency component of the sinusoid. Thus, we hear only the pure tone pitch corresponding to the frequency of the sine wave and not the clicking sound from the presence of other frequencies.

TRANSIENTS

In studying hearing, it is useful to present a very brief acoustic signal, such as the stimulus shown in Figure 4.4a. This brief pulse is referred to as a *click* or a *transient*. The stimulus comes on at some peak value and goes off after a brief duration, D. The spectra of a click

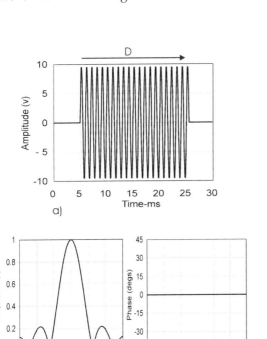

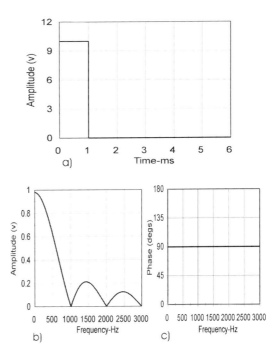

FIGURE 4.3 The amplitude (**b**) and phase (**c**) spectrum of a sinusoid that is turned on and off with duration, D (**a**). Note that this pulsed sinusoid is a complex sound with a continuous amplitude spectrum with zero amplitude at frequencies equal to integer multiplies of $1/2D$.

FIGURE 4.4 The time-domain waveform (**a**) and spectra (amplitude in **b** and phase in **c**) are shown for a single transient of duration D ($D = 1$ ms).

are depicted in Figure 4.4b, which shows the amplitude spectrum, and Figure 4.4c gives the phase spectrum. Notice that the amplitude spectrum is a continuous spectrum, and the amplitude goes to zero at frequencies equal to $1/D$ when D is expressed in seconds. Thus, if D is short, the spectrum is broad. For instance, if D is 0.0001 seconds (0.1 msec), then the spectrum has frequencies with non-zero amplitudes up to 10,000 Hz (10,000 Hz = $1/0.0001$ sec). The phase spectrum is constant at 90°.

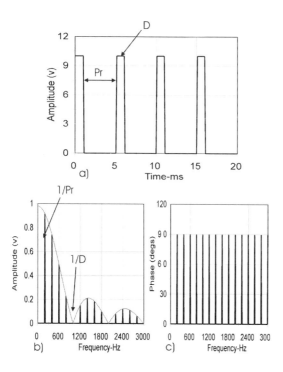

FIGURE 4.5 The time-domain and amplitude spectrum of a transient with duration D, repeated periodically with a period of Pr. (**a**) The time-domain plot; (**b**) the amplitude spectrum plot, and (**c**) the phase spectrum. The lines in the spectrum are at the harmonics of the fundamental frequency of the repetition ($1/Pr$), and the amplitudes are determined by the shape of the spectrum of a single transient of duration D (see Figure 4.4b).

The transient may be repeated at a periodic rate (Pr = period), as shown in Figure 4.5. The spectra of this waveform are also shown in Figure 4.5. For the repeated transient, the spectra are line spectra with the lowest frequency equal to the frequency at which the click is repeated, with each successive harmonic present in the spectrum. The amplitudes of the harmonics change with the same function as shown in Figure 4.4b. That is, the amplitude spectrum has a general shape (sometimes called the *spectral envelope*) determined by the duration, D, of the individual click. If the spectra in Figure 4.5 were based on a 1-msec (0.001 sec) duration-click repeated at a 200-Hz rate, then each harmonic of 200 Hz would be present in the spectra and the amplitudes would decrease to zero at 1000 Hz (i.e., 1000 Hz = $1/0.001$ sec). Again, the phase spectrum would be flat at 90°.

In Figures 4.4 and 4.5, the transients have been displayed as positive-going changes in amplitude. It is assumed that such a stimulus would produce a condensation pattern of activity in the air (see Chapter 3). Thus, this type of click is sometimes referred to as a *condensation click*. If the click were negative-going, then it could be called a *rarefaction click* (again, see Chapter 3). A rarefaction click has the same amplitude spectrum as a condensation click, but its phase spectrum contains frequency components all with 270° starting phases (180° phase shift relative to a condensation click).

BEATS AND AMPLITUDE MODULATION

Perhaps the most obvious way to change a simple stimulus into a complex one is to add two sinusoids of different frequencies (f_1 and f_2). Figure 4.6 shows two examples of such an

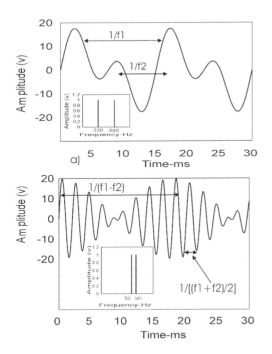

a)

FIGURE 4.6 Time-domain waveform for the addition of two frequencies (f1 and f2). (**a**) Large frequency difference between the tones. (**b**) Small frequency difference between the tones, which also shows the beating in amplitude. The insets in each panel describe the amplitude spectrum of each complex sound.

addition. In the top part of the figure, the two tones are of very different frequencies (the frequencies can be seen as the time separations $1/f1$ and $1/f2$, shown in Figure 4.6a), whereas in the bottom of the figure the tones are very close together in frequency. In the bottom portion of Figure 4.6, the time-domain waveform appears as a single tone with an overall amplitude that changes in a sinusoidal manner. Such a stimulus is called a *beating* stimulus because it is perceived as a tone with a loudness that beats, or waxes and wanes (see Chapters 11 and 13 for a discussion of the beating sensation). If the frequencies of the two tones that are added are *f1* and *f2*, then the

frequency of the sinusoid that appears to beat is equal to $(f1 + f2)/2$. The change in the beating amplitude has a frequency of $(f2 - f1)/2$, and the loudness or beats appear at a rate of *f2* − *f1*. If a waveform consists of 560-Hz (*f1*) and 660-Hz (*f2*) sinusoids of equal amplitudes and starting phases, then the sinusoid that is beating has a frequency of 610 Hz, the beating envelope changes at a rate of 50 Hz, and the loudness changes would occur at the rate of 100 times per second (100 Hz). The spectra consist of two tones with equal amplitudes and starting phases.

Another type of waveform appears similar (*but is not identical*) to the beating tone. An example of this waveform, referred to as an *amplitude-modulated* stimulus, is shown in Figure 4.7. An amplitude-modulated sinusoid is generated by having the amplitude of a sinusoid change over time. Consider the sinusoidal function (equation (2.1))

$$D(t) = A \sin (2\pi F_c t), \qquad (4.1)$$

where F_c is sinusoidal frequency. Now let A change as a sinusoid, $A(t)$:

$$A(t) = (1 + m \sin (2\pi F_m t)), \qquad (4.2)$$

where F_m is the frequency of the change in amplitude, and m is the magnitude of this amplitude change. Thus, combining equations (4.1) and (4.2):

$$D(t) = A[1 + m\sin (2\pi F_m t)] \sin (2\pi F_c t). \quad (4.3)$$

This yields a sinusoid with frequency, F_c (called the *carrier frequency*), whose amplitude is changing with frequency F_m (called the *modulation frequency*). The spectra of a sinusoidally amplitude-modulated tone shown in Figures 4.7b,c consist of three frequencies: the center frequency is equal to the carrier frequency (F_c), and the two side frequencies

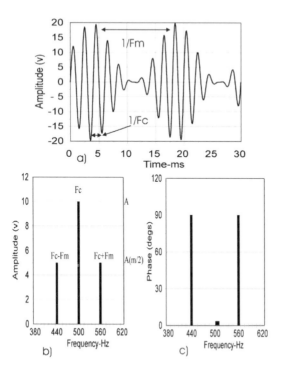

FIGURE 4.7 (a) The time and frequency domain descriptions of a sinusoidally modulated amplitude of a sinusoid. A sinusoid with a carrier frequency (Fc) is amplitude modulated (AM) at a modulation frequency (Fm). The amplitude spectrum of an AM signal contains the carrier frequency, and two sidebands at the carrier frequency plus and minus the modulation frequency: (b) Amplitude Spectrum and (c) Phase Spectrum. The amplitude of the sidebands are $A(m/2)$, where A is the amplitude of the carrier and m is the modulator amplitude.

(called *side bands*) have frequencies of $F_c + F_m$ and $F_c - F_m$. The side bands are at the carrier frequency plus and minus the modulation frequency, and the side band amplitudes are each equal to $A(m/2)$. The phase spectrum is also shown in Figure 4.7. Notice the *differences* in the spectra between a beating stimulus (insert in Figure 4.6b) and an amplitude-modulated sinusoid (Figure 4.7b).

In the examples given above, the amplitude has been modulated with a sinusoidal frequency (equation (4.2)). Other functions can be used to vary the amplitude of a stimulus. Therefore, for sinusoidal modulation the stimuli are sometimes referred to as *sinusoidal amplitude modulation* (SAM) stimuli.

SQUARE WAVE

Figure 4.8 describes the time waveform and the spectrum of a square wave with a period of 2 msec. The resulting amplitude spectrum has a fundamental frequency at the reciprocal of the period (500 Hz in Figure 4.8, 1000/2 msec = 500 Hz), and the spectra contain the odd harmonics of this fundamental frequency (e.g., 1500, 2500, 3500). The amplitudes of the harmonics equal 1 over their harmonic number (e.g., the 3rd harmonic's amplitude is 1/3, the fifth harmonic's amplitude is 1/5). For the square wave shown in Figure 4.8, the phase of all harmonics is 90°. A square wave is equivalent to turning a sound on and off in a periodic manner such that the off time is the same as the on time. A square wave can be used as a modulator just like a sine wave. However, the spectra of a square-wave–modulated sound is different from that of a sinusoidally modulated sound. If the off and on times are not equal, the spectra of the sound changes. If the off time is very long and the on time very short, then the sound might be very similar to a repeated transient as shown in Figure 4.5.

FREQUENCY MODULATION

Rather than varying the amplitude of a signal, one might wish to vary its frequency. Figure 4.9 shows two time-domain waveforms in which the frequency of the signal is changing over time. In the top panel the frequency of the waveform decreases from high frequency

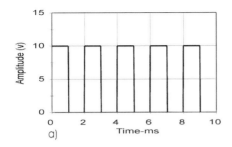

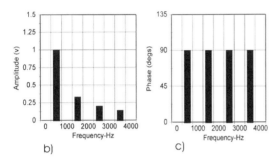

FIGURE 4.8 Time domain (**a**), amplitude spectrum (**b**), and phase spectrum (**c**) descriptions of a 2-ms square wave. The square wave has a spectrum with the greatest level at 500 Hz, as well as a 2-ms period.

scribes the range of frequency modulation. The sinusoidal frequency-modulated equation can then be rewritten as

$$D(t) = A \sin [(2\pi F_c t) + (m/F_m)\sin (2\pi F_m t)]. \quad (4.4)$$

In considering the spectra of such stimuli, we must realize that their frequency content is changing over time. In these cases, two types of spectra can be considered: (1) *long-term spectra*, which relate to the entire waveform over its complete duration, and (2) *short-term spectra*, which consider the spectrum at any

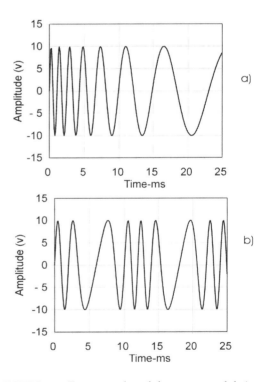

FIGURE 4.9 Two examples of frequency modulation (FM) are shown. (**a**) The frequency of the waveform increases from high to low frequency from the beginning to the end. (**b**) the frequency of the waveform varies through two cycles of a sinusoid, staring with a high frequency going to lower frequencies and back to a high frequency.

to low frequency over the duration of the stimulus. In the bottom panel of Figure 4.9, the change in frequency over time is sinusoidal: the frequency starts high, then goes low, then goes back to a high frequency, and so on. For sinusoidal frequency modulation the time-domain waveform can be written as

$$D(t) = A \sin (2\pi F_c t).$$

Because F_c varies as a sinusoidal function of time, we write this frequency term as

$$(F_c + m \sin (2\pi F_m t)),$$

where F_c is the base or carrier frequency described above; m is the magnitude of the sinusoidal change in frequency, that is, m de-moment in time during stimulus duration. The rules for determining the long-term spectra of frequency-modulated (FM) stimuli are too complex for discussion in this text, but, like amplitude modulation, frequency modulation produces side-band components spaced at the combinations of $F_c \pm nF_m$ (where $n = 1, 2, \ldots, \infty$), along with a spectral component at the carrier frequency, F_c. Such spectral considerations are crucial for understanding the sensitivity of the auditory system to frequency-modulated signals.

The changes in frequency that occur over time for the waveforms shown in Figure 4.9 can be graphed, as shown in Figure 4.10.

These are simple descriptions of changes in frequency for these frequency-modulated signals. They display part of the information required for a short-term amplitude spectrum. Such displays do not contain any information about the amplitude of the various frequency components. For the cases shown in Figure 4.9, the amplitudes of all the frequency components were the same, so the amplitude information may not be needed.

Suppose that the waveform shown in Figure 4.11a was displayed in some sort of short-term spectral graph. This time-domain waveform is a signal whose frequency increases over time, and the amplitude of each fre-

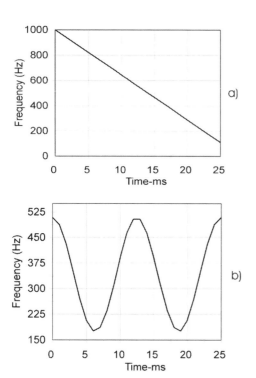

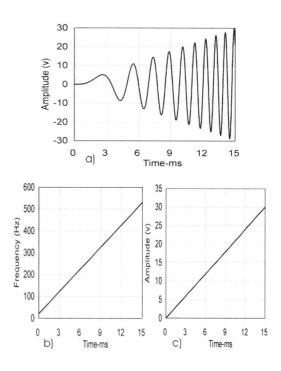

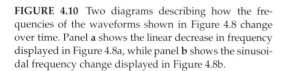

FIGURE 4.10 Two diagrams describing how the frequencies of the waveforms shown in Figure 4.8 change over time. Panel **a** shows the linear decrease in frequency displayed in Figure 4.8a, while panel **b** shows the sinusoidal frequency change displayed in Figure 4.8b.

FIGURE 4.11 (**a**) A time-domain waveform in which both the amplitude and frequency increase over time. (**b**) The change in frequency as a function of time for the time-domain waveform shown in Figure 4.11a. (**c**) Shows the change in amplitude for the waveform in Figure 4.11a.

quency also increases over time. Figures 4.11b,c show the changes in frequency and amplitude over time. This information can be combined into one figure called a *spectrogram* or *spectrograph*, as shown in Figure 4.12, where time is shown on the horizontal axis, frequency on the vertical axis, and amplitude as the darkness of the bar. This is a type of short-term spectral display.

Many common stimuli change both in frequency and amplitude over time: they are both amplitude and frequency modulated. Music and speech are two such stimuli. Time-domain representations, such as those shown in Figure 4.11, are useful means for describing these stimuli, but in many cases a spectrograph such as that shown in Figure 4.12 provides additional information valuable in understanding auditory processing.

NOISE

Noise, as used here, means a sound with an instantaneous amplitude that varies over time

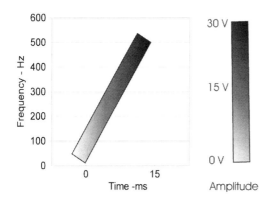

FIGURE 4.12 A spectrogram of the waveform in Figure 4.9 is shown. Time is displayed along the horizontal axis, frequency along the vertical axis, and amplitude as the darkness of the bar. Note that the bar rises, indicating a rise in frequency over time, and the darkness of the bar increases indicating an increase in amplitude over time.

in a random manner. When a noise is "Gaussian," the instantaneous amplitude varies in its probability of occurrence according to the "normal," or Gaussian, distribution. A normal distribution of amplitude fluctuations is shown in Figure 4.13. The distribution shows that the mean amplitude of Gaussian noise is zero (on average, the instantaneous amplitude of the noise at any moment in time is zero) and that the higher or lower the amplitude, the less probable it is to occur at any moment in time.

"White" noise indicates that all frequencies between some limits (these frequency limits define the *bandwidth* of the noise) are present at the same average intensity or pressure, that is, white noise has a continuous and flat power spectrum over its bandwidth. This spectrum is shown in Figure 4.14, with *average power* for each sinusoidal component in the spectrum shown as a function of frequency for the white Gaussian noise. If this is Gaussian noise, then it has a random distribution of starting phases as a function of frequency.

There are two measures of intensity for a white noise: (1) *total power* and (2) *noise power per unit bandwidth*, often called *spectrum level*, abbreviated N_o. Most measuring instruments calculate total power. Total noise power (TP) can be viewed as the sum of the amplitudes of the sinusoids in the spectrum of the noise (as explained in Figure 4.14). This is approximately equal to the area (height × width or, bandwidth [BW] × intensity not in dB) of the spectrum shown in Figure 4.14. The energy at frequencies far from the frequency region of interest will often have no bearing on auditory perception. In this case, the spectrum level of the noise is used. The spectrum level is the average intensity measured in a band of noise 1 Hz wide. Because the instantaneous noise intensity varies, the spectrum level is the *average* noise power in a band of noise 1 Hz wide.

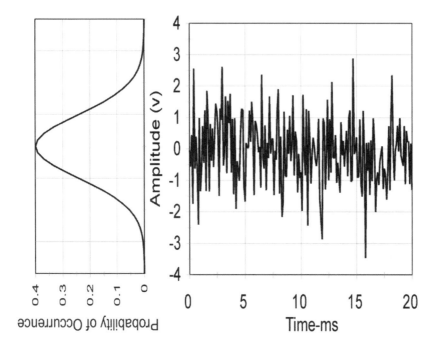

FIGURE 4.13 The time domain of Gaussian noise. The value of the instantaneous amplitude (displacement) has a normal or Gaussian probability of occurrence, as shown on the vertical axis.

This can be represented by the height of the spectrum in Figure 4.14. The procedure is analogous to finding the height of a rectangle (N_o) when the area (TP) and width (BW) are known: $N_o = TP/BW$ or, in decibels, N_o in dB = TP in dB minus 10 log BW. (See Figure 4.14 for a description of the use of TP and N_o.) For example, if a white noise has a total power of 80 dB SPL and a bandwidth of 1000 Hz (e.g., noise has energy from 500 to 1500 Hz), then N_o is 50 dB SPL:

$$N_o \text{ in dB} = TP \text{ in dB} - 10 \log 1000;$$

therefore,

$$N_o \text{ in dB} = 80 \text{ dB} - 30 \text{ dB} = 50 \text{ dB},$$

since 10 log 1000 = 30 dB (recall that dB is simply 10 times the log of the ratio of two quantities, in this case the quantities are frequencies).

Another type of noise, *pink noise*, with the spectrum shown in Figure 4.15, is often used in audition. Note that the x-axis of Figure 4.15 is plotted with frequency on a logarithmic (log) scale, that is, each equal distance along the axis is a doubling of frequency. Such log plots will be used in later chapters.

For pink noise, spectrum level (N_o) decreases with increasing frequency such that with each doubling of frequency the amplitude is halved. The result is that the total power within any band of pink noise remains constant if the bandwidth of this band of pink

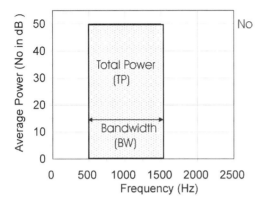

FIGURE 4.14 Schematic diagram of amplitude spectrum of a white noise. The total power (*TP*) of the noise is the summed intensity of all the sinusoids in the noise. N_o is the spectrum level or the average noise power per unit bandwidth (BW), that is, the average power in a band of noise one cycle wide. The bandwidth of the noise is 1000 Hz (500 to 1500 Hz).

noise is proportional to the center frequency of the noise band. For instance, a band of pink noise centered at 1000 Hz with a 500-Hz bandwidth would have the same total power as a band of pink noise centered at 2000 Hz with a 1000-Hz bandwidth.

NARROWBAND NOISE, ENVELOPE, AND FINE STRUCTURE

The noise stimuli discussed above are broadband, containing a large range of frequency components. Noises can also be *narrowband*, containing a limited number of frequency components. Figure 4.16 shows two samples of narrowband noise. In Figure 4.16a the bandwidth of frequencies is approximately 10 Hz (e.g., between 500 and 510 Hz),

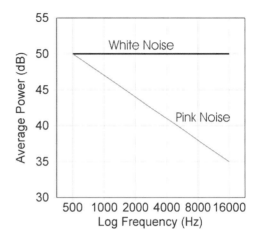

FIGURE 4.15 The amplitude spectrum of pink noise. Pink noise has a high-frequency roll-off of 3 dB for each doubling of frequency. That is, the noise is 3 dB less at 1000 Hz than it is at 500 Hz, etc. This property of pink noise means that bands of pink noise whose bandwidths are proportional to the center frequencies of the band will always have the same total power. Note the log-frequency axis.

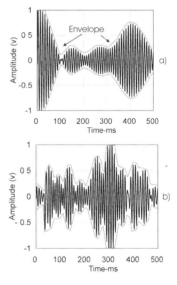

FIGURE 4.16 Two examples of narrowband noise and the resulting envelopes (thin lines): (**a**) a 10-Hz wide band of noise, and (**b**) a 25-Hz wide band of noise.

while in Figure 4.16b the bandwidth is 25 Hz (e.g., between 500 and 525 Hz). In both examples, there is an overall slow change in amplitude (outlined by the thin curves). The rate of this overall amplitude change is slower for the 10-Hz wide noise than for the 25-Hz wide noise. This slow overall amplitude change is often called the *temporal envelope* of the waveform. The rate of envelope fluctuation increases as the bandwidth of the noise increases.

Many signals can be characterized by an envelope and the *fine-structure* waveform that falls under the envelope. The SAM signal mentioned earlier in this chapter has an envelope defined by the modulator frequency, and a fine structure defined by the carrier frequency. In fact, most complex waveforms can be described with the following formula:

$$x(t) = e(t) f(t), \qquad (4.5)$$

where $e(t)$ is the envelope function, $f(t)$ is the fine-structure waveform, and $x(t)$ is the complex waveform. For the hearing sciences, $e(t)$ is most useful when $e(t)$ changes much more slowly than $f(t)$. Note that equation (4.3) for a SAM tone is in the same form as equation (4.5). For a SAM tone, $e(t) = [1 + m \sin (2\pi F_m t)]$ and $f(t) = \sin (2\pi F_c t)$ (see equation (4.3)). The amplitude of other carrier stimuli, such as white noise, can also be amplitude modulated, as shown in Figure 4.17. In this case the modulator was a sinusoid, generating SAM noise, and,

$$x(t) = [1 + m \sin (2\pi F_m t)] \, n(t), \qquad (4.6)$$

where $n(t)$ is the noise waveform. Thus, the envelope $[1 + m \sin (2\pi F_m t)]$ changes more slowly than the fine structure of the noise $[n(t)]$.

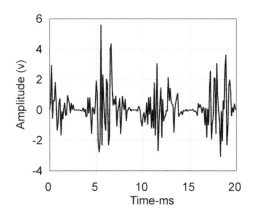

FIGURE 4.17 Sinusoidally amplitude modulated (SAM) white Gaussian noise (rate of modulation is 200 Hz, modulation period 5 msec).

SUMMARY

Complex stimuli are represented in the frequency domain by their amplitude and phase spectra. These spectra can be either line or continuous spectra. White Gaussian noise, narrowband noise (with a characteristic envelope), pink noise, rarefaction, and condensation clicks (transients), beating tones, amplitude-modulated signals, square waves, and frequency-modulated signals are a few of the complex stimuli used to study the auditory system. Sound spectrographs are used to display the spectral information of signals with frequencies and amplitudes that change over time.

SUPPLEMENT

The ability to define any complex stimulus depends on an understanding of Fourier analysis. Appendix C introduces the student to Fourier analysis. Chapter 2 in the book by Rosen and Howell (1991) explains many of these concepts and their applications. Any serious student of hearing with a background in math, physics, or engineering must read *Signals, Sounds, and Sensation* by Hartmann (1998).

A sinusoid turned on and off does not have a single-line spectrum but rather a continuous spectrum with the spectral shape described by the function $|(\sin x)/x|$. This is the function shown as the spectral envelope in Figure 4.3. As duration decreases, there is a *spread of energy* to other frequency regions, which is often audible.

The operations of AM and FM are well-known calculations because of their use in radio and other communications broadcasting. β is sometimes used to denote the ratio m/F_m in equation (4.4) for sinusoidally frequency-modulated tones.

The spectrum level (N_o) of noise has units of energy because it is noise power divided by frequency. Because frequency is the reciprocal of time, N_o is the same as power divided by one over time and will have units of energy.

The *normal equation* may be used to describe the probability of an instantaneous amplitude of Gaussian noise:

$$P(a_i) = 1/(\sigma \sqrt{(2\pi)}) \exp [(\frac{1}{2})(a_i/\sigma)^2],$$

where a_i is instantaneous amplitude, $P(a_i)$ is the probability of an instantaneous amplitude, and σ^2 is variance of the distributions of instantaneous amplitudes and is proportional to the average total (rms) power of the noise. Gaussian noise may be generated by adding sinusoids with frequencies spanning the bandwidth of the noise to be generated. To obtain true Gaussian noise, the amplitudes of the sinusoids should be sampled from a Rayleigh distribution of amplitudes and the starting phases for each sinusoid should be sampled from a rectangular distribution of starting phases. However, if more than 20 sinusoids are being added, then a close approximation to a Gaussian noise can be obtained by adding sinusoids with amplitudes all equal to N_o and starting phases sampled from a rectangular distribution of phases. The article by Hartmann (1987) describes these calculations.

In explaining the derivation of spectrum level (see Figure 4.14), a rectangle was used to approximate the shape of the noise power spectrum. However, in most situations the power spectrum of the noise does not cut off abruptly at the noise bandwidth, but rather noise power decreases gradually as frequency increases above or below the edges (cutoff frequencies) of the bandwidth. In these cases, the *equivalent rectangular bandwidth* (ERB) is sometimes used to define noise bandwidth. The ERB computation assumes that the noise does have a rectangular power spectrum as shown in Figure 4.14. The total power of the noise (which is easily measured) is divided by a direct estimate of spectrum level, with the resulting ERB being the bandwidth of a rectangle that has the total power of the noise and the spectrum level of the noise.

In Chapter 2, the term rms was defined. Its definition depends on determining the period of the sound (T in equation (2.4)). Some sounds like noise are not periodic, and it would be difficult for a measuring instrument to determine the period of a sound before calculating its rms amplitude. Thus, rms measuring meters usually compute the rms over a long period of time like one second. While one second may not be near the period of a sound, most sounds have periods that are much shorter than 1 second, so the error is extremely small.

The derivation of the envelope and fine structure and other aspects of narrowband stimuli are found in the book by Hartmann (1998).

CHAPTER 5

Sound Analysis

RESONATORS

In Chapter 2, the concept of free vibration was described as a damped sinusoidal vibration that occurred once an object is set into motion. In Chapter 3, in discussing standing waves, a continuous driving force was necessary to establish standing wave motion. The vibratory system in this second case is called *forced vibration* because the vibrating system is being forced to vibrate by some external object (for example, a hand vibrating a string). Two vibratory properties are involved: the vibration of the driving object and that of the object being vibrated. Most real-world acoustic situations are accurately described in terms of forced vibrations.

In the simplest case, the object being vibrated has its own natural vibratory frequency, as described by its free vibration properties. The closer the frequency of the driving force is to the natural frequency of the receiving object, the easier it is for the driving force to vibrate the receiving object. The natural vibratory frequency of the receiving object is called its *resonant frequency*.

Chapter 3 stated that, if two objects differed in their characteristic impedance, total transfer of vibratory intensity between the two objects would not occur. However, transfer of vibration from one object to another object is as large as it can be when the driving force has a frequency at or near the resonant frequency of the receiving object. A maximal amount of amplitude transfer will occur when the frequency of the driving object is equal to the resonant frequency of the receiving object. If the driving frequency is less than or greater than the resonant frequency of the receiving object, the receiving or resonating object (called a *resonator*) will vibrate less than its maximal possible amplitude. In Figure 5.1, a resonator with a 200-Hz resonant frequency is driven by driving an object at three different driving frequencies. Notice that the maximal response occurs for the resonating object when the driving frequency is 200 Hz (equal to the resonant frequency of the resonator).

The reactance components (both mass and spring reactance) of the characteristic impedance of simple objects determine their resonant frequencies. In general,

$$f_r = \sqrt{(s/m)}/(2\pi), \qquad (5.1)$$

where f_r is the resonant frequency, s is a measure of stiffness, and m is a measure of mass.

The resistance component of impedance also reduces the motion. The amount of resistance determines the sharpness of the peak in

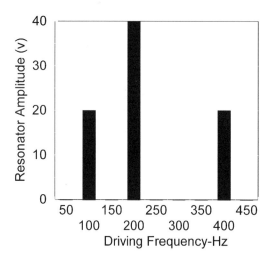

FIGURE 5.1 Relative amplitudes of a resonator driven by vibrators with differing driving frequencies. The 200-Hz driving frequency provides the greatest amplitude of the resonator, most likely indicating that the resonance frequency is 200 Hz.

the resonant function (Figure 5.1). That is, with little resistance the peak is very sharp, and thus it is difficult for driving frequencies not close to the resonant frequency to drive the resonator. The greater the friction, the broader the resonant peak and the easier for drivers with different frequencies to cause significant vibrations in the resonator.

If a driving force creates standing waves within a closed tube (see Chapter 3), the standing wave will oscillate at the tube's resonant frequency. It is possible to drive a tube with a vibration of small amplitude and produce within the tube a vibration with a large amplitude at the tube's resonant frequency. This point will become significant when we consider sound waves entering the ear canal. Does the ear canal resonate and increase the sound amplitude coming in? If so, what are the resonance frequencies of the ear canal and other parts of the auditory system?

FILTERS

Resonators may be especially designed to determine or to modify the frequency of vibrating objects; such resonators are called *filters*. We refer to those frequencies to which the filter vibrates as the ones the filter will *pass*. If a filter passes sinusoids with frequencies between 200 and 400 Hz, and if the filter is driven with these sinusoids, the amplitudes of these sinusoids at the output of the filter will be relatively unaffected by the filter. The amplitudes of sinusoids of other frequencies will be attenuated (reduced) as a result of the filter. Thus, if the filter passes sinusoids with frequencies between 200 and 400 Hz with little or no attenuation, and if the driving force has a frequency of 200 to 400 Hz, these sinusoids will have amplitudes at the output of the filter that equal their driving-force amplitudes. If an object with a driving frequency of 100 Hz and a level of 80 dB SPL drives a 200- to 400-Hz filter, then at the output of the filter the 100-Hz sinusoid *might* have its level reduced to 20 dB SPL. Thus, a filter produces sinusoids with no reduction in amplitude or with attenuated amplitudes.

There are four types of filters: *low pass, high pass, bandpass,* and *band reject*. A low-pass filter will pass all sinusoids with frequencies below a particular value; sinusoids with frequencies above that value will have their amplitudes attenuated. A high-pass filter passes sinusoids with frequencies above a particular value. A bandpass filter passes all sinusoids with frequencies between two particular values. The frequency values above, below, or between which the filter passes the sinusoids without reducing their amplitudes are called the *cutoff frequencies* of the filter. A band-reject filter attenuates the amplitudes of all frequencies between its cutoff frequencies (i.e., a band-reject filter is the opposite of a bandpass filter). Thus, a low-pass filter with a cutoff of 2000 Hz will pass all sinusoids with frequencies below 2000 Hz and will attenuate the amplitudes of all sinusoids with frequencies above 2000 Hz. A 2000-Hz cutoff high-pass filter will do just the opposite. A bandpass filter with cutoffs at 1000 and 4000 Hz will pass only sinusoids with frequencies between 1000 and 4000 Hz, while a band-reject filter with these same cutoff frequencies does the opposite, that is, it will attenuate the amplitudes of all sinusoids with frequencies between 1000 and 4000 Hz and will pass all sinusoids with frequencies less than 1000 Hz and greater than 4000 Hz.

Figure 5.2 shows the four types of filters (for the bandpass filter notice the similarity to the resonance curve shown in Figure 5.1). The amount of attenuation beyond the cutoff frequency is expressed in decibels of attenuation per *octave of frequency*. An octave indicates a doubling of frequency. The octave is expressed relative to the cutoff frequency of the filter. The first octave of 200 Hz is 400 Hz, the second octave is another doubling, or 800 Hz, the third octave is 1600 Hz, and so on. Table 5.1 shows the harmonics and octaves of a 1000-Hz sinusoid. Octaves should not be con-

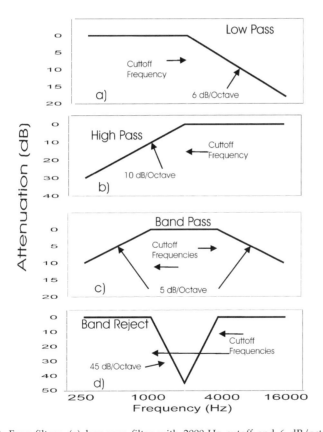

FIGURE 5.2 Four filters. (**a**) low-pass filter with 2000-Hz cutoff and 6 dB/octave roll-off; (**b**) high-pass filter with 2000-Hz cutoff and 10 dB/octave roll-off; (**c**) bandpass filter with 1000- and 4000-Hz cutoffs and 5 dB/octave roll-off. (**c**) band-reject filter with 1000- and 4000-Hz cutoffs and 45 dB/octave roll-off. The amount the filter attenuates the sound is shown as a function of frequency, which is plotted on a log axis. The frequency axis is a log axis.

TABLE 5.1 Harmonic and Octave Relations of a
1000-Hz Tone

Tone	Harmonics	Octaves
Fundamental frequency	1000 Hz	1000 Hz
2nd harmonic & 1st octave	2000 Hz	2000 Hz
3rd harmonic		3000 Hz
4th harmonic & 2nd octave	4000 Hz	4000 Hz
5th harmonic	5000 Hz	
6th harmonic	6000 Hz	
7th harmonic	7000 Hz	
8th harmonic & 3rd octave	8000 Hz	8000 Hz

fused with harmonics; octaves express a ratio
scale, with the ratio being 2 to 1, and harmonics represent integer multiples. If a low-pass
filter has a 2000-Hz cutoff and an attenuation
rate of 6 dB per octave, then the filter passes
all sinusoids with frequencies below 2000 Hz,
and the amplitudes of sinusoids with frequencies above 2000 Hz are attenuated at the rate
of 6 dB for each doubling of the cutoff frequency, 2000 Hz. Thus, the amplitude of a
4000-Hz sinusoid would be attenuated 6 dB
below its amplitude at the input, an 8000-Hz
sinusoid is attenuated 12 dB, a 16,000-Hz
sinusoid is attenuated 18 dB, and so forth. In
Figure 5.2a the *roll-off*, or rate of attenuation,
for the various filters is 6 dB per octave, in
Figure 5.2b it is 10 dB per octave, in
Figure 5.2c it is 5 dB per octave, and in Figure 5.2d it is 45
dB per octave.

Filters may be used to determine the frequency spectra of unknown signals. That is,
filters can be used to perform a type of spectrum analysis of a complex waveform. Chapter 4 described computing the time-domain
waveform from the amplitude and phase
spectra, but not how to determine the spectra
from the time-domain waveform. To estimate
the spectral components of a waveform we
can establish a series of bandpass filters with,

for instance, the following cutoff frequencies
(see Figure 5.3): 95–105 Hz, 195–205 Hz, 295–
305 Hz, and 395–405 Hz. If we found, after
analyzing an input waveform with these filters, that the amplitude at the output of the
first (95–105 Hz) and third filters (295–305 Hz)
was greater than that for the other two filters,
then the input waveform must have a greater
amplitude in the 100- and 300-Hz regions of
the spectrum than in the 200- and 400-Hz regions. The actual amplitude at the output of
each filter can be used to estimate an amplitude spectrum for the unknown input waveform. In this case, we plot the amplitude at the
output of each filter as a function of the center
frequency of each bandpass filter, and this

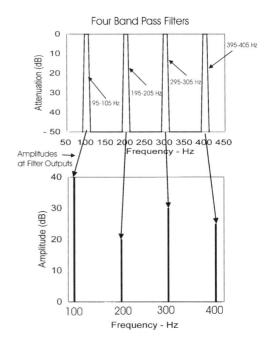

FIGURE 5.3 A stimulus with an unknown amplitude
spectrum is analyzed by four bandpass filters: 95–105,
195–205, 295–305, and 395–405 Hz. The amplitudes of the
outputs of these filters can be used to form an estimate of
the sound's amplitude spectrum.

plot forms an estimate of the waveform's amplitude spectrum.

Another use of filters involves "shaping" the spectra of signals. Imagine that a complex waveform with the amplitude spectrum shown Figure 5.4a is passed through the filter shown in Figure 5.4b. The complex wave will have at the output of the filter the amplitude spectrum shown in Figure 5.4c. Because the filter is a bandpass filter, the amplitudes of the sinusoids within the band of frequencies passed by the filter (the *passband*) have not been attenuated, whereas those outside the passband have been attenuated by 12 dB per

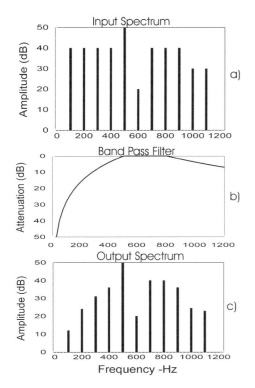

FIGURE 5.4 An input stimulus with the amplitude spectrum shown in (**a**) is passed through the filter described in (**b**); the output stimulus has the amplitude spectrum shown in (**c**). The output spectrum is obtained by subtracting the attenuations (dB) shown in **b** from the input amplitudes in decibels (**a**).

octave. If the amplitude spectrum and attenuation of the filter are expressed in decibels, then the amplitudes in decibels in the output spectrum are obtained by subtracting the attenuation values introduced by the filter from the amplitudes in the input spectrum. Thus, a filter has been used to alter the amplitude spectrum of a complex waveform.

Filters also introduce a phase delay to signals that pass through them. To describe a filter fully, we must indicate not only the type of filter, the cutoff frequencies, and the roll-off rate, but also how the phase of each sinusoid is altered. Although the phase shift introduced by a filter is not as easy to describe as the attenuation rate, both of these values must be known if the waveform at the output of the filter is to be determined.

As described in Chapter 4, if the amplitude or the phase spectrum of a complex wave is modified, then the time waveform representing the complex wave will also be altered. Because filters modify spectra, they will also alter time waveforms. Thus, a filter will change the time-domain representation as well as the spectrum of a signal. An example is shown in Figure 5.5.

Any object that has mass and is moved by a driving force can act as a filter. The cone of a loudspeaker, fluids, and the eardrum all act as filters. When the driving force pushes on the loudspeaker cone, the cone moves outward. When the driving force draws in, the speaker cone does likewise. When the frequency of the driving force is low, the speaker cone can follow the driving vibration with no loss in amplitude. As the driving frequency increases, the loudspeaker cone cannot follow the driving force, and the amplitude of the speaker cone movement will be reduced. This happens because the cone has mass, and mass requires time to move. When the driving force has a high frequency, the speaker cone responds to the outward push after a small time delay due to the force attempting to move some mass. Thus, by the time the speaker cone begins to move outward, the driving force is moving inward, so the speaker cone also tries to move inward. As a result, the cone does not move as far outward as it did at lower frequencies, creating a lower amplitude at this high frequency. The speaker cone thus acts as a low-pass filter. The eardrum and other structures within the ear have mass. They can react like low-pass filters. These effects will be discussed in greater detail in the next few chapters.

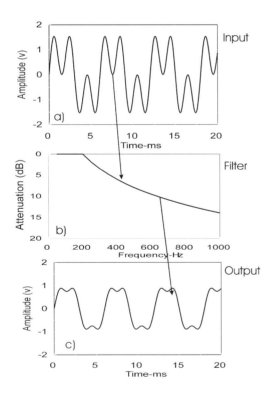

FIGURE 5.5 A complex sound (**a**) is passed through a low-pass filter with a 200-Hz cutoff and a 6-dB-per-octave roll-off (**b**). As can be seen, the complex time-domain waveform at the filter's output is different from that at the input because of the attenuation of the level of sinusoids with frequencies greater than 200 Hz (**c**).

NONLINEARITIES

So far we have discussed situations in which the sound *input* to some object, such as a filter, has been modified by an increase or decrease in the magnitude of the input value. If a waveform consisting of 100, 200, and 300 Hz is the input to a filter, the filter will alter only the amplitudes and phases of these sinusoids. There are, however, passive objects and systems that not only *modify* the values of the input but also *add* sinusoids at other frequencies. For instance, a complex input waveform to a system might have frequency components of 100, 200, and 300 Hz, but at the output of the system the measured frequency components might consist of the original 100-, 200-, and 300-Hz components and perhaps the additional frequencies of 400, 500, and 600 Hz. That is, sinusoids are present in the output of this system that are not present in the input. A passive system that adds sinusoids to the input waveform is called a *nonlinear system*, whereas a system that changes only the amplitudes and phases of the input signal is referred to as a *linear system*. For instructive purposes, the interaction of a nonlinear system and waveforms will be discussed in terms of two waveforms: one, a simple waveform consisting of only one frequency $f1$, and the other, a complex waveform consisting of the sum of two sinusoids with frequencies $f1$ and $f2$. If one sinusoid (with $f1$) is an input to a nonlinear device, then the output contains sinusoids with $f1$ and its higher *harmonics*: $2f1$, $3f1$, $4f1$, and so on. Thus, if $f1$ is 1000 Hz, the nonlinear device would yield the following nonlinear frequency components: 2000, 3000, 4000 Hz, and so on.

If the two sinusoids with frequencies $f1$ and $f2$ are inputs, then the nonlinear outputs consist of frequency components at the harmonics of each frequency—$2f1$, $2f2$, $3f1$, $3f2$, and so on—as well as frequency components consisting of the combination of $f1$ and $f2$. These frequencies, called *combination tones*, consist of both *summation tones* and *difference tones*. Typical summation tones are $f1 + f2$, $2f1 + f2$, and $f1 + 2f2$, and typical difference tones are $f1 - f2$, $2f1 - f2$, $2f2 - f1$. If the input frequency components were 100 and 250 Hz, the nonlinear combination tones could be 350, 450, and 600 Hz (summation tones), and 150, 50, and 500 Hz (difference tones). Thus, the total nonlinear output contains such harmonics and combination tones as 350, 200, 150, 50, 450 Hz, and so on. In general, a nonlinear device pro-

duces harmonics and combination tones according to this equation:

$$mf_1 \pm nf_2, \qquad (5.2)$$

where $m = 0, 1, 2, 3, \ldots$ and $n = 0, 1, 2, 3, \ldots$

Table 5.2 displays combination tones for sinusoids with input frequencies of 100 and 250 Hz. The amplitudes and phases of the nonlinear components present in the output depend on the frequencies, amplitudes, and phases of the input *and* the nature or type of nonlinearity.

TABLE 5.2 Some (for $n = m = 3$) Combination (Summation and Difference) Tones and Harmonics Resulting from Two Input Frequencies, 100 Hz (f_1) and 250 Hz (f_2)

Symbol	Combination tones		
	Harmonic tones	Summation tones	Difference tones
f_1	100 Hz		
f_2	250 Hz		
$2f_1$	200 Hz		
$3f_1$	300 Hz		
$2f_2$	500 Hz		
$3f_2$	750 Hz		
$f_1 + f_2$		350 Hz	
$f_1 + 2f_2$		600 Hz	
$f_1 + 3f_2$		850 Hz	
$2f_1 + f_2$		450 Hz	
$3f_1 + f_2$		550 Hz	
$2f_1 + 2f_2$		700 Hz	
$2f_1 + 3f_2$		950 Hz	
$3f_1 + 2f_2$		800 Hz	
$f_1 - f_2$			150 Hz
$f_1 - 2f_2$			400 Hz
$f_1 - 3f_2$			650 Hz
$2f_1 - f_2$			50 Hz
$3f_1 - f_2$			50 Hz
$2f_1 - 2f_2$			300 Hz
$2f_1 - 3f_2$			550 Hz
$3f_1 - 2f_2$			200 Hz

So far, a nonlinear device has been discussed in terms of changes in the frequency domain. *Time-domain* waveforms are also changed by nonlinear devices, as shown in Figure 5.6. A simple linear device is one that changes only the amplitude or the phase of the input. The change in the time-domain representation resulting from a nonlinear device is sometimes called *distortion*. Thus, the output time waveform of a nonlinear device is distorted and differs from the input time waveform. A linear system, such as a filter (a

complex linear system), also modifies the input time-domain waveform as shown in Figure 5.5. Therefore, we cannot simply investigate the difference between the input time-domain waveform and the output time-domain waveform to decide whether a system is nonlinear; we must investigate changes in the frequency domain. If there are frequencies present in the output spectrum that are not present in the input spectrum, then a passive system is nonlinear. One question of great interest to hearing scientists is whether the auditory system is linear or nonlinear. If it is a nonlinear, what types of nonlinearities are produced?

A linear system is one in which the output of the system (*out*) is a *linear function* of the input (*in*), that is, *out* = a(*in*) + b, where a and b are constants, and *in* and *out* are time-domain descriptions. An equation of the form

$$y = ax + b$$

is a linear equation and is defined by a *straight-line* relationship between x(*in*) and y(*out*). A nonlinear relationship can be defined as:

$$out = a(in) + b(in)^2 + c(in)^3 + d(in)^4 + \text{etc.}$$

This equation represents a nonlinear or *curvilinear relationship* between *in* and *out*. The presence of the terms with a power greater than one (e.g., *in*^{2}) produces harmonics and combinations tones (see Appendix A).

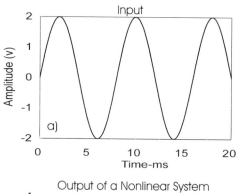

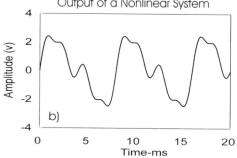

FIGURE 5.6 A nonlinear system changes (*distorts*) a time-domain waveform. A simple sinusoid (**a**) becomes a complex sound (**b**) when a nonlinear system adds harmonics. Figure 5.6b is a complex wave because the nonlinearity added tones of other frequencies to the original sinusoid. Note that both a filter (Figure 5.5) and a nonlinear system (Figure 5.6b) change a time-domain waveform.

SOUND AND ITS ANALYSIS

Sounds can be described by the frequencies, amplitudes, and phases of sinusoidal components. Thus, to describe hearing we need to describe how the auditory system determines the frequencies, amplitudes, and starting phases of the various sinusoids that make up any sound. We will use the analogy of bandpass filters (similar to Figure 5.3) to describe how the auditory system determines the "neural" spectrum of a sound and, from such a neural spectrum, how the system estimates frequency and amplitude.

Our everyday hearing experience reveals that we often perceive sound in terms of its frequency-domain representation. For instance, when you listen to music, you often hear the many pitches that make up any one musical sound. The perception of these pitches means you are sensitive to the frequencies that make up the complex musical sound. As such, your auditory system has analyzed the musical sound in terms of the frequency domain.

Thus, we will show that the auditory system determines the frequencies and amplitudes of sound with processes that are analogous to bandpass filters. As indicated above, the many structures of the auditory system alter sound and its analysis (often in a nonlinear manner). We will use the physical concepts of resonance, filtering, and nonlinearity to describe these alterations.

SUMMARY

In a forced vibration system, the receiving or resonant object has a resonant frequency that depends on its characteristic impedance. Filters are resonators that can be used to determine the amplitude spectrum of an unknown waveform. There are four types of filters (low pass, high pass, bandpass, and band reject), and each is characterized by its cutoff frequency and the attenuation roll-off in decibels per octave. A linear system changes only the phase and amplitude of an input waveform, whereas a nonlinear system adds sinusoids to the input. These non-linear additional sinusoids are harmonics and combination tones (summation and difference tones) of the input sinusoids. Although filters and nonlinear devices are usually described in terms of altering the frequency domain of waveforms, both types of systems also alter the time domain of the waveform. Resonance, filtering, and nonlinearity are concepts that will be used when we describe how the auditory system processes sound.

SUPPLEMENT

Filtering and nonlinearity are *independent* aspects of any system that analyzes sound. These concepts are described in any introductory physics or electrical engineering textbook. The book by Rosen and Howell (1991, chapters 6–8) presents the material at a slightly more advanced level. The more advanced reader should consult the book by Hartmann (1998).

The cutoff frequency of a filter is usually defined as that frequency at which the power of the signal has been attenuated by half its maximal power. Because decreasing the power by one-half yields 3 dB of attenuation, the cutoff frequency is determined at that frequency for which the intensity of the sound has been decreased by 3 dB from that existing in the *passband* of the filter. This definition of the cutoff frequency is referred to as the half-power or 3-dB cutoff. The bandwidth of a bandpass filter is usually computed as the frequency difference between the half-power cutoff frequencies. When comparing the bandwidths from one filter to another, a term called the Q of the filter is used:

$$Q = \frac{\text{(center frequency of the filter)}}{\text{(bandwidth of the filter)}}.$$

Although the Q of a filter can be determined at the 3-dB cutoff frequencies, it can also be determined at other cutoff frequencies, for instance, at the 10-dB cutoff frequencies. The bandwidth is referred to as Q_{10} in this case.

Attenuation in decibels ($Atten_{dB}$) for any particular frequency (f), cutoff frequency (cut), and roll-off rate ($roll$) is expressed as follows:

$$Atten_{dB} = roll \times \log_2(cut/f)$$

for $f < cut$, and

$$Atten_{dB} = roll \times \log_2 (f/cut)$$

for $f > cut$, and $\log_2$ is the logarithm to the base 2 ($\log_2 = 3.3219 \times \log_{10}$).

Filters are used to modify stimuli that are presented to listeners as well as to describe the type of processing that might take place within the auditory system. The phase shifts introduced by the filters mean that certain frequencies relative to other frequencies may be delayed in reaching the output of the filter. For instance, if a filter introduces a 180° phase shift to a 1000-Hz tone, then a sound passing through that filter will have its 1000-Hz component delayed by 0.5 msec (a 1000-Hz tone has a 1 msec period and 180° is one-half of that period, or 0.5 msec) relative to the time delay introduced to other frequency components.

A more general description of a linear system involves the relationship between the input to and the output from the system. The change the stimulus undergoes while "passing through" the system can be described by some function $F[x(t)]$. If $x(t)$ is the input, then the output, $y(t)$, can be expressed as

$$y(t) = F[x(t)].$$

For the system to be linear, the function F must meet certain requirements. The major property is *superposition*. If the function F meets the concept of superposition, then the following equation would hold true for a situation involving more than one input: $x1(t)$, $x2(t)$, $x3(t)$, and so on:

$$y(t) = F[x1(t) + x2(t) + x3(t) + \text{etc.}];$$

then if superposition holds:

$$y(t) = F[x1(t)] + F[x2(t)] + F[x3(t)] + \text{etc.}$$

The equations state that the function F can be distributed across the various inputs treated separately. This, in turn, means that, once we know the function F, we can always predict how the system will act when a new, but unknown, input is presented. Notice, for instance, that superposition does not allow for more output functions to occur than there are input functions. This was the primary concept we used in defining linear and nonlinear systems. A linear system produces the same frequency at the output as in the input, whereas a nonlinear system produces sinusoids with different frequencies at the output than were in the input.

PART II

PERIPHERAL AUDITORY ANATOMY AND PHYSIOLOGY

CHAPTER 6

The Outer and Middle Ears

The previous chapters have discussed the nature, transmission, and analysis of sound. Chapters 6 through 9 will investigate how acoustic energy is collected by the outer ear and transmitted to the fluids of the inner ear, how the inner ear transforms this energy into neural impulses, and how these neural impulses code for or analyze the basic acoustic information dealing with amplitude, frequency, and time/phase.

Figure 6.1 summarizes the structure, mode of operation, and function of four gross divisions of the auditory system: *outer ear*, *middle ear*, *inner ear*, and *central auditory nervous system*. The central auditory system includes all of the complex interconnections in the auditory system from the auditory branch of the *eighth cranial* (VIIIth) *nerve* to and including the *auditory cortex*. (The complex central auditory nervous system is not diagrammed in Figure 6.1; see Chapter 15). This chapter deals with the outer and middle ears and Chapters 7–9 with the inner ear.

Acoustic pressure is transmitted to the fluids of the inner ear via the outer and middle ears in a variety of ways. The outer and middle ears help overcome middle and inner ear impedances (Chapter 3) and thus allow for very efficient transmission of the acoustic stimulus to the inner ear. The outer and middle ears also provide protection to the inner ear against excessive changes in the environment.

Appendices E and F will assist the reader in understanding the general areas of anatomy (the structure of the body and the interrelation of its parts) and physiology (the function of an anatomical system).

STRUCTURE OF THE OUTER EAR

The changing acoustic pressures that constantly impinge upon us are collected by the outer ear. The outer ear consists of the visible part of the ear (*pinna*) and a canal (*external auditory canal*) leading to the eardrum. The human pinna (illustrated in Figure 6.1) is formed primarily of cartilage without useful muscles, and it has many small superficial bumps and grooves. The deep center portion of the pinna is called the bowl, or *concha* (cave). The concha has a diameter of 1 to 2 cm (1 inch = 2.54 cm = 25.4 mm) and leads to an opening with a diameter of about 5 to 7 mm. This opening (or *meatus*) is called the *external auditory meatus*; it opens into a canal 2 to 3 cm in length called the external auditory canal.

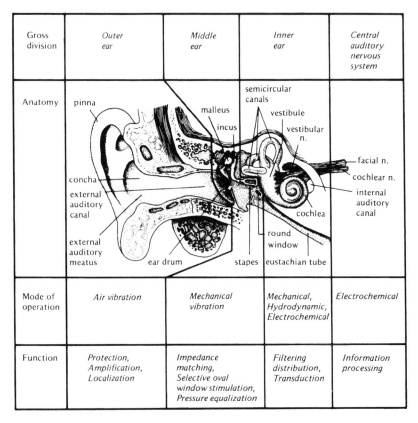

Gross division	Outer ear	Middle ear	Inner ear	Central auditory nervous system
Anatomy				
Mode of operation	Air vibration	Mechanical vibration	Mechanical, Hydrodynamic, Electrochemical	Electrochemical
Function	Protection, Amplification, Localization	Impedance matching, Selective oval window stimulation, Pressure equalization	Filtering distribution, Transduction	Information processing

FIGURE 6.1 Cross-section of human ear, showing divisions into outer, middle, and inner ears and central nervous system. Below are listed the predominant modes of operation of each division and its suggested function. Adapted with permission from Ades and Engstrom (1974) and Dallos (1973).

The lateral (toward the pinna) third of the canal consists of cartilage containing glands and lined with hairs; the rest of the canal is bony, with a tight skin lining close to the *eardrum* or *tympanic membrane*.

STRUCTURE OF THE MIDDLE EAR

The tympanic membrane is held in place by fibers and cartilage situated in a bony groove.

The tympanic membrane denotes anatomically one boundary of the large cavity known as the *middle ear cavity* or *tympanum*. The main portion of the middle ear cavity lies between the tympanic membrane and a bony wall (*promontory*), which forms the outer boundary of the inner ear. The middle ear cavity is about 2 cubic centimeters (cm^3) in volume and includes a smaller upper cavity called the *epitympanum* or *epitympanic recess*. The middle ear cavity is connected to the *nasopharynx* (nose cavity) by a long (35 to 38 mm) tube known as the *eustachian tube*.

The tympanic membrane is a cone-shaped, relatively transparent membrane, 55 to 90 mm^2 in area. It is constructed of layers of tissue, of which the central fibrous layers are structurally the most important. The tympanic membrane consists of two sets of fibers: one set radiates from the center to the outside of the membrane, and the other set is composed of rings of fibers. The fibers are very sparse in the upper (superior) portions (toward the epitympanum) of the tympanic membrane called the *pars flaccida* (Figure 6.2). The region of maximum concavity of the tympanic membrane is the *umbo* (Latin for the knob on a warrior's shield). The tympanic membrane is attached to the three middle ear bones or *ossicles*; first to the *manubrium* (handle) of the *malleus*, the outermost (lateral) of the ossicle. The head (upper portion) of the malleus is the rounded part that occupies half the epitympanic recess. The head of the malleus is connected to the next ossicle, the *incus*. The ossicles and their relation to one another can be seen in Figure 6.3. The incus lays medial (toward the center of the head) to the malleus, and the head of the malleus and the body of the incus are connected at a double saddle joint occupying most of the epitym-

panic recess. The *inferior process* (also called the *long process*) of the incus projects downward and then bends toward the inner ear to form the *lenticular process*, which joins the third ossicle, the *stapes*.

The stapes, which is the smallest bone in the body, consists of the head (or *capitulum*), two bony struts called *crura* (singular *crus*), and a flat oval bone called the *footplate*. The footplate is implanted in the *oval window* (part of the inner ear) and is attached to its head by the crura. The stapes's footplate is surrounded at its rim by a ring-shaped ligament situated along the edge of the oval window. This ligament, called an *annular ligament*, assists in supporting the footplate in the oval window. Thus, the tympanic membrane is directly joined or "coupled" to the inner ear by the three ossicles. The three ossicles are suspended in the tympanic cavity by means of ligaments. Those ligaments most important for suspension are the *axial ligaments*. The axial ligaments consist of the *posterior ligaments* of the incus and the *anterior ligaments* of the malleus, as shown in Figure 6.4. Also shown in Figure 6.4 are the middle ear muscles. One of the middle ear muscles is the *tensor tympani muscle*, which is about 25 mm in length. Most of this muscle is enclosed within a bony canal that runs parallel to and above the eustachian tube. A tendon emerges from the bony canal to connect the muscle to the upper part of the manubrium of the malleus. The other middle ear muscle, the *stapedial muscle*, also originates within a bony canal. Only about 6 mm in total length, it is the smallest muscle in the body. The tendon of the stapedial muscle completes the attachment to the head of the stapes.

Thus, the inner ear communicates with the acoustic environment outside of the body by means of a funnel (the pinna), a short tube (the external auditory canal), a thin membrane (the tympanic membrane), and three small bones. We will now consider the functions served by the pinna, the external canal of the

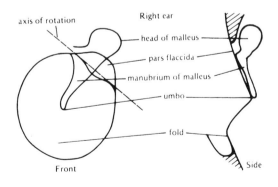

FIGURE 6.2 Front and side views of a right tympanic membrane and its connection to the malleus.

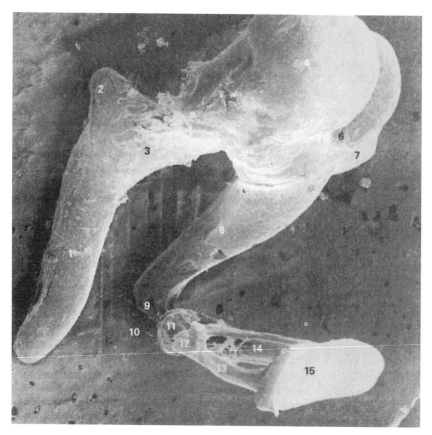

FIGURE 6.3 Middle ear ossicles. (**a**, above): from the front (anterior aspect); (**b**, below): from within the inner ear (medial aspect). Parts of the *malleus*: 1 = manubrium; 2 = anterior process; 3 = neck; 4 = head. Parts of the *incus*: 6 = body; 7 = short process; 8 = long process; 9 = lenticular process. Parts of the *stapes*: 11 = head; 12 = neck; 13 = anterior crus; 14 = posterior crus; 15 = footplate. Also, 5 = incudomalleal articulation; 10 = incudostapedial articulation. Photographs courtesy of Dr. Ivan Hunter-Duvar, Hospital for Sick Children, Toronto

outer ear, the tympanic membrane, and the ossicle of the middle ear.

FUNCTION OF THE OUTER EAR

As sound travels from its source to the outer ear, it passes over the torso and head (including the pinna). These parts of the body provide obstacles to the sound, and thus change the sound before it reaches the outer ear. One way to describe the changes to sound transmission that take place is to measure the changes in the amplitude and phases of the spectral components of the sound due to the influences of the various parts of the body and the outer and middle ears. That is, the structures of the torso and head attenuate and slow

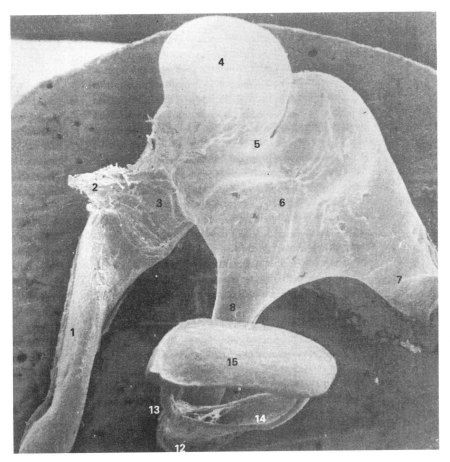

FIGURE 6.3 Continued.

sound in a frequency-dependent manner (due to the interaction of structure size and wavelength, see Chapter 3) as sound travels from its source to the outer ear. To measure these spectral changes, the spectrum of the sound source (input spectrum) is first determined, and then the spectrum at the outer ear (output spectrum) is measured. The difference between the input and output spectra describes how the structures of the torso and head alter the amplitudes and phases of the sinusoidal components that make up the input stimulus. Sound pressure measurements in the outer ear can be accomplished by inserting either very small microphones into the outer-ear canal or by placing a tube into the outer-ear canal and measuring the sound pressure in the tube. The combination of the amplitude and phase spectra that describe sound pressure changes due to the intervening structures is called a *transfer function* (e.g., the attenuation and phase shifts provided by a filter describe the filter's transfer function). This process is sometimes called *real-ear measurement*.

Figure 6.5a describes the amplitude spectra of the outer-ear transfer function for one hu-

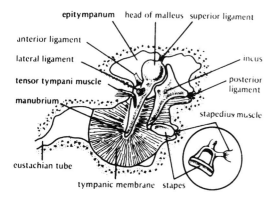

FIGURE 6.4 Schematic drawing of the human middle ear (right side) seen from within: medial view. Adapted with permission from Moller (1970).

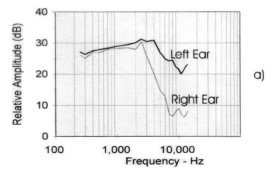

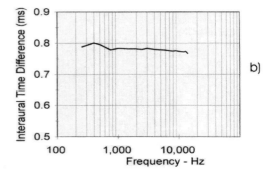

FIGURE 6.5 The HRTFs for a human when presented a wideband noise source directly opposite the left ear. Panel **a** shows the amplitude spectra at each ear (measured within the ear canal) in an arbitrary decibel scale. The sound at the right ear is attenuated relative to that at the left, especially at high frequencies. Panel **b** shows the interaural time difference between the sound arriving at the two ears. The interaural time difference remains approximately constant across frequency. Adapted with permission from Wightman and Kistler (1989a).

man listener. Figure 6.5b describes the difference in the time between the sound arriving at the right and left ears of a listener (*interaural time difference*) when the sound is placed directly opposite the left ear. The interaural time differences are derived from the phase spectra of the transfer function for each ear. These transfer functions describing changes between the source and outer ear are called *Head-Related Transfer Functions* (HRTFs). Thus, the HRTF describes how the torso and head change the amplitudes (attenuate the amplitudes of the spectral components of the originating sound) and phases (add phase shifts to those of the spectral components of the originating sound) of a sound as it travels from a source toward the outer ear. The HRTF has important consequences for sound localization (see Chapter 12).

Many of the changes in the HRTFs at high frequencies are due to the pinnae. Only mammals have pinnae, but among mammals there is great diversity in their form. In general, only animals with relatively high-frequency hearing have mobile pinnae. Because the pinnae of humans and other primates have no

useful muscles, they are relatively immobile. Mobile, and to some extent immobile, pinnae help in localizing high-frequency sounds by funneling them toward the external canal, in distinguishing noises originating in front of the head from those behind the head, and in providing other types of filtering of an incoming sound wave.

In addition to the spectral changes such as those shown in Figure 6.5, the outer ear causes an increase in level of about 10 to 15 dB in a frequency range from roughly 1.5 to 7 kHz (kHz means 1000 Hz; thus, 7 kHz is 7000 Hz). Experiments have shown that this frequency-dependent increase in sound pressure level between the free field measurement and the measurement at the tympanic membrane is due mainly to the effects of the concha and the external auditory canal, as shown in the transfer functions of Figure 6.6. In Chapter 5, we discussed the concept of resonance. The resonant frequency of the external auditory canal is about 2.5 kHz, and the resonance frequency of the concha is closer to 5 kHz. The resonance of the concha and external canal complement each other to produce a gain in acoustic pressure for the frequency components (in the range from 1.5 to 7 kHz) of the originating sound.

The only other known function of the external ear is that of protection of the tympanic membrane against foreign bodies and changes in humidity and temperature. Be-

cause it completely seals off the external auditory canal, the tympanic membrane provides some protection for the middle ear against foreign bodies.

The tympanic membrane vibrates as a result of sound waves traveling in the external auditory canal, and this vibration is passed along to the ossicular chain. The vibratory pattern of the tympanic membrane has been the subject of much research since Helmholtz published his first experiments in 1868. Figure 6.7 shows that the membrane vibrates maximally at a point below the umbo and directly about the fold (see Figure 6.7, point 15). Modern investigative methods show the complicated vibratory patterns illustrated on the bottom of Figure 6.7. It is generally agreed that the vibratory pattern of the tympanic membrane is the most complicated at higher frequencies and higher levels. The complicated vibratory response illustrated by time-averaged holography shows variations with frequency of the stimulus. At the low frequency used in Figure 6.7b, the point of maximum displacement is in the superior posterior (top rear, the 8.2 contour) section of the membrane. The vibratory pattern of the tympanic membrane, although complicated, allows for an efficient transfer of the acoustic stimulus from the outer ear to the middle ear and the middle ear structures.

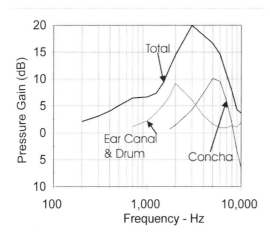

FIGURE 6.6 Estimated change in sound pressure level from free field to site of measurement—concha, ear canal, and ear drum—and the total transfer function from the free field to the tympanic membrane (total). The total curve is essentially the sum of the other two curves. Adapted with permission from Shaw (1974).

FUNCTION OF THE MIDDLE EAR

Once the acoustic stimulus reaches the tympanic membrane, it can be transmitted through the middle ear to the inner ear via three methods: (1) bone conduction (i.e., sound could travel via the bones of the skull, bypassing the middle ear, going directly to the

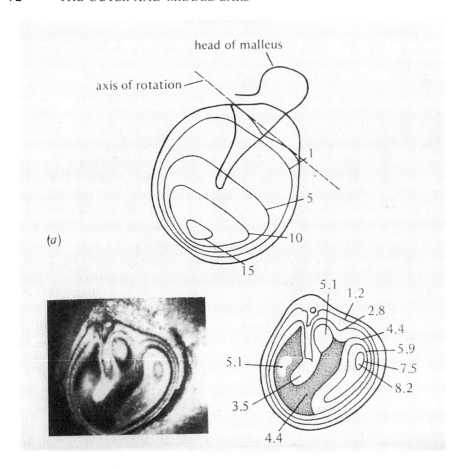

FIGURE 6.7 (a) The vibratory pattern of the right tympanic membrane to a 2-kHz tone. The closed curves represent contours of equal displacement amplitude on a relative scale; 15 is the maximum amplitude. This is von Bekesy's (1941) stiff plate model. (b) Tonndorf and Khanna's 1972 time-averaged holograph of the left tympanic membrane vibration. The stimulus was a 525-Hz 121-dB SPL tone. In the explanatory drawing, each isoamplitude contour must be multiplied by 10^{-5} cm to obtain the actual displacement. Thus, the maximum displacement of 8.2 is actually 8.2×10^{-5} cm, the minimum displacement is 1.2×10^{-5} cm, etc. Adapted with permission from Tonndorf and Khanna (1972).

inner ear), (2) the air in the middle ear cavity, and (3) across the middle ear cavity by means of the ossicular chain (the malleus, incus, and stapes) to the inner ear. The ossicular chain, which vibrates in response to tympanic membrane vibration and passes this vibration onto the fluids and structures of the inner ear, is the

most effective method of transmitting sound to the inner ear.

For the ossicular chain to vibrate efficiently, the middle ear must not be a closed cavity. If the middle ear cavity were completely closed, then changes in atmospheric pressure would cause air pressure to build up or be reduced in the middle ear without any release. Thus, changes in air pressure that occur frequently in elevators, airplanes, or under water would cause the tympanic membrane to move more in one direction than in the other. If the tympanic membrane is already stretched due to unequal pressure between the middle and outer ears, then pressure changes caused by a sound wave will not be as successful in vibrating the tympanic membrane. The eustachian tube allows for equalization of pressure differences across the tympanic membrane by providing another path for the pressure, via the nasal passages. That is, the eustachian tube allows the tympanic membrane to operate efficiently in a variety of atmospheric pressures since the pressure on its outside (in the outer ear connected to outside pressures) is the same as that on the inside (in the middle ear connected to outside pressures via the eustachian tube and nasal passages). When the eustachian tube or nasal cavity is blocked, sound transmission is not very efficient, and one can experience a hearing loss and perhaps pain (due to stretching of the tympanic membrane and the tissues around it).

The ossicular chain vibrates a membrane (*oval window membrane*) of the inner ear, which causes the fluids of the inner ear to move. If this oval window membrane were pushed or driven by air, then air pressure would be the driving force. Because the inner ear fluid is denser than air, air is not very efficient in moving the fluid (i.e., the fluid has a higher characteristic impedance than air). Thus, the auditory system would lose some of its sensitivity due to the impedance (Chapter 3) of the fluids and structures of the inner ear. The difference is referred to as an impedance mismatch, meaning that more pressure is required for a stimulus to be propagated in the inner ear fluids than in air. Nature has compensated for this mismatch, principally by the size difference between the areas of the tympanic mem-

brane and the stapes footplate and, to a more limited extent, by the lever action of the ossicular chain.

When stimulated by high sound pressure levels, the tympanic membrane operates as a stretched membrane, and only part of the pressures acting on it is transferred to the manubrium of the malleus. Measurements show that only two-thirds of a total area of 85 mm^2 of the tympanic membrane is stiffly connected to the manubrium, and thus vibrates at high levels. Therefore, the effective surface area of the tympanic membrane is approximately 55 mm^2. The stapes, which is the last part of the middle ear chain, makes contact with the oval window and the fluids of the inner ear. The area of the stapes footplate is about 3.2 mm^2, which is considerably less than the effective area of the tympanic membrane. This difference in surface area between the effective area of the tympanic membrane and the stapes footplate acts to increase the pressure exerted on the tympanic membrane. Consider that, if all of the force that impinges on the tympanic membrane is transferred to the stapes footplate, then the force per unit of area ($p = F/A$) must be greater at the footplate because it is smaller than the tympanic membrane. The increase in pressure can be expressed as the ratio of the effective area of the tympanic membrane to that of the stapes footplate—that is, 55 mm^2/3.2 mm^2 = 17. Thus, the pressure at the footplate is 17 (i.e., 25 dB greater) times greater than at the tympanic membrane, and, therefore, the air pressure can stimulate the fluid-filled inner ear.

The transfer of the force from the tympanic membrane to the stapes footplate depends on the action of the ossicles. The ossicles work as a lever system because the length of the manubrium and neck of the malleus are longer than the long process of the incus (see Figure 6.3). The lever action of this system is 1.3 to 1, that is, the force at the tympanic membrane is increased by a factor of 1.3 at the stapes. In addition, the tympanic membrane tends to buckle as it moves, due to its conical shape, causing the malleus to move with about twice the force. Thus, the pressure increase of a factor of 17 due to the area difference of the tympanic membrane and stapes footplate is multiplied by a factor of 1.3 due to the lever action, and then the buckling action causes an additional multiplication by 2. These calculations result in a theoretical maximum total pressure increase of $17 \times 1.3 \times 2$, or 44.2 (a pressure increase of 44 to 1 corresponds to 33 dB) at the stapes footplate. The actual pressure transformation depends on the frequency of the acoustic stimulus, since the structures of the middle and inner ear offer both resistive and reactive impedance (see Chapter 3). Figure 6.8 shows that the increase in pressure between the eardrum and inner ear is 30 dB or more in the region of 2500 Hz. Above this frequency, the ratio decreases. In addition, as we showed earlier, the resonances of the concha and external canal are in the 2–5-kHz range. These various pressure gains cause a significant pressure increase across a wide frequency range, which is important for the perception of sounds like speech. Nature has designed an ingenious system to match the impedance of air to that of the inner ear fluids. The losses and gains in amplitude provided by the outer and middle ears help define the sensitivity of the auditory system (see Chapter 10 for further discussion).

Previously we described the suspension system of the ossicles and stated that there were several ligaments and muscles that performed this function. The muscles are normally in a state of tension, but when the ear is excited by sound they exert an increased pull, which is a reflex action. The sound that elicits the reflex must be about 80 dB SL. Contraction of the middle ear muscles reduces the transmission of pressure through the ossicular

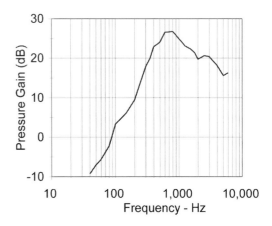

FIGURE 6.8 Transfer function for the middle ear showing that the pressure in the fluids of the inner ear is increased over that at the tympanic membrane by the decibel value shown as a function of the frequency of the stimulus. Adapted with permission from Nedzelnitsky (1980).

chain. The reduction in transmission of sound due to middle ear muscle contraction has a maximum value of about 0.6 to 0.7 dB per decibel increase in stimulus level above the threshold of the reflex (approximately 80 dB SL). The reduction amounts to approximately 10 to 30 dB for loud sounds and is frequency dependent, having more effect for low frequencies below 2 kHz. The time for the reflex action to occur is a minimum of 10 msec for high-intensity sounds and can be as long as 150 msec for relatively low-intensity sounds. Therefore, short-duration sounds of sudden onset are not greatly attenuated or reduced by the middle ear reflex. The muscles can provide a type of protection because they can reduce the amount of fatigue and damage that might exist in the inner ear from exposure to high-intensity, low-frequency, steady sound.

Another possible function of the middle ear muscles is to limit distortion (nonlinearities). If the contact pressure between two vibrating bodies is smaller than the driving pressure acting during the vibration, the two bodies (such as the ossicular bones) will separate, and as a result of the separation a distortion of the transmitted signal could easily occur at the joint. In the middle ear when the vibrations are small, the elastic ligaments can exert a pressure sufficiently great that the bones do not separate. For larger vibrations, however, the pressure of the ligaments may not be sufficient, and the muscles work against one another to press the stapes against the incus, thus limiting the separation. Anatomically, it is also interesting that each of the middle ear muscles is enclosed throughout its length in a long narrow canal. This permits the muscles to produce only a pull and not to be set into vibration by the sound pressure in the middle ear. If the muscles did vibrate due to acoustic stimulation, they would produce harmonics (nonlinearities) that could be translated to the fluids of the inner ear and could become audible.

The ossicular chain does not vibrate in a simple manner for all frequencies and levels. Due to the size of the heads of the malleus and incus, the mass of the ossicle is distributed evenly around an axis through the two large ligaments of the malleus and the short process of the incus. At moderate intensities, the ossicular chain moves so that the footplate of the stapes swings about an imaginary axis drawn vertically through the posterior crus much like a swinging door pivoting about its hinges. The anterior portion of the footplate pushes into and out of the cochlea like a piston. This pivotal action is possible because of the asymmetric fiber length of the annular ligament. At very low frequencies (below 150 Hz) and at extremely high intensities, this rotation of the footplate is thought to change dramatically. Under these conditions, the axis of rotation is through the crura, becoming perpendicular to the previous vertical axis. The motion of the stapes becomes a rocking motion around the axis much like that of a seesaw, as seen in Figure 6.9. At very high levels the resultant displacement of the inner ear fluids is zero. Because of this complex form of rotation, further increases in level results in very little motion of the fluids of the inner ear. This complex motion is thought to help protect the inner ear from being overstimulated at very high levels not often encountered in nature.

Despite the claims that the middle ear reflexes and stapes footplate motion may provide protection, deafness caused by exposure to manmade environmental noises (see Chapter 16), such as some industrial noises, might be positive evidence that the ear does not really have an adequate protective mechanism against our present levels of acoustic stimulation.

MIDDLE EAR IMPEDANCES

The membranes, fluids, bones, muscles, and ligaments of the middle ear all contribute to its impedance, and the differences in the impedance of various tissues and fluids lead to most of the pressure changes described in this chapter. Remember from Chapter 3 that impedance is composed of resistance and reactance. The main resistive component of the middle ear is the resistance of the inner ear fluid against the motion of the stapedial footplate. Friction caused by the moving parts of the middle ear also contributes to the resistance, but to a lesser degree. This resistive component of the impedance does not vary with the frequency of stimulation as the reactive components do. Two types of reactance are involved in middle ear impedance. One is the mass reactance that is due to the mass of the middle ear structures. This mass reactance affects high frequencies more than low fre-

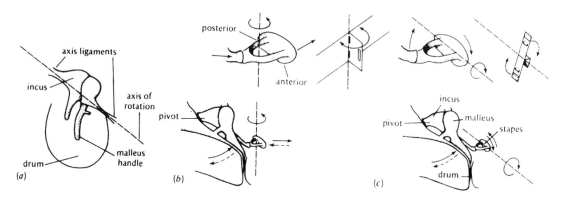

FIGURE 6.9 Motions of the middle ear ossicles. (**a**) The motion of the middle ear ossicles is a rocking motion about an axis drawn through the axial ligaments and the short process of the incus. Since the large heads of the malleus and incus offset the rest of the mass of the ossicular chain, this axis is also the center of gravity. (**b**) The normal stapes motion is like that of a swinging door with the posterior section as the hinge and the anterior section pushing in and out of the oval window like a plunger. (**c**) At very high levels, the stapes is thought to rock around an axis drawn through the crura. This seesaw motion affects the inner ear fluid only slightly. Adapted with permission from von Bekesy (1936).

quencies, so the impedance due to mass reactance increases with increasing frequency of stimulation. The other reactance of the middle ear is the springlike property of the ligaments and muscles. When we are referring to the ease with which a spring may be stretched, we speak of its *compliance*; but if we are speaking of the difficulty with which we can stretch a spring, we refer to its *stiffness*. Thus, stiffness and compliance are reciprocally related. The springlike properties affect low frequencies of stimulation (below 1 kHz), and the springlike reactance dominates impedance at low frequencies. In the middle frequencies (about 1–2 kHz), these two reactive components (mass and springlike) partially cancel each other, and the resistive component dominates the middle ear impedance.

One tool used to measure the impedance of the middle ear is the *electroacoustic impedance bridge*. The outer ear cavity must be sealed for this measurement. A very small loudspeaker presents a low-frequency tone (usually 220 Hz) into this sealed outer ear cavity. A small microphone (probe microphone) is used to record the sound pressure created by the small stimulating loudspeaker in the sealed outer ear cavity. If the middle ear is stiff, the pressure in the cavity will be higher than if the middle ear is compliant (that sound would be absorbed at the tympanic membrane for a compliant middle ear). Thus, the sound pressure recorded by the probe microphone can be used to measure the springlike qualities of the middle ear.

Often the middle ear impedance measured with the electroacoustic bridge is expressed in terms of compliance. This is because the impedance is usually measured at low frequency (220 Hz), and most often only the springlike quality of the middle ear is being measured. The higher the compliance, the lower the impedance and the easier the system works. The electroacoustic bridge provides a convenient method for measuring abnormal function of the tympanic membrane and middle ear.

When the middle ear muscles contract (e.g., due to a loud sound), the middle ear becomes less compliant (stiffer). This increased stiffness due to middle ear muscle contraction, or the *acoustic reflex*, as it is often called, can be measured by the electroacoustic bridge. Measuring the acoustic reflex is useful for measuring the action of the middle ear muscles. Because the acoustic reflex is a neural reflex involving the inner ear, the auditory nerve, and parts of the central nervous system, measuring the acoustic reflex can help in determining the function of these parts of the auditory system.

From the point of view of evolution, it is not surprising that, as animals develop from living in water to living on land, a middle ear system evolved that matched the high impedance of the fluid-filled inner ear with airborne stimulation. However, what about those pathways to the inner ear other than the middle ear ossicle—that is, direct air conduction and bone conduction? If the middle ear ossicle were missing, the inner ear could be stimulated directly by air pressure variations in the middle ear cavity. Again, since the impedance difference between air and the structures and fluids of the inner ear are so great that the inner ear fluids are driven by negligible amounts by the direct influence of sound pressure changes in the middle ear.

There is another reason why air pressure alone is not an efficient way to vibrate the fluids and structures of the inner ear. We will see in the next chapter that the round and oval windows lead to the inner ear on opposite sides of the basilar membrane. Positive pressure on the round window will cause the basilar membrane to move in one direction, and positive pressure on the oval window will cause it to move in the opposite direction. Thus, with only air pressure changes in the middle ear, the oval and round windows would be stimulated simultaneously. Applying the same pressure to both windows simultaneously results in a very inefficient system because the basilar membrane would not move very much, and its movement is crucial for hearing. A hearing loss of 60 dB is often found in patients who have no ossicles unless the round window is shielded from air pressure variations. Thus, another function of the middle ear system is to direct the stimulation to the oval window, allowing the round window to move according to pressure vibrations within the inner ear.

The other way the inner ear can be stimulated is by bone conduction, in which acoustic pressure changes cause the bones of the head (primarily the temporal bone) to vibrate, and this vibration is passed on directly to the inner ear fluids. Because of the enormous difference in characteristic impedance between air and bone, stimulation of the inner ear by bone conduction does not occur as an important part of normal auditory function. If, however, a vibrator is applied to the skull, the inner ear can be stimulated (vibrated) via direct conduction of sound through the bones of the skull. The impedance difference between the skull and the fluids of the inner ear is small enough to allow for some sound transmission to the inner ear. Measuring hearing by vibrating the skull may be employed to practical advantage in the event of damage to the middle ear. That is, stimulation of the skull by a vibrator can be used during diagnosis to test the viability of the middle and inner ears (see Supplement to Chapter 10).

SUMMARY

The outer ear consists of the pinna and the external auditory canal and ends at the lateral border of the tympanic membrane. The three ossicles (malleus, incus, stapes) in the middle ear couple the tympanic membrane to the inner ear. Two muscles and several ligaments help support these ossicles. Transfer functions, such as the Head-Related Transfer Function, describe the spectral changes that take place as a sound travels from its source to the tympanic membrane. The resonance of the external auditory canal and the tympanic membrane, along with the lever action of the ossicular chain and the area difference between the tympanic membrane and oval window membrane, help increase the air pressure at the external auditory meatus so that the air pressure can drive the dense fluids of the inner ear. In some limited fashion, the muscles may help protect the auditory system. For higher levels, motion of the ossicular chain varies as the frequency and level of the input stimulus change. The structures of the middle ear provide sources of impedance for transmission of sound to the inner ear.

SUPPLEMENT

Chapter 2 in the book by Pickles (1988) covers many of the topics of Chapter 6. The early research on the function of the external ear was conducted by von Bekesy (1941, 1989/1960), Wiener (1947), and Wiener and Ross (1946). The vibratory pattern of the tympanic membrane and of the middle ear ossicles has been the subject of research ever since 1868, when Helmholtz published his work (see Warren and Warren, 1968). Tonndorf and Khanna suggested the complicated vibratory pattern shown in Figure 6.7b. Their results are consistent with those of Helmholtz and Kirikae. Those interested in the use of the middle ear impedance measurements in the clinic as a diagnostic tool should read Jerger and Northern's (1980) book, especially the chapter by Jerger and Hayes. The books by Geisler (1998) and Moller (2000) should also be consulted for a description of outer and middle ear function.

It is often stated that the middle ear muscles protect against high-intensity stimulation. The middle ear reflex has a threshold of 80 dB SL and a minimum latency of 10 msec. It attenuates primarily stimulation below 2 kHz and probably not by more than about 10 dB. Thus, it protects only against gradual onset low-frequency sounds. Von Bekesy (1989/1960) suggested that the middle ear muscles may reduce distortion by tightening the joint between the malleus and stapes at high levels. The middle ear muscles might also help reduce sounds produced by one's own speech and mouth movements, because these actions stimulate the muscles. Guinan and Peake's (1967) work suggests, however, that a stimulus of 150 dB SPL is needed before a rocking motion occurs. Therefore, we can conclude that, while the external and middle ears collect, amplify, and transmit acoustic information to the inner ear, the high incidences of deafness due to overstimulation of the inner ear make it evident that none of the suggested protective mechanisms can defend us completely against present levels of stimulation. Rosowski (1991) discusses the changes that take place in the middle ear when the ossicles are missing.

Summary of Human Outer and Middle Ear Measurements

Most of the data listed here come from Wever and Lawrence's *Physiological Acoustics* (1954). Their Appendix D is a compilation of many authors' works, from which we have chosen the data to illustrate the range of measures. When available, a mean value is given. Additional data from recent investigations have also been added.

Pinna (MALE)

length: 60–75 mm, mean = 67 mm
breadth: 30–39 mm, mean = 34.5 mm
angle that length axis is inclined to head: 15°
concha volume: 2.5 cm^3
concha resonance frequency: 4.5 kHz

External Auditory Meatus

cross-section: 0.3–0.5 cm^2

External Auditory Canal

cross-section: 0.3–0.5 cm^2
length: 2.3–2.97 cm
diameter: 0.7 cm
volume: 1.0 cm^3
resonance frequency: 2.6 kHz

Tympanic Membrane

diameter along the manubrium: 7.5–9 mm
diameter perpendicular to the manubrium:
    7.5–9 mm
area: 0.5–0.9 cm^2
effective area: 42.9–55 mm^2
inward displacement of umbo: 2 mm
thickness: 0.1 mm
weight: 14 mg
breaking strength: 0.4–3.0 × 10^6 dyne/cm^2,
    mean = 1.61 × 10^6 dyne/cm^2

displacement amplitude for low-frequency tones
    at threshold of feeling: 10^{-2} cm
displacement amplitude for a 250-Hz tone: 12.5 Å
    at 75 dB SPL, 7.5 Å at 70 dB SPL, 5 Å
    at 65 dB SPL (1 Å = 10^{-10} m, thus,
    1 µm = 10,000 Å)

Middle Ear Cavity

total volume: 2.0 cm^3
volume of ossicles: 0.5–0.8 cm^3

Malleus

weight: 23–32 mg
length from end of manubrium to end of lateral
    process: 5.8 mm
total length: 7.6–9.1 mm

Incus

weight: 25–32 mg
length of long process: 7.0 mm
length of short process: 5.0 mm

Stapes

weight: 2.05–4.34 mg, mean = 2.86 mg
height: 2.50–3.78 mm, mean = 3.26 mm
length of footplate: 2.64–3.36 mm, mean =
    2.99 mm
width of footplate: 1.08–1.66 mm, mean =
    1.41 mm
area of footplate: 2.65–3.75 mm^2, mean = 3.2 mm^2
width of elastic ligament: 0.015–0.1 mm
amplitude of displacement for a constant
    eardrum pressure of 1 dyne/cm^2:
    125 Hz: 75 × 10^{-8} cm
    200 Hz: 28 × 10^{-8} cm
    400 Hz: 20 × 10^{-8} cm
    750 Hz: 18 × 10^{-8} cm
    1500 Hz: 10 × 10^{-8} cm
    2000 Hz: 6 × 10^{-8} cm
    2500 Hz: 2 × 10^{-8} cm
maximum displacement: 0.1 mm

CHAPTER 7

Structure of the Inner Ear and Its Mechanical Response

*I*n Chapter 6 we considered the course of the acoustic stimulus as it traveled from the environment toward the inner ear. The present chapter describes the remaining anatomy of the inner ear and how this anatomy relates to the vibratory stimulation it receives from the stapes.

In general, the motion of the stapes moves the fluid and other structures of the inner ear. This motion causes the haircells of the inner ear to be stimulated and to elicit neural discharges in the auditory nerve. Thus, the mechanical energy of sound vibration is changed into neural information within the inner ear. The inner ear provides the nervous system with information about the frequency, intensity, and temporal content of acoustic stimulation. Part of the spectral analysis of sound is provided by the mechanics of the inner ear in a way that can be described as filtering.

STRUCTURE OF THE INNER EAR

The inner ear can be divided into three parts: the *semicircular canals*, the *vestibule*, and the *cochlea*, all of which are located in the *temporal bone*. Three semicircular canals open into the vestibule—*superior*, *posterior*, and *lateral*—and the *utricle* and *saccule*. Under normal circumstances, these structures affect the sense of balance (the *vestibular system*) rather than hearing. They are, however, part of the total system to which acoustic disturbances are delivered. The sensory receptor cells of the vestibular system, as in the auditory system, are *haircells*, and the two systems are often discussed together. However, we will not cover the structure and function of the vestibular system.

The vestibule is the central inner ear cavity (see Figure 7.1). It is about 5 mm front to back and top to bottom and about 3 mm in width. The vestibule is bounded on its lateral side by the *oval window*, which is located in its wall facing the *middle ear cavity* (*tympanic wall*). The footplate of the stapes connects to the oval window. The vestibule contains the utricle and the saccule, which are sense organs of the vestibular system.

The cochlea, a small shell-shaped part of the bony labyrinth, shown in Figure 7.2 and illustrated schematically in Figure 7.1, con-

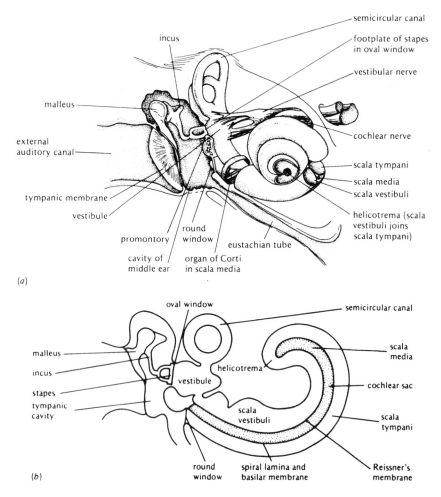

FIGURE 7.1 (a) Main components of the inner ear in relation to the other structures of the ear. Adapted with permission from Dorland (1965). (b) Schematic diagram of middle ear and partially uncoiled cochlea, showing the relationship of the various scalae. Adapted with permission from Zemlin (1981).

tains the primary auditory organ of the inner ear. It resembles a tube of decreasing diameter, which is coiled increasingly sharply upon itself, approximately 2⅝ times in humans. The cochlea terminates blindly in its third turn at the *apex*. Its central axis is called the *modiolus*, which acts as an inner wall. The spiral canal of the cochlea is about 35 mm long and is par-

tially divided throughout its length by a thin spiral shelf of bone, known as the *osseous spiral lamina*, projecting from the modiolus (Figures 7.2 and 7.3). Across this shelf, the basilar membrane connects to the outer wall of the bony cochlea at the *spiral ligament* and completes division of the canal into two passages, except for a small opening at the apex called the *heli-*

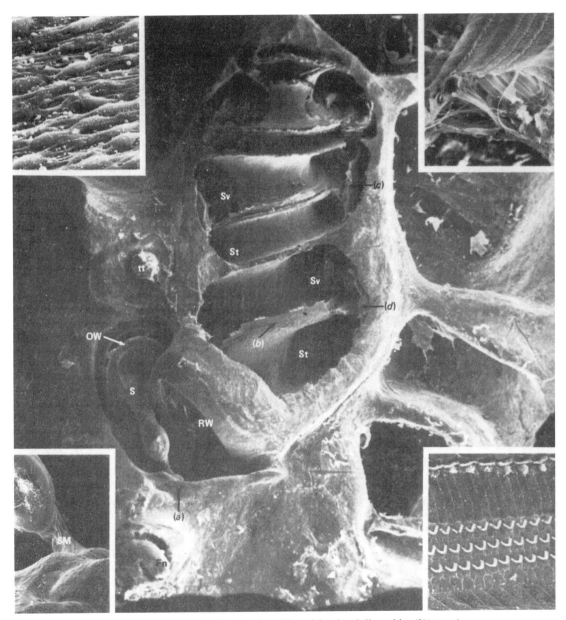

FIGURE 7.2 Scanning electron micrograph showing the coiling of the chinchilla cochlea (3¾ turns). Also shown: S = Stapes; OW = oval window; RW = round window. The inserts show details at the points indicated on the central photograph (a = lower left, b = upper left, c = upper right, and d = lower right): (**a**, lower left): SM = stapedial muscle; (**b**, upper left): tympanic layer below the basilar membrane; (**c**, upper right): cross-section of the organ of Corti. (**d**, lower right): cilia of the inner and outer hair cells. Sv = scala vestibuli; ST = scala tympani; tt = attachment of tensor tympani; Fn = facial nerve. Photograph courtesy of Dr. Ivan Hunter-Duvar, Hospital for Sick Children, Toronto.

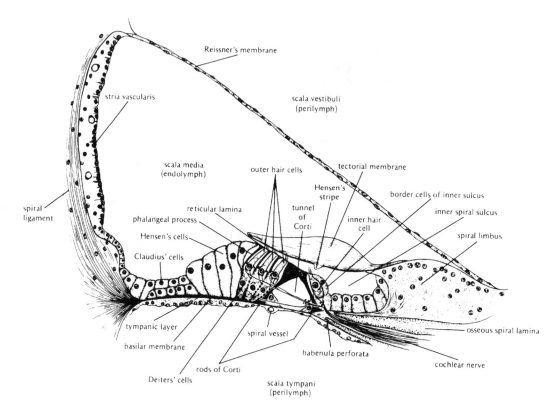

FIGURE 7.3 Drawing of a cross-section of the cochlea showing the organ of Corti situated in the scala media on the basilar membrane. Drawing by Sara Crenshaw McQueen, Henry Ford Hospital, Detroit.

cotrema. The lower passage of the canal (*scala tympani*) has an opening, known as the *round window*, which is covered with a thin membrane (*round window membrane*) separating the scala tympani from the tympanic cavity. A delicate membrane, *Reissner's membrane*, extends upward diagonally from the osseous spiral lamina to a region of the outer wall slightly above the *basilar membrane*; it extends the length of the cochlea to the apex, where it joins the basilar membrane at the helicotrema. Thus, there is a completely sealed sac within the middle of the cochlea, called the *cochlear sac* (or *cochlear duct*), which runs the length of

the cochlea except for a small portion at the apex. The cochlear sac is bounded "above" by Reissner's membrane, "below" by the basilar membrane, and on its one "side" by the stria vascularis; surrounded by a watery fluid called *perilymph*, it contains its own fluid called *endolymph*. The stapes pushes on the oval window, causing perilymph to move through the scala vestibuli around the helicotrema to the scala tympani and then to push on the round window.

Cross-sections of the cochlea, as in Figure 7.3, show the canal divided into three parts, ducts, or scala. These three ducts are the *scala vestibuli, scala tympani,* and *scala media.* The scala vestibuli extends from the oval window in the vestibule to the helicotrema. In cross-sections (Figure 7.3) of the cochlea, it is usually shown as the upper scala. The scala tympani extends from the round window to the helicotrema. In these sections, it is often shown at the bottom. The scala media is the central duct, which is bounded by the basilar membrane, a portion of the outer wall of the cochlea, and Reissner's membrane. The outer wall of the scala media is covered by the *stria vascularis,* which has a dense layer of blood capillaries and specialized cells. The stria vascularis provides the basic metabolic control of the cochlea. The scala vestibuli and scala tympani contain perilymph, and the scala media contains the endolymph. It is easy to remem-ber the names of the different scalae, since the scala vestibuli lies opposite the vestibule, the scala tympani communicates with the tympanic cavity via the round window, and the scala media lies in the middle. The two peripheral scalae, the scala tympani and scala vestibuli, contain perilymph, and the middle or enclosed scala, the scala media, contains endolymph.

Whereas the bony cochlea becomes smaller and smaller in cross-sectional area as the apex is approached, the basilar membrane lying within the cochlea becomes progressively wider as it approaches the apex, as shown in Figure 7.4. The osseous spiral lamina is broadest at the vestibular (basal) end, where the basilar membrane is about 0.16 mm wide; near the helicotrema the basilar membrane

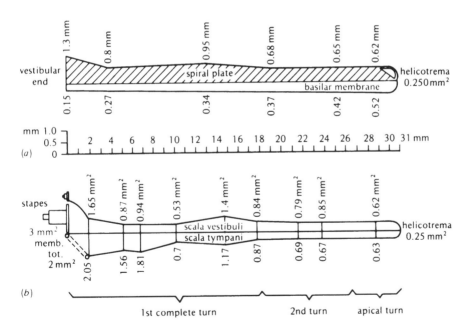

FIGURE 7.4 Schematic diagrams. (**a**) Dimensions of the basilar membrane. (**b**) Dimensions of the scalae of the human cochlea. The basilar membrane is wider at the apical end than at the basal end, but the scalae are smaller at the apex than at the base. Adapted with permission from Fletcher (1953); data are from Wrightson and Keith (1918). Additional inner-ear measurements are listed in the Supplement to this chapter.

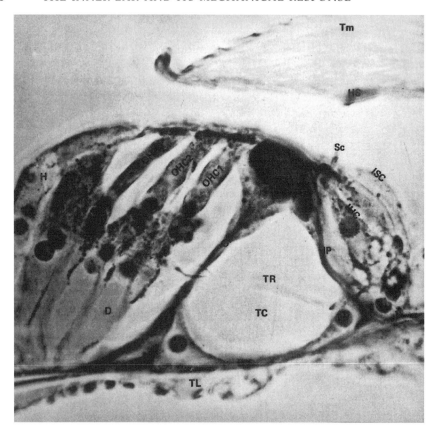

FIGURE 7.5 (a) Light micrograph of a cross-section of a chinchilla organ of Corti. Clearly shown are: IHC = inner hair cells; OHC = the three rows of outer hair cells; OP, IP = outer and inner pillars of Corti; TC = tunnel of Corti; BM = basilar membrane; TL = tympanic layer of cells below BM; D, H = supporting cells of Deiters and Hensen; TM = tectorial membrane; HS = Hensen's stripe; ISC = inner sulcus cells; TR = a tunnel radial nerve fiber. (b) Scanning electron micrograph of a cross-section of a chinchilla organ of Corti. Many of the same parts of the organ of Corti are shown, but with the advantage of added dimensionality. This view also shows the head of the IP cells between the stereocilia (Sc) of the inner and outer hair cells. The TM is pulled back to expose the stereocilia. Photographs courtesy of Dr. Ivan Hunter-Duvar, Hospital for Sick Children, Toronto.

has broadened to 0.52 mm. Measuring the various characteristics of the basilar membrane was of primary interest to von Bekesy, who later built a model of the membrane and described its complicated pattern of motion.

From von Bekesy's work it may be concluded that the human basilar membrane is a stout layer of closely attached fibers about 34 mm long from base to helicotrema. It is wider, more flaccid, and under no tension at the apical end. The base end is narrower and stiffer than the apical end and may be under a small amount of tension. These facts become important in considering the vibratory pattern of the membrane in response to acoustic stimulation.

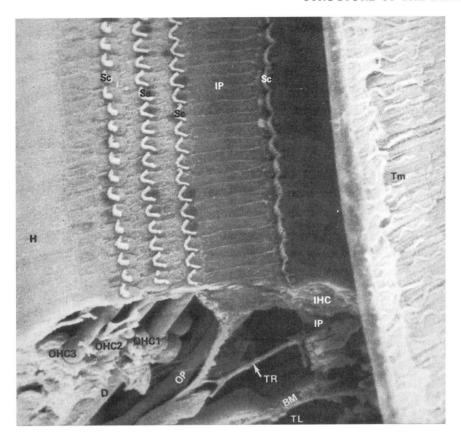

FIGURE 7.5 Continued.

On the scala media surface of the basilar membrane lies the *organ of Corti* (see Figures 7.3 and 7.5). The organ of Corti is divided into inner and outer portions by the *pillars* or *rods of Corti*. The rods of Corti form a tunnel that is almost triangular in cross-section and runs the length of the cochlear sac. The space between the rods is called the *inner tunnel of Corti*, which contains a fluid called *cortilymph*. The inner portion of the organ of Corti is on the modiolar side (inside) of the rods, and the outer portion lies toward the bony cochlear wall (toward the stria vascularis).

On the inner side of the inner rods is a single row of haircells, called the *inner hair-cells*, and on the outer side of the outer rods are three or four rows of smaller haircells, called the *outer haircells*, with various supporting cells. The inner and outer haircells slant toward each other and are held in place in three ways. First, cells of *Deiters* and *Hensen* serve as lateral buttresses for the outer haircells, and the *Border Cells* of the inner sulcus serve the same purpose for the inner haircells. Second, a *reticular membrane* (or *lamina*) holds the upper ends of the haircells and assists in maintaining their alignment. Third, the Deiters cells are found below the haircells, with one Deiters cell per haircell. The upper end of the cylindrical Deiters cell is a cup-shaped structure that encloses the base of the haircell. From the cupped end of the cylindrically shaped Deiters cell, a slender process passes to the surface of the organ of Corti, forming a *phalangeal process* and a part of the reticular membrane (see Figures 7.5 and 7.6). These Deiters cells and phalangeal processes form the main vertical supportive mechanism for the haircells.

Figure 7.7 shows the detailed structure of the inner and outer haircells and the difference in their shape. The upper surface of each

inner haircell consists of a membrane with a cuticle into which the bases of the *stereocilia* (cilia, or tiny hairs) are rooted. On each inner haircell are about 40 cilia, arranged in two or more parallel rows that form a very shallow "U" (Figures 7.8 and 7.9a). One small area of the surface of the haircell is free of cuticle, and in this area a basal body or modified *kinocilium* is usually found. A kinocilium is an extremely long cilium found in other haircell receptor systems. During embryonic life, the human cochlea haircells also have a long,

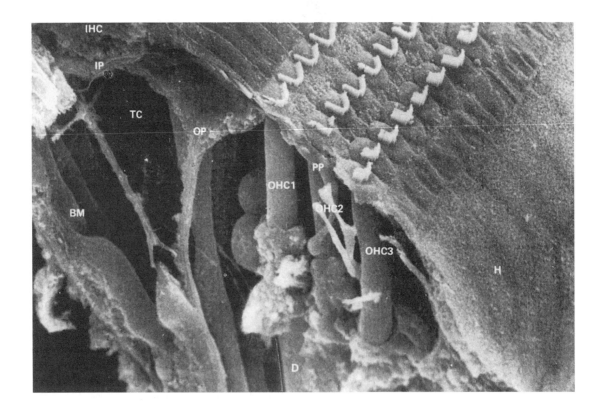

FIGURE 7.6 Scanning electron micrograph showing the supporting structure of the organ of Corti provided by Deiters cells (D) and their phalangeal processes (PP) in relation to the hair cells (OHC1, OHC 2, OHC3). The phalangeal processes arise from the top of the Deiters cells and form part of the reticular lamina at the top of the hair cells. The stereocilia are above the reticular lamina; the body of each hair cell sits in the cup-shaped top of a Deiters cell. Chinchilla photograph courtesy of Dr. Ivan Hunter-Duvar, Hospital for Sick Children, Toronto.

coarse kinocilium. After birth, the kinocilium generally disappears and only a basal body remains, although a rudiment of the kinocilium may also remain. The basal body of the inner haircell is located on the outer edge of the cell, that is, away from the modiolus, toward the inner rods of Corti.

The upper surface of the outer haircell contains about 150 stereocilia arranged in three or more rows on each cell in the shape of a "V" or "W" (Figures 7.8 and 7.9b). The basal body lies at the bottom of the W, toward the stria vascularis. The tips of the tallest row of cilia of each outer haircell are in contact with a struc-

ture of colorless fibers, known as the *tectorial membrane* (Figure 7.10). The tectorial membrane is a soft, ribbon-like structure, attached along one edge to the spiral limbus and perhaps attached along the other edge to the outer border of the organ of Corti (near the supporting cells).

The stereocilia are arranged in an elegant geometrical form that provides a great deal of strength. Figure 7.11 shows the detailed structure of the stereocilia with the strength-enhancing *cross bridges* or *tip lengths*. These cross bridges help strengthen the stereocilia, and they play a crucial role in transduction of the

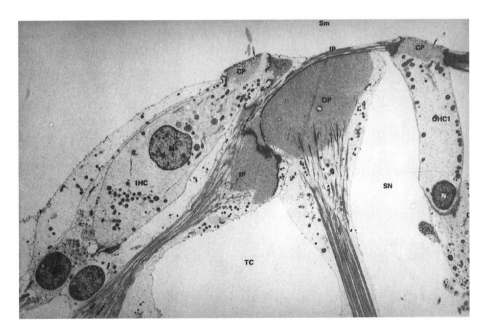

FIGURE 7.7 (a) Transmission electron micrograph showing the detailed structure of an inner hair cell (IHC) and an outer hair cell (OHC1). The stereocilia (Sc) of the hair cell project into the scala media (Sm) and are rooted in the cuticular plate (CP). The nucleus (N) of the inner hair cell is clearly seen. At the base of the hair cell, nerve fibers of the inner spiral bundle (IS) are visible. Also shown: inner and outer pillars (IP, OP) of the tunnel of Corti (TC). (b) Transmission electron micrograph of the outer hair cells (OHC). Note the difference in shape between IHC and OHC. The OHC is seen sitting in the cup-shaped Deiters cell (D). At the base of OHC an efferent nerve fiber (E) is also seen. The space between OHCs is the space of Nuel (SN); within SN parts of the phalangeal processes (PP) are seen. The tops of the PP form part of the reticular lamina and separate and hold in place the hair cells. Chinchilla photographs courtesy of Dr. Ivan Hunter-Duvar, Hospital for Sick Children, Toronto.

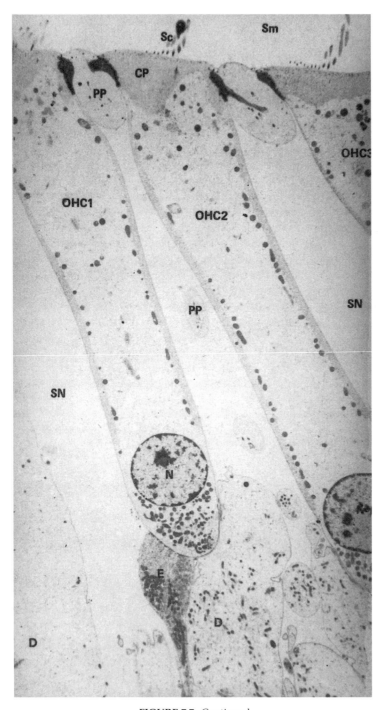

FIGURE 7.7 Continued.

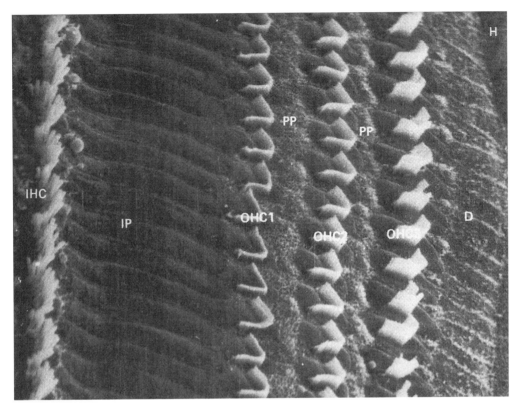

FIGURE 7.8 Scanning electron micrograph of the top of the organ of Corti, with the tectorial membrane removed to expose the stereocilia of the inner and outer hair cells. One row of inner hair cells and three rows of outer hair cells are clearly seen. Between the inner and outer hair cells, the heads of the inner pillar cells (IP) are seen. Between the tops of the rows of outer hair cells, the tops of the phalangeal processes (PP) are seen. The supporting cells of Deiters (D) and Hensen (H) are at the outer edge. Note that the "W" formation of the stereocilia of the outer hair cells is slanted in relation to the inner hair cells and is slightly different for each row. Chinchilla photograph courtesy of Dr. Ivan Hunter-Duvar, Hospital for Sick Children, Toronto.

movement of the stereocilia into the chain of chemical events that take place within the haircells that eventually leads to neural impulses in the auditory nerve (see Chapter 8).

MECHANICAL RESPONSE OF THE INNER EAR

The vibratory patterns representing the acoustic message reach the inner ear via the stapes. As we have seen, the stapes moves the oval window, which causes the fluid-filled system (cochlea) that contains the cochlear sac to vibrate. These vibratory undulations contain the information that must be coded into neural information. One key factor in this process is the mechanical response of the basilar membrane and the organ of Corti. Here we will refer to the basilar membrane instead of the whole cochlear sac, because its vibratory patterns are more easily illustrated.

FIGURE 7.9 (**a**, left) Scanning electron micrograph showing the stereocilia (Sc) of an inner pillar cell (IP) and some small cilia called microcilia (m). (**b**, right) Scanning electron micrograph showing the head of an inner hair cell (IHC) in detail. Also shown: the stereocilia (Sc) of an outer hair cell (OHC) in detail. In both photographs, the graduation in size of the stereocilia is clearly seen. Chinchilla photographs courtesy of Dr. Ivan Hunter-Duvar, Hospital for Sick Children, Toronto.

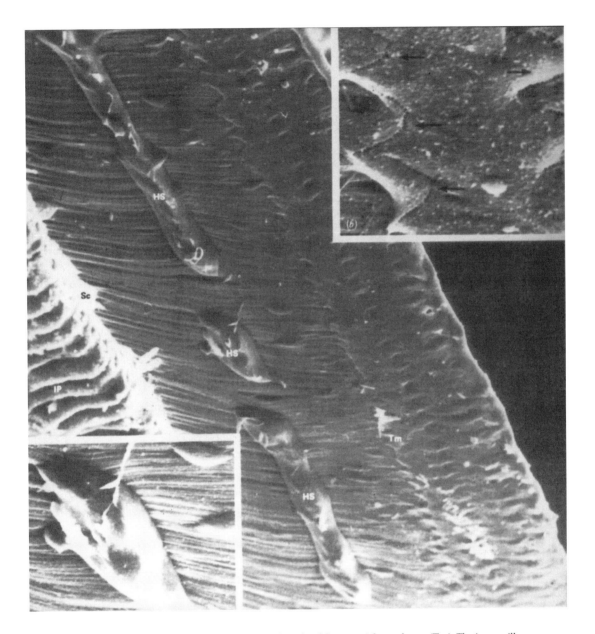

FIGURE 7.10 Scanning electron micrograph of the underside of the tectorial membrane (Tm). The inner pillar cells (IP) of the tunnel of Corti and the stereocilia (Sc) of the inner hair cells are shown at left for reference. The inserts, further magnified, show: (**a**) Hensen's stripe (HS) and (**b**) the "W" imprint of the tallest row of stereocilia for each outer hair cell. These imprints are evidence that the stereocilia of the outer hair cells are in contact with the tectorial membrane. There are no imprints in the tectorial membrane corresponding to the stereocilia of the inner hair cells. Chinchilla photograph courtesy of Dr. Ivan Hunter-Duvar, Hospital for Sick Children, Toronto.

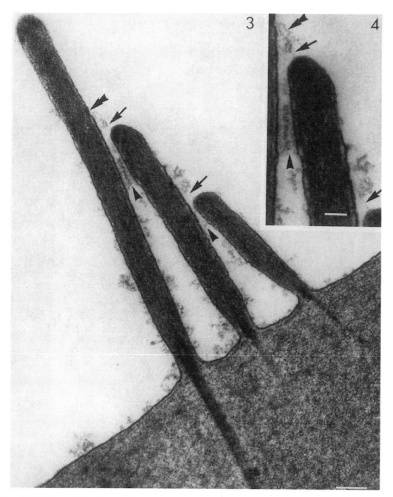

FIGURE 7.11 A high magnification of the stereocilia and the cross bridges, or tip links, shown at the arrows. The insert is a view of the tip links at higher magnification. Reprinted with permission from Osborne *et al.* (1988).

The general function of the cochlea, and hence of the basilar membrane, is to translate the mechanical vibrations of the stapes and the inner ear fluids into neural responses in the auditory branch of the VIIIth cranial nerve. The auditory nerve interacts with the haircells by way of synaptic junctions, and the haircells are attached to the basilar membrane by the supporting cells. Thus, vibration of the fluids causes the basilar membrane to move, which in turn causes the cilia of the haircells to bend. The bending of the cilia causes the nerve at the base of the haircell to initiate a neural potential, which is sent along the auditory nerve. Thus, the haircells, in connection with the basilar membrane, translate mechanical information into neural information.

Most of the early work that demonstrated the importance of the vibration of the basilar membrane was done by von Bekesy, using both direct observation of the cochlea and cochlear models. These early measures have been confirmed and refined by several modern methods (see Appendix F). Changing the elasticity of the round window or the length of the cochlear canals does not affect the vibratory pattern of the basilar membrane. Even altering the position of the stapes and changing the fluid does not affect the vibratory pattern of the basilar membrane. Thus, the vibratory pattern of the basilar membrane remains very stable under most conditions. This implies that the vibratory pattern must depend heavily on the characteristics of the basilar membrane itself—that is, the changes in elasticity and width discussed previously. The exact nature of the vibratory pattern of the entire cochlea (not just the basilar membrane) depends on the condition of the inner ear. Even moderate damage or alterations of the inner ear can cause significant abnormal changes in the biomechanics of the cochlea.

Each point along the basilar membrane that is set in motion vibrates at the same frequency as the stimulus. However, the amplitude of the membrane vibration is different at different locations, depending on the frequency and level of the input stimulus. In a sense, a wave motion is set up along the membrane as the fluids in the inner ear are driven by the stapes.

The basilar membrane becomes wider as the distance from the stapes to the helicotrema increases, and its flaccidity also increases toward the helicotrema; thus, the natural frequency of vibration of the basilar membrane decreases toward the helicotrema (i.e., larger, flexible membranes have lower resonant frequencies than smaller, stiffer ones). Because of the variation in width and stiffness, different frequencies will cause a maximum vibration amplitude at different points along the membrane. If two different frequencies are received by the cochlea simultaneously, they will each create maximum displacement at

different points along the basilar membrane. This separation of a complex signal into different maximal points of displacement along the basilar membrane, corresponding to the sinusoids of which the complex signal is composed, means that the basilar membrane is performing a type of spectral analysis (see Chapters 3 and 5, and Appendix C).

The response of the membrane is a *traveling wave*. The wave motion of the membrane appears to travel toward the helicotrema. The pressure of the sound is distributed immediately throughout the cochlea because the stapes is transmitting the pressure variations to a relatively incompressible fluid environment. An example of the vibratory pattern along the membrane that is "frozen" or stopped at an instant in time is called an *instantaneous basilar membrane pattern*.

Figure 7.12 shows a representation of two instantaneous patterns of the traveling wave along the basilar membrane. The bottom figure presents a more realistic representation than the usual display (top of Figure 7.12) because the basilar membrane is shown attached at its two edges and is displaced or bent in response to sound in a transverse (medial or crosswise) direction as well as in the longitudinal direction. Most instantaneous patterns of basilar membrane displacement are shown as if Figure 7.12a were viewed from the side, the basilar membrane being represented by a single line (see Figure 7.13). Keep in mind that the condition shown in the bottom of Figure 7.12b is more accurate. Figure 7.13 shows several instantaneous patterns in successive temporal order (1–4) for a sinusoidal input. That is, curve 1 shows the basilar membrane displacement for the first instant in time, curve 2 shows the next instant in time, and so on. These instantaneous patterns indicate that one part of the basilar membrane may be displaced toward the scala tympani, while at the same time an adjacent part may be deflected

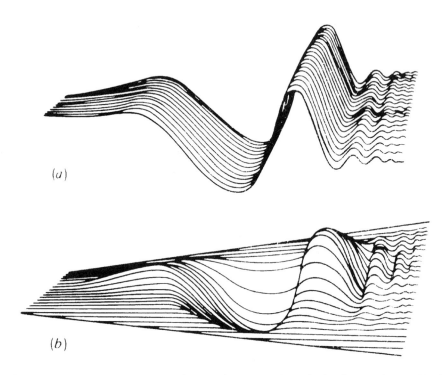

(a)

(b)

FIGURE 7.12 Instantaneous pattern of a traveling wave along the basilar membrane. (**a**) The pattern that would result if the membrane were ribbon-like. (**b**) The membrane vibration illustrated more realistically; since the basilar membrane is attached along both edges, it must vibrate in a radial or transverse direction, as in Figure 7.11b. Adapted with permission from Tonndorf (1960).

in the opposite direction. This is also obvious in Figure 7.12. This difference in direction of displacement can be viewed as a local phase difference between the vibratory pattern of two adjacent portions of the membrane. This local phase difference is greater in the apical portion (ahead of the peak of the displacement pattern) of the wave than in the basal portion.

If a line is drawn through the points of maximum displacement for a specific traveling wave, the resulting curve is the *envelope of the traveling wave*. The envelope of the maximum points of positive and negative displacement of the traveling wave of Figure 7.13 is illustrated as the solid line; many illustrations

of the envelope of the traveling wave show only the positive envelope. Figure 7.14 is a schematic representation of the cochlea and the envelope of the traveling wave that would occur for stimuli of three different frequencies. One instantaneous waveform is also shown for each frequency. Notice that the maximum point of the envelope is different for each frequency. For the lowest frequency (60 Hz) the maximum displacement of the envelope is near the apical end, for the highest frequency (2 kHz) the maximum displacement is near the base, while the intermediate frequency has its maximum between the other two. Figure 7.14 illustrates another important point—that is, lower frequency stimulation will stimulate

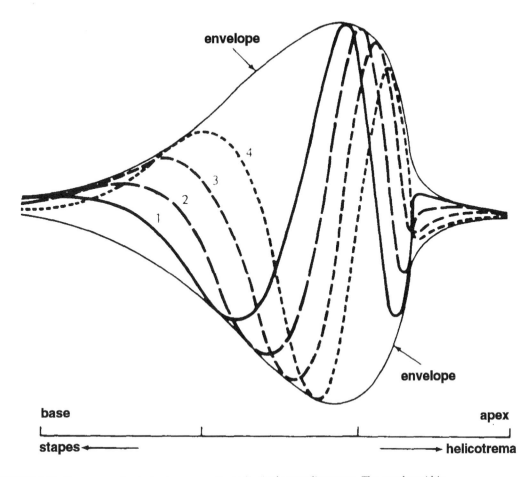

FIGURE 7.13 Four successive instantaneous patterns (1–4) of a traveling wave. The envelope (thin solid line) of the traveling wave formed by connecting all the points of maximum amplitude along the membrane is also shown. Adapted with permission from Ranke (1942).

not only the basal end of the membrane but also the apical end, where the point of maximum displacement occurs. Higher frequencies, however, stimulate only the basal end of the cochlea. As shown by the envelopes of the traveling waves created by high-frequency stimulation, the displacement apical to the point of maximum displacement is reduced rapidly. In contrast, the displacement basal to the point of maximum displacement is re-

duced slowly in a basal direction for all frequencies and extends completely to the base. We may conclude then that the traveling wave always travels from the base toward the apex. How far toward the apex it travels depends on the frequency of stimulation; lower frequencies travel farther. Figure 7.15 depicts traveling wave motion for a complex sound input consisting of two tones. The two peaks in the basilar membrane vibratory pattern are

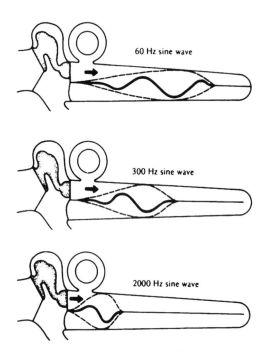

FIGURE 7.14 Instantaneous patterns and envelopes of traveling waves of three different frequencies shown on a schematic diagram of the cochlea. Note that the point of maximum displacement, as shown by the high point of the envelope, is near the apex for low frequencies and near the base for higher frequencies. Also note that low frequencies stimulate the apical end as well as the basal end, but that displacement from higher frequencies is confined to the base. Adapted with permission from Zemlin (1981).

caused by the two different stimulating frequencies.

If we measure the phase relationship between the vibration of the stapes and that at locations along the basilar membrane for different frequencies of stimulation, we can calculate the time it takes for the traveling wave to "travel" to a particular location on the membrane. Figure 7.16 shows the phase shift between a point on the membrane (horizontal axis) and the motion of the stapes. For in-

stance, a 200-Hz tone will set up a vibration that moves in phase with the stapedial motion at 20 mm from the stapes. At about 27 mm from the stapes, the vibration from the 200-Hz tone is 180° out of phase with stapes vibration. Because a 200-Hz tone has a 5-msec period, the 180° phase shift indicates that the vibration at 27 mm from the stapes occurs 2.5 msec (that is, one-half period) after the stapes moves. Thus, the time delay between displacement at the base (or stapes) and that at the apex results in a phase shift along the cochlear partition. Because the time delay from base to apex is relatively constant, the phase shift depends on the frequency of stimulation. In addition, membrane displacement increases in magnitude as the level of sound increases.

In trying to visualize these traveling-wave patterns, we should keep in mind some basic characteristics of the basilar membrane. First, *the basal end of the basilar membrane responds best*

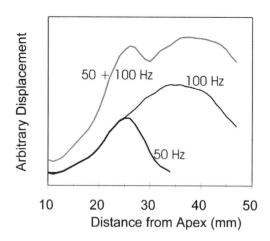

FIGURE 7.15 Individual traveling wave envelopes to 50-Hz and 100-Hz tones, and the envelope to a complex sound consisting of the sum of the two tones, showing two displacement peaks. Adapted with permission from Tonndorf (1962).

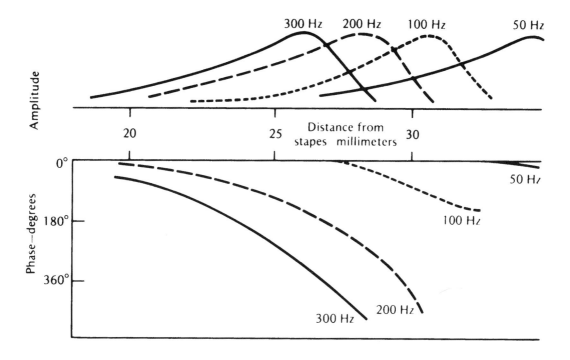

FIGURE 7.16 (**a**, top) Envelopes of traveling waves of four different frequencies. (**b**, bottom) The phase shift in degrees between the motion of the stapes and a point on the basilar membrane. By calculating the period of the sinusoid and the amount of phase shift, one can calculate the time for the traveling wave to "travel" to a particular point on the membrane. See text for an example. Adapted with permission from von Bekesy (1947).

to high-frequency stimulation, but it can also re-spond to low-frequency stimulation. *The apical end of the basilar membrane vibrates only to low-frequency stimulation.* Furthermore, *there is a time lag between the stapes movement and the movement of the apical end of the basilar mem-brane.* Thus, high-frequency stimulation causes maximum displacement in the basal end of the membrane. Low-frequency stimu-lation causes the basal end to vibrate first, with a small displacement, while high fre-quencies cause the base to vibrate with the greatest amplitude and there is essentially no vibration at the apex. *The greater the stimulus level, the greater the amount of basilar membrane*

displacement. And, finally, *the temporal pattern of basilar membrane displacement follows that of the stimulating sound.* That is, the membrane moves up and down in synchrony with the vibrating stimulus, more at one point than at others due to the traveling wave properties. Thus, basilar membrane displacement pro-vides potentially useful information about the frequency, level, and temporal pattern of acoustic stimulation.

With a constant, moderately intense input, a particular location along the basilar mem-brane will be displaced maximally by only one frequency. It is also true that for that par-ticular location the basilar membrane will be

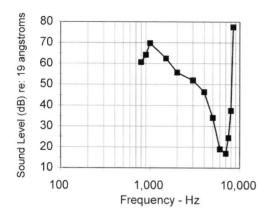

FIGURE 7.17 The amplitude required to maintain the basilar membrane at a constant displacement (1.9×10^{-8} cm) as a function of the frequency of the tonal input is shown using the Mössbauer technique for measuring basilar membrane motion in the chinchilla. The particular location at which the measurement was made had its maximum displacement when the frequency of the input was approximately 8350 Hz (8.35 kHz). Based on data from and adapted with permission from Robles, Ruggero, and Rich (1986).

displaced by frequencies lower than the one that displaces it maximally. In contrast, higher frequencies will displace that location on the membrane only a little, if at all, even at high levels. From our discussion of filters in Chapter 5, it can be seen that a given location along the basilar membrane acts as a filter with a sharp high-frequency roll-off. That is, a location along the basilar membrane will pass (or vibrate best) to certain frequencies. Higher frequencies cause very little displacement at that point, and because lower frequencies progressively displace the membrane less and less at that location, that location also has a gradual low-frequency roll-off. Thus, a particular location along the basilar membrane acts as a bandpass filter of the vibrating motion. This filter has a sharp high-frequency roll-off for a constant, moderately intense input. We might expect that the neural output of

the cochlea would reflect this filtering characteristic of the basilar membrane. This neural output and the bandpass filtering characteristics described above will be seen in the discharges of the individual nerves that leave the cochlea (Chapter 9).

Measurements made using the Mössbauer technique and with the technique of laser interferometry (see Appendix F) allow one to study the details of basilar membrane displacement and acoustic stimulation. Figure 7.17 shows the results of basilar membrane motion with the Mössbauer technique. These data were collected by presenting different frequencies to the inner ear and measuring the level for each tone required to displace the basilar membrane a fixed amount (1.9×10^{-8} cm). Thus, at the point of measurement along the cochlear partition of the chinchilla, an 8350-Hz (8.35 kHz) tone required the least level to displace the basilar membrane. This curve, therefore, represents the responsiveness of the basilar membrane at this point (near the base) of the cochlear partition. An 8.35-kHz tone causes the most vibration at this place along the basilar membrane, and as such this tonal frequency is called the *critical* or *center frequency* (CF). For this location along the basilar membrane, tones with frequencies higher and lower than 8.35 kHz had to have higher levels in order to vibrate the basilar membrane to the same extent as that caused by the 8.35-kHz tone, as is consistent with the traveling wave motion described above.

Measurement of the biomechanical response of the inner ear can take on a variety of forms. Displacements at different points along the basilar membrane can be measured for a *fixed stimulus*, as was done in Figures 7.12–7.16. In Figure 7.17, the tonal level required to *displace the membrane a fixed amount* (the *isosensitivity* measure) was measured as a function of varying the tone's frequency and level. The displacement at one point along the membrane for tones of different frequencies but of constant level could also be measured. In these measures, the indices of tonal level must take into account the transfer function of the outer and middle ears.

The motion of the basilar membrane is not entirely linear. One consequence of the nonlinear motion (see Chapter 5) is that basilar membrane displacement may not be linearly related to stimulus level. Figure 7.18 shows *input–output* functions for basilar membrane velocity as a function of stimulus level for tones of different frequencies. The amount of basilar membrane velocity was measured at one point along the membrane for 9000- and 1000-Hz tones presented at different levels. A tone with a frequency of 9000 Hz causes maximal displacement at the point along the basilar membrane where the measurements were made (i.e., 9000 Hz is the CF for this basilar

membrane location). The 9000-Hz CF tone leads to a nonlinear input–output relationship between stimulus level and basilar membrane velocity, whereas the 1000-Hz tone produces a nearly linear input–output relationship (compare the data curves with the light straight line). The form of nonlinearity for CF tones is *compressive*, in that, from about 30 to 80 dB SPL, increases in tonal level produce less and less change in membrane velocity. At very low (below 30 dB SPL) and very high (above 90 dB SPL) levels, the relationship appears more linear. Tones above and below the CF (e.g., the 1000-Hz tone in Figure 7.18) produce nearly linear relationships.

There are several consequences of this nonlinearity near the CF. First, the range over which the membrane moves is less than the range over which stimulus level varies, leading to compression of the *dynamic range* of basilar membrane motion near the CF. Thus, *changes* in low-level sounds cause a greater *change* in basilar membrane motion than *changes* at higher stimulus levels. That is, note that for the 9000 Hz tone a change in level from 20 to 30 dB (at a low level) resulted in a change by about a factor of 4–5 in velocity, while a change from 50 to 60 dB (at a higher level) resulted in a change of only about a factor of 2–3 in velocity. Second, the nonlinearity can produce distortion products such as harmonics and difference tones (see Chapter 5). As we will learn in Chapter 13, listeners can detect nonlinear harmonic and difference tones. Thus, the nonlinear motion of the basilar membrane is probably partially responsible for generation of audible harmonic and difference tones.

Knowing the exact motion of the organ of Corti is crucial to understanding how the neural response relates to the biomechanical response. If the neural response is substantially different from the biomedical response, another mechanism that helps transform the

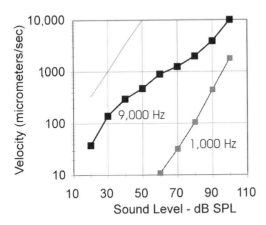

FIGURE 7.18 Input–output basilar membrane velocity functions showing basilar membrane velocity as a function of sound level for a 9000- and a 1000-Hz tone. The CF for this place is 9000 Hz. The response shows a compressive nonlinearity for 9000 Hz, which is linear at 1000 Hz. A straight line is also shown for comparison. Based on data and adapted with permission from Ruggero and Rich (1991).

biomechanical response into the neural response must be present. If the biomechanical and neural responses are similar, then the neural response is probably reflecting faithfully the biomechanical responses of the organ of Corti. In addition, damage to these inner ear structures are likely to have dire consequences for how well the auditory nervous system can process sound (see Chapter 16).

SUMMARY

The auditory part of the three inner ear structures is the cochlea, which is divided by the basilar membrane and Reissner's membrane into three sections: scala vestibuli, scala media, and scala tympani. The scala vestibuli and scala tympany contain the fluid perilymph, and the scala media contains endolymph. The basilar membrane is supported on the modiolar side by the osseous spiral lamina and on the stria vascularis side by the spiral ligament. The basilar membrane is wider, more flaccid, and under no tension at the apical end. The base end is narrower and stiffer than the apical end and may be under a small amount of tension. In the scala media is the organ of Corti. On the inner (modiolar) side of the tunnel of Corti are the inner haircells. On the outer side of the tunnel of Corti are the three rows of outer haircells; their cilia are in contact with the tectorial membrane. The haircells and nerve fibers are held in place by supporting cells. The basilar membrane has a traveling-wave motion when excited by the stapes. The traveling wave yields maximal displacement, after a time delay, at the apex for low-frequency stimulation and at the base for high-frequency stimulation. Level variations cause variations in the amount of dis-placement of the basilar membrane, and the temporal pattern of vibration follows that of the stimulus. The motion of the basilar membrane is nonlinear, especially near the CF.

SUPPLEMENT

Chapter 3 in the textbook by Pickles (1988) covers most of the topics of this chapter. In addition, the books edited by Altschuler *et al.* (1989) and those by Fay and Popper (1992) and Dallos, Popper, and Fay (1996) provide an in-depth overview of the anatomy and physiology of the inner ear. Von Bekesy pioneered our understanding of the vibratory characteristics of the basilar membrane. Much of his work is summarized in his book *Experiments in Hearing* (reprinted by the Acoustical Society of America, 1989).

To study the micromechanics of the cochlea, we must know as much as possible about the organ of Corti. However, because it is small, inaccessible, and fragile, the organ of Corti is extremely hard to investigate systematically. As we will learn in Chapter 16, the responses of the organ of Corti vary significantly if the inner ear is damaged. Thus, much of the original biomechanical work of von Bekesy does not agree in detail with that obtained with modern measures. It is now known that most of the differences are because von Bekesy worked with damaged inner ears.

Modern measurements also suggest that the responses of the auditory nerve fairly faithfully represent basilar membrane motion. Thus, the haircells (inner haircells) appear to behave as true biological transducers, changing basilar membrane vibration into neural impulses without significantly altering the information about sound provided by basilar membrane motion.

The primary distortion product generated by nonlinear basilar membrane motion is the cubic difference tone. That is, if two tones with frequencies $f1$ and $f2$ are presented with frequencies near the CF, then the basilar membrane will vibrate at frequencies of $f1$ and $f2$, *and* at a frequency equal to the cubic difference tone of $2f1–f2$ (see Chapter 5), although the magnitude of vibration at the frequency of $2f1–f2$ is less than that occurring at $f1$ and $f2$ (see Ruggero *et al.*, 1992).

Some scientists have taken a comparative (studying a variety of different animals) approach to understanding the inner ear in an attempt to relate structure with function. This work demonstrates that less complex haircell sensory systems of other animals can aid in drawing conclusions about the more complex mammalian cochlea (see Fay and Popper, 1994).

Summary of Inner Ear Measurements

Since most of the experimental work on the inner ear has been performed with animals, both animal and human data are listed. The principal sources for these data are Angelborg and Engstrom (1973), Rauch and Rauch (1974), Spoendlin (1966), and Wever and Lawrence (1954). These sources are often secondary, so the data given here actually represent the results of measurements by many authors.

Oval Window

dimensions: 1.2×3 mm to 2.0×3.7 mm *human*
area: $1.12–1.27$ mm^2, mean = 1.20 mm^2 *cat*
mean: 1.4 mm^2 *guinea pig*

Round Window

dimensions: 2.25×1.0 mm *human*
area: 2 mm^2 *human*
   3 mm^2 *cat*
   1 mm^2 *guinea pig*

Cochlea

number of turns: $2\frac{5}{8}$ *human*
volume: 98.1 mm^3 (including the vestibule proper) *human*
length of cochlear channels: 35 mm *human*

Helicotrema

area: 0.08-0.04 mm^2 *human*

Basilar Membrane

length: $25.3–35.5$ mm
   mean = 34 mm *human*
$19.4–25.4$ mm, mean = 22.5 *cat*
   mean = 18.8 mm *guinea pig*
   mean = 20.7 mm *squirrel, monkey*
   mean = 18.4 mm *chinchilla*
width, basal end: $0.08–0.16$ mm *human*
   0.062 mm *guinea pig*
width, apical end: $0.423–0.651$ mm, mean = 0.5 *human*, $0.194–0.240$ mm, mean = 0.209 mm *guinea pig*

Organ of Corti

cross-sectional area, basal end:
   0.00053 mm^2 *human*
   0.0055 mm^2 *cat*
cross-sectional area, apical end:
   0.0223 mm^2 *human*
   0.0201 mm^2 *cat*

Outer Haircells

number: 12,000 *human*
cell body length:
    20 μm (basal end)
    50 μm (apical end) *human*
cell body width: 5 μm *human*
cilium width: 0.05 μm at its base to 0.2 μm at tip
    human
cilium length: 2 μm in basal turn, 6 μm in apical
    turn *human*
arrangement: 100–150 stereocilia per outer
    haircell *human*
cilia 6–7 rows in **W** or **V** pattern per outer haircell
    human
cilia 3 rows in **W** or **V** pattern per outer haircell
    cat and *guinea pig*
cilia of outermost row have tips embedded in the
    tectorial membrane
angle in **V** shape 120° in the basal turn and 60° in
    the apical turn

Inner Haircells

number: 3500 *human*
    2600 *cat*
arrangement: 40–60 cilia per cell arranged in a
    shallow **U** shape
2–4 rows of cilia per cell, length of cilia longer in
    apical turn than in basal turn
lengths and diameters vary among individual
    coils of the cochlea, and within a single cell

Spiral Ligament

cross-sectional area: 0.543 mm^2 near basal end,
    0.042 mm^2 at the apex *human*

Scala Vestibuli

volume: 54 mm^3 including vestibule proper *human*

Scala Tympani

volume: 37.4 mm^3 *human*

Cochlear Duct

volume: 6.7 mm^3 *human*
    length: 35 mm *human*

Perilymph

volume: scala vestibuli, 10–15 μL *human*
scala tympani, 6–8 μL *human*
    total, 16-23 μL *human*
K$^+$ concentration: 4 meq/liter* (scala vestibuli)
    human
Na$^+$ concentration: 139 meq/liter (scala vestibuli)
    human
pH: 7.4–7.8 *human*, 7.9 *guinea pig*, 7.3–7.87 *cat*
viscosity: 1.030–1.050, in relation to H$_2$O at 27°C
    = 1 *human*
surface tension: 49.6 dyne/cm *guinea pig* (at 23°C)
protein: 70-100 mg/100 ml *human*
    142 mg/100 ml *cat*
    70–107 mg/100 ml *guinea pig*

Endolymph

volume: scala media, 2.7 μL *human*
K$^+$ concentration: 144 meq/liter *human*
Na$^+$ concentration: 13 meq/liter *human*
pH: 7.5 *human* (postmortem), 7.4 *guinea pig*, 7.82
    cat
viscosity: saccule, 1.030–1.050 in relation to H$_2$O
    at 27°C = 1 *human*
surface tension: 52 dyne/cm *guinea pig* (at 23°C)
protein: 20–30 mg/1000 ml *human*, 118 mg/100
    ml *cat*, 25 mg/100 ml *guinea pig*

---

*meq/liter = milliequivalent per liter.

CHAPTER 8

Peripheral Auditory Nervous System and Haircells

*I*n Chapter 7, we emphasized the importance of the haircell—the biological transducer for the auditory system. Within the inner ear are many electric potential differences that play a role in transduction of mechanical energy into neural energy. Our purpose in this chapter is to describe these cochlear potentials, to explain the function of the haircells as biological transducers, and to describe the anatomy of the auditory nerve that is connected to the haircells and sends neural signals to the auditory brainstem, and also receives neural input from the brainstem.

COCHLEAR POTENTIALS

The intricate motions and interactions of the various cochlear structures generate electric potentials that are relatively easy to study. Whether or not all of these potentials actually play an important role in the transduction process is an unanswered question. Even if they are nonfunctional byproducts of the mechanics of the inner ear, they are of interest because they yield information about the co-chlear transduction process. Four potentials can be recorded from the cochlea:

1. The *resting potential*, which is a dc (direct-current or nonalternating) potential, and exists without acoustic stimulation (see Appendix E).

2. The *summating potential*, which is also a dc potential, but appears only during acoustic stimulation.

3. The *cochlear microphonic*, which is an ac (alternating-current) potential difference that appears only during acoustic stimulation.

4. The *action potential*, which is also an ac potential difference, but is generated by the nerves rather than in the structures of the inner ear.

In general, the potentials in the cochlea as well as those of the auditory nerve are measured with very small wire or glass micropipettes, as shown in Figure 8.1. The electrode measures the electric potential at the site of its penetrations *relative* to some reference site. The reference site might be another part of the auditory system or, more often, a neutral location such as a neck muscle. Thus, the potentials are the differences in electrical charge between two points, for instance, between points A and B in Figure 8.1.

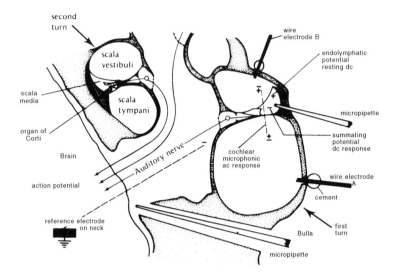

FIGURE 8.1 Drawing of a mid-modiolar section of the first and second turns of the guinea pig cochlea. The wire electrodes produce differential recordings of the CM, AP, and SP from the first turn. One glass micropipette records the EP, CM, and SP from the scala media of the first turn; the other records single neurons of the auditory nerve as they exist from the modiolus. Adapted with permission from Davis (1956).

Resting Potentials

Three dc potentials can be observed in the resting and responding cochlea. The one of greatest significance is the *endocochlear* or *endolymphatic potential* (EP). The EP is located in the endolymph of the scala media (SM). The EP is about +80 millivolts (mV), that is, the electrical charge in the scala media of a normal resting cochlea is 80 mV above 0 mV. This is the highest positive resting potential found in the body, and it is not found in the endolymph of the vestibular system despite the fact that the endolymph is continuous in both structures. This can be explained if the source of the EP is the stria vascularis, which is located on the wall of the cochlear duct but is not present in the other endolymphatic spaces (e.g., vestibular system) of the inner ear. Because the haircells have an *intracellular* resting potential of about –70 mV, the +80 mV of the

external endolymph makes the potential difference across the top of the haircells about 150 mV, an extremely high potential difference to be found in the body. Because of this large potential difference, it is thought that the EP could serve to increase the size of the electrical response of the haircells, making it a more effective biological transducer. Damage to the stria vascularis will result in the loss, or decrease, of the EP, and without the EP the inner ear will not perform the mechanical to electrical transduction process. Figure 8.2 shows the resting potentials of the three scalae of the cochlea.

Summating Potential (SP)

The summating potential (SP) is a stimulus-related dc electrical response recorded from the cochlea. There is a baseline shift in

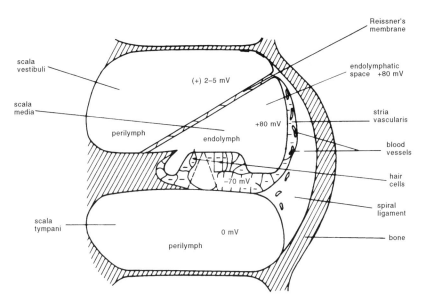

FIGURE 8.2 Schematic cross-section of the cochlea showing potentials. The scala vestibuli is 2 to 5 mV more positive than the scala tympani. The scala media has a +80 mV potential called the endocochlear or endolymphatic potential (EP). In sharp contrast to the EP, the intracellular potentials of the organ of Corti are about –70 mV. Thus, the potential difference across the top of the haircells is about 150 mV. Adapted with permission from Tasaki *et al.* (1954).

the potentials recorded from the cochlea whenever a stimulus is present, as shown in Figure 8.3. This baseline shift can be either in the positive or negative (voltage) direction. Experimenters who have attempted to quantify the SP have found it to be composed of different potentials, which interact in a complicated manner. Thus, we are not certain of the exact source of the SP, although six sources have been suggested.

Cochlear Microphonic (CM)

The Cochlear Microphonic (CM) is an ac potential that occurs only during the presentation of an acoustic stimulus. For instance, if the ear is stimulated with a 500-Hz pure tone of moderate level, the CM will appear as a

500-Hz electrical sine wave. In other words, the electrical potential difference in the cochlea, the CM, oscillates in the same manner as the driving stimulus. From observations of its growth and shape, it appears that over a fairly wide range of sound levels the CM faithfully reflects the intensity and frequency components of the sound input. If the sound intensity becomes too high, however, the voltage increase of the CM stabilizes and then decreases in magnitude. Figure 8.4 is one example of the growth in the level of the CM output as a function of the level of the input sinusoid. Such intensity curves are sometimes called *input–output functions*. Repeated or lengthy stimulation at extreme sound levels results in either temporary or permanent impairment of haircell function (see Chapter 16), which is clearly manifest in a decline in CM output.

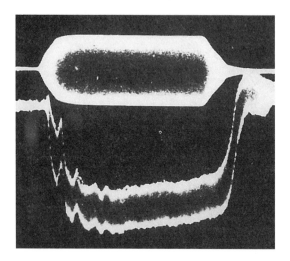

FIGURE 8.3 The trace above records the stimulus, a 21.5-kHz tone burst. The lower trace is a round window electrode recording the response to the stimulus. The downward dc shift of the trace during presentation of the stimulus is the SP. The section of the recording that replicates the input signal is the CM. The wiggle at the beginning of the response as it begins its downward shift is the AP. The round-window recording technique, therefore, yields a composite response composed of SP, CM, and AP. Reprinted with permission from Pestalozza and Davis (1956).

Consequently, understanding the decline in the input–output function at high levels may provide clues to preventing impairment of haircell function caused by over stimulation.

The source of the CM is thought to be the cilia-bearing end of the outer haircells. This can be inferred when an electrode approaches the top of a haircell. The CM becomes larger and then reverses its electrical signal phase as the electrode passes through the organ of Corti. That is, if in the scala tympani the CM potential was positive, then the CM becomes negative when the electrode enters the scala media. The growth and polarity reversal of the CM indicates that its generation is at the boundary of the haircells and the scala media, near the level of the reticular lamina and the roots of the cilia of the haircells.

The *level of the CM is proportional to the displacement of the basilar membrane*. This fact, together with the ability to record from small areas along the cochlear partition by the differential electrode technique (recording the CM with a technique that allows for cancellation of the SP and the ability to record from local regions of the organ of Corti), led to the electrophysiological investigation of the motion of the basilar membrane by means of the CM. That is, the CM has been used to measure the displacement action of the basilar membrane. The resultant space–time pattern of the CM reveals a traveling wave that moves toward the apex to a location dependent on the frequency of the stimulus. This pattern agrees well with the mechanical movement observed by von Bekesy in his models. That is, the apical end of the cochlea only responds to low frequencies, but the basal end responds to all

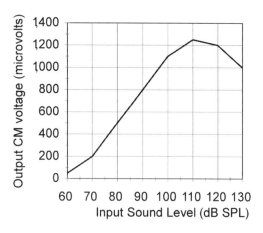

FIGURE 8.4 Input–output function (or intensity function) for the cochlear microphonic response recorded differentially from the first turn of the guinea pig cochlea (see Figure 8.1 for electrode placements) to an 8-kHz tone burst. As the sound pressure level (SPL) increases to 90 dB, the CM increases linearly. From 90 to about 100 dB SPL, the CM increases at a slower rate. For this stimulus the CM is largest at about 100 dB SPL and then decreases as the input intensity increases beyond 100 dB SPL.

frequencies. Figure 8.5 shows that a low-frequency (500-Hz) tone generates a CM in all turns of the cochlea, along the entire basilar membrane. However, the amplitude of the CM is greatest at the apex (third turn), and a phase shift occurs (180° at 500 Hz) between the first-turn (base) recording and the third-turn recording. A high-frequency tone (8000 Hz), however, generates a CM only in the first turn (base) of the cochlea. This agrees exactly with the characteristics of the traveling wave discussed in Chapter 7. Thus, even with the recent developments of laser light and Mössbauer techniques, the CM remains (with certain limitations) a useful tool for studying the

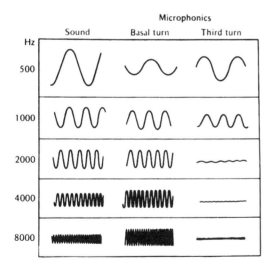

FIGURE 8.5 Cochlear microphonics recorded differentially from the first and third turns of the guinea pig cochlea. Rows: responses to the various frequencies. Column 1: acoustic stimulus. Columns 2 and 3: CM from turns I and III. For 500 Hz, the CM is larger in turn III than in turn I. Also, the 500-Hz CM shifts 180° out of phase between turns I and III because of the time taken by the traveling wave of displacement to reach turn III. For the highest frequency stimulus used (8000 Hz), the response only in turn I indicates that the traveling wave does not reach turn III. Adapted with permission from Davis (1960).

traveling-wave motion of the cochlear sac to a wide range of levels and frequencies of stimulation.

Action Potential (AP)

The whole nerve action potential (AP), unlike the other potentials discussed, is not a true cochlear potential, although it can be recorded from the cochlea. The AP is the sum of the action potentials of many individual auditory neurons that are firing nearly simultaneously within the bundle of auditory nerves (see Chapter 9). Because stimulation of the individual neurons depends on the displacement of the basilar membrane, neurons that innervate the base of the cochlea will be stimulated at an earlier point in time than those at the apex. The time difference is the length of time it takes for the traveling wave to move from the base to the apex (i.e., 2.5 to 4 msec, depending on the species and the measuring technique). This situation creates a general asynchrony of discharge among the individual neurons, which terminate at different points along the cochlea. Because each neuronal discharge consists of potentials with both positive and negative values that vary over time (see Appendix E), the potentials tend to cancel each other or to sum only partially when added. At the onset of a click, or of a high-frequency tone burst with a sudden onset, however, only the basal end of the cochlea is stimulated initially. The near simultaneous discharge of a large number of neurons in the base is then followed by less synchronous nerve impulses from more apical locations of the cochlear partition. The sum of the initial synchronous discharge of the nerve impulses to an abrupt high-frequency tone burst or click is the part of the AP usually measured. The AP is negative at first and then positive, and it is usually followed by a later negative component arising from the nerves that innervate more apical portions of the cochlea and possibly other, more central, sources of potentials. Figure 8.3 shows the waveform of an AP superimposed on the CM. The most widely used stimulus for AP studies is the click (Chapter 4), because a click has a sudden onset and a broadband amplitude spectrum.

Within the limitations we have mentioned (i.e., the overwhelming contribution of the basal turn), the AP reflects grossly the output of the cochlea. The input–output function and waveshape of the AP from a normal cochlea are known, and so deviations from normative data can be used as indications of cochlear dysfunction. This is particularly important because of our present ability to record the AP for humans.

HAIRCELLS, STEREOCILIA, OUTER HAIRCELL MOTILITY, AND NEURAL TRANSDUCTION

The haircells are stimulated via bending of the stereocilia. Situated at the top of the organ of Corti is the tectorial membrane, which runs parallel to the basilar membrane. When the basilar membrane vibrates, the tectorial membrane must also move, but there is an important difference in the motion of the two membranes due to differences in their support. As shown in Figure 8.6, the tectorial membrane is hinged on one end at the spiral limbus, whereas the modiolar edge of the basilar membrane is thought to be hinged on the osseous spiral lamina. As the basilar and tectorial membranes are displaced, they pivot about the two different hinging points, and the indicated shearing forces are created. The actual method of shearing is probably different for inner and outer haircells, since the stereocilia of outer haircells appear firmly attached to the tectorial membrane, while the stereocilia of inner haircells are probably not

attached to the tectorial membrane, and fluids trapped between the stereocilia and tectorial membrane probably cause inner haircell shearing. Shearing is a particular form of bending in which in this case the top moves more than the bottom.

Thus, mechanical energy is transduced into electrochemical activity by shearing of the stereocilia. In order to function well for a long period of time and to provide a connection between haircell and tectorial membrane, the stereocilia need to be strong, resilient, and resistant to breakage. The stereocilia are arranged in an elegant geometrical form that provides a great deal of strength (see Figure 7.12). It is believed that transduction of neural information takes place near the top of the haircells in the region of the stereocilia. In experiments in which a haircell is excised from the organ of Corti (see Appendix F) and the stereocilia are carefully displaced in small amounts with a micropipette, recordings of electrical changes either within the haircell (*intercellular recordings*, again see Appendix F) or near the haircell (*extracellular recordings*) implicate the top of the haircell or a region near the bottom of the shortest stereocilia as the sites where neural transduction is most likely occurring.

In order for the haircell to transduce stereocilia shearing into a neural response, the permeability of the haircell membrane must change to allow for neural transduction (see Appendix E). It is believed that the tip lengths (see Figure 7.12) aid in causing "channels" to open and close near the top of the haircell. These channels allow for ionic transport into and out of the haircell that is necessary for neural transduction. As the tip lengths stretch and contract during the shearing of stereocilia, they may act to open and close a type of "trapdoor" on the stereocilia, and the trapdoor allows for ions to flow in and out of the haircell starting the neural transduction process.

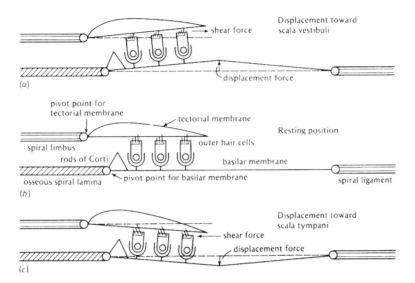

FIGURE 8.6 Schematic diagrams of shearing forces created between the haircells and the tectorial membrane as a result of basilar membrane displacement. (**a**) Shearing force that results from displacement of the basilar membrane toward the scala vestibuli. (**b**) Relationship between haircells and tectorial membrane with no stimulation. (**c**) Shearing forces in the direction opposite to that in panel **a** for displacement in the opposite direct. The shearing of the cilia is thought to activate the neurons at the base of the haircells. Adapted with permission from Zemlin (1981).

There are two types of haircells—inner and outer—and they differ in several ways as we have described. Why are there two types of haircells, and how do their differences help in transducing vibrations into a neural code? The next section explains additional differences between the two types of haircells that lead to a theory as to how the two types work together, allowing for the exquisite sensitivity and frequency-resolving capacity of the inner ear.

Outer Haircell Motility

So far we have described the changes that occur within the organ of Corti when sound is present in terms of the passive reaction of tissue to vibration. The outer haircells exhibit a type of active response in that they change their size in response to stimulation. That is, when stimulated the outer haircells expand and contract. Inner haircells do not exhibit this *motility*.

The changes in haircell size, primarily in length, appear to take place on two different time scales. There is a fast motility (changes that occur within 0.15 to 0.2 msec after stimulation) that probably can take place on a cycle-by-cycle basis for all frequencies of stimulation. The slower change that takes place in haircell length after stimulation (changes that take place after several seconds) might reflect some slower biochemical action such as that which occurs when muscles contract. It is known that the outer haircells contain some of

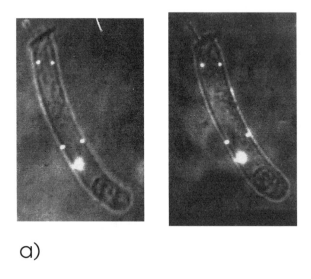

a)

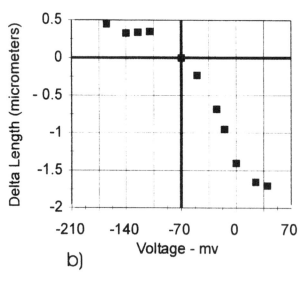

b)

FIGURE 8.7 (**a**, top): An example of the change in the length of a outer haircell due to electrical stimulation. The haircell was excised from the cochlea in an *in vitro* experiment. The picture on the left shows the haircell before chemical stimulation with potassium, and the picture on the right shows the decrease in length and increase in width after stimulation. Reprinted with permission from Zajic and Schacht (1991). (**b**, bottom): Change in the length of an outer hair cell as a function of applying hyperpolarizing and depolarizing voltages to the hair cell. Reprinted with permission from Ruggero and Santos-Sacchi (1997).

the crucial proteins (*actins*) for muscle-like contractions and, therefore, might contract and expand based on similar principles. Figure 8.7 shows changes in outer haircell length as a function of changing the voltage on the excised haircell in a patch-clamp experiment (see Appendix F). Note that the length decreases with increasing voltage (depolarization, see Appendix E) more than it increases with decreasing voltage (hyperpolarization, see Appendix E). Not only do the outer haircells themselves change, but, since they are attached to the tectorial membrane by means of the stereocilia, such changes could affect the connection between the basilar and tectorial membranes and hence the exact way in which the stereocilia are bent as described above.

Outer haircell motility is also influenced by efferent fibers that come from the brainstem (the olivocochlear bundle, see below) and directly synapse on the outer haircell. This efferent control of outer haircells is a slow process because of the time it takes for a neural signal to be sensed, travel to the brainstem, and then travel back down to the outer haircell. Since the innervation pattern of the olivocochlear bundle fibers comes from brainstem nuclei on both sides of the brain, efferent control of outer haircell function is bilateral. The exact role of efferent control of outer haircell function is not well understood.

The Roles of the Inner and Outer Haircells

As Chapter 16 will explain in more detail, the ability of the inner ear to efficiently transduce vibration into a neural response is seriously compromised when either the inner or outer haircells are damaged, but outer haircell damage causes a significant loss in sensitivity and frequency resolution. As we will

discuss below, most of the fibers in the auditory nerve connect with the inner haircells. Thus, it appears almost certain that the inner haircells are the biological transducers for sound and that the outer haircells help provide for the high sensitivity and high-frequency resolution of the cochlea so that inner haircell transduction will also be highly sensitive and exhibit high-frequency resolution. The general theory is that the stereocilia of the outer haircells sense the vibrations within the cochlea that cause a change in the length (and width) of the outer haircell. This motility effects a change in the mechanical coupling between the basilar and tectorial membranes. Presumably, this coupling is changing synchronously with the vibrations within the cochlea, even when these vibrations occur at very high frequencies. In so doing, the changes in the coupling make the vibratory pattern of the cochlear partition very sensitive to small vibrations and highly frequency selective. This vibratory pattern also reflects significant nonlinear responses, which are crucial for proper function of the inner ear. The stereocilia of the inner haircells are sheared in response to the sensitive cochlear vibrations, and the neural response generated within the inner haircell is transmitted to the auditory nerve fibers that innervate the inner haircell and sends a neural signal up the auditory nerve bundle to the brainstem. Thus, the outer haircells seem to feedback energy into the cochlea via its motility, and as a consequence the sensitivity of the cochlea is amplified, leading to the concept that the outer haircells help provide a *cochlear amplifier* for transducing small vibrations into neural impulses.

COCHLEAR EMISSIONS

If a click is delivered to the ear, then approximately 5–10 msec after the click has ceased sound may occur within the sealed outer ear as if there were an echo from the inner ear to the click. Although there is no evidence that this "echo" is a cochlear potential, the data show that the echo comes from the cochlea and not from the middle ear or from neural activity in the central nervous system. In addition to the recording of an acoustic echo in the outer ear, there is also evidence of tonal-like emissions measurable within the outer ear that appear to originate from the cochlea in the absence of any acoustic stimulation. Both the *acoustic echo* (sometimes referred to as an *evoked otacoustic emission*, EOAE or OAE) and the *cochlear emission* (sometimes referred to as *spontaneous otacoustic emissions*, SOAEs) are very low in level (near or below the thresholds of hearing), and the nature of the phenomenon varies greatly from person to person and from animal to animal. However, almost every person with normal hearing shows evidence of some form of otacoustic emission, although the variability from person to person (and even ear to ear) can be large.

Two types of evoked emissions, or echoes, are typically measured: *transiently evoked otacoustic emissions* (TEOAEs) and *distortion-product otacoustic emissions* (DPOAEs). The TEOAE is evoked by a click or other short-duration sounds delivered to the ear and the echo's time-domain waveform or its amplitude spectrum are measured in the sealed outer ear cavity. If tonal stimuli are used, the emission contains frequencies near that of the stimulating tone. If broadband stimuli (such as clicks or noise) are used, then the emissions often have several narrow spectral peaks that are not a direct consequence of the stimulus but appear to reflect how the stimulus is being affected by the biomechanical actions of the cochlea.

The DPOAE is measured when a two-tone complex ($f1$ and $f2$) is used for stimulation. The evoked emission is then spectrally ana-

lyzed in the frequency region of the cubic-difference tone (see Chapters 5 and 13). The cubic-difference tone is equal to $2f1 - f2$ (e.g., if the two-tone stimulating complex consisted of 1333.33 and 1666.66 Hz, then $2f1 - f2 = 2 \times 1333.3$ Hz $- 1666.66$ Hz $= 1000$ Hz, so that 1000 Hz is the cubic-difference tone and the frequency of the recorded DPOAE). The echo evoked in the frequency region of the cubic-difference tone has a relatively high level and can be used to measure both the frequency dependence of the OAE and some of its nonlinear properties. Figure 8.8 displays a TEOAE, the spectrum of an SOAE, and the spectrum of a DPOAE.

It is generally believed that spontaneous emissions (SOAEs) are a consequence of the feedback system involving the motile outer haircells. Indeed, such feedback systems in electrical circuits often become unstable under several different conditions, and this instability generates sinusoidal oscillations. Thus, by analogy, spontaneous emissions may be a result of such instabilities in the outer haircell feedback system, producing sinusoidal oscillations that are picked up in the outer ear as spontaneous emissions. Evoked acoustic emissions are also presumably a direct result of the active cochlear system, especially the nonlinearities of this system. The prevalence of distortion product emissions underlies the presumption that otacoustic emissions reflect cochlear processes, especially the nonlinear properties of these processes.

The motility of the outer haircells, acoustic echoes, and cochlear emissions suggest that there are active processes within the inner ear that are set off by stimulation and that haircell motility might be related to acoustic echoes and/or cochlear emissions. That is, when sound enters the inner ear, the structures do not only vibrate in direct response to the sound, but some structures are stimulated to produce their own unique changes and act to

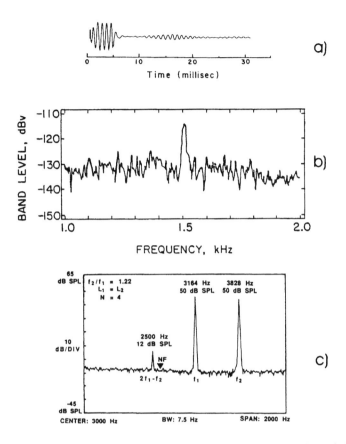

FIGURE 8.8 (**a**, top): An acoustic echo (*TEOAE*). The eliciting stimulus (sound impulse) is reflected at the beginning of the recording made in the closed outer ear. The small waveform shown around 10–20 msec later is the small echo. Adapted with permission from Wilson (1980). (**b**, middle): An example of a spontaneous cochlear emission (*SOAE*) showing a spectral analysis of the pressure in a closed outer ear that has not been stimulated. The amplitude spectrum shows peaks at 1.5 kHz, which represents the frequency of the cochlear emissions. Adapted with permission from Zurek (1985). (**c**, bottom): The spectrum of the outer ear waveform for a measure of the distortion product otacoustic emission (*DPOAE*). The two primary tones used to elicit the distortion product had levels (*L1* and *L2*) of 50 dB SPL and frequencies of 3164 Hz (*f1*) and 3828 Hz (*f2*), with an *f2/f1* ratio of 1.22. The cubic-difference tone (*2f1 – f2*) is at 2500 Hz, and the distortion product measured in the outer ear at this frequency has a level of 12 dB SPL, which is clearly above the measurement noise floor (NF). The measurement was based on averaging the outer ear waveform for four (*N* = 4) presentations. Reprinted with permission from Martin *et al.* (1990).

add energy to that which occurs in the cochlea in reaction to sound stimulation. Such *active processes* are probably necessary to explain the very low sensitivity of the inner ear to sound stimulation and the very fine ability of the inner ear to determine the frequency of sound (frequency selectivity). The motility of the outer haircells could be the active process, and

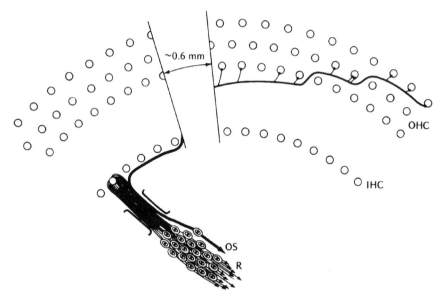

FIGURE 8.9 Horizontal schema of the afferent innervation of the organ of Corti. Reprinted with permission from Spoendlin (1978) in Naunton and Fernandez (1978).

echoes and emissions could be measurable consequences of this motility.

STRUCTURE OF THE AUDITORY NERVE

In our discussion of the anatomy and physiology of the auditory system in Chapter 7, we have established that the basilar membrane performs a type of frequency selectivity via the traveling wave in such a manner that haircells at different locations along the cochlear partition can respond differentially to frequency. Electrophysiological recordings of cochlear potentials reflect this frequency selectivity. Intensity has been associated with the amount of basilar membrane displacement and the resultant shearing of haircells. Level changes are also reflected in the various cochlear potentials. Brief reference has been made to the nerves that synapse with haircells and carry information about the acoustic stimulus centrally, toward the auditory cortex. If we are to understand how the auditory nerve preserves the frequency, intensity, and timing information existing in the cochlea, we must know the pattern of *innervation* between the nerve fibers and the haircells as well as the nature of the neural discharges. The patterns made by the auditory nerve communicating with or innervating the haircells along the basilar membrane limit the types of encoding possible in auditory nerve fibers. We shall describe two basic types of nerve fibers: afferent and efferent. *Afferent fibers* are sensory nerves carrying information from the peripheral sense organ (in this case, the organ of Corti) to the brainstem and brain. *Efferent fibers* typically bring information from higher neural centers (e.g., a nucleus in the brainstem) to the periphery.

Afferent Fibers

There are two types of auditory afferent fibers: (1) *radial fibers* (R) or *type I fibers* and (2) *outer spiral fibers* (OS) or *type II fibers*. The radial fibers comprise about 85–95% of the total number of afferent fibers. As illustrated in Figure 8.9, type I fibers innervate the inner haircells exclusively, each fiber innervating the base of only one or two inner haircells and then leaving the organ of Corti through an

opening in the osseous spiral lamina called the *habenula perforata* to travel to the modiolus. Figure 8.10 is a transmission electron micrograph showing the path taken by the radial fibers. There are about 45,000 or 50,000 afferent fibers in the cat, and 95% of them are thought to innervate the 2600 inner haircells. Thus, each inner haircell may be innervated by 16 to 20 radial fibers, as shown schematically in Figure 8.9. Similar calculations for the human cochlea estimate that an average of eight radial fibers would innervate one or two inner haircells. Thus, the radial or type I fibers are said to have a many-to-one connection to the inner haircells. The type I fibers tend to be

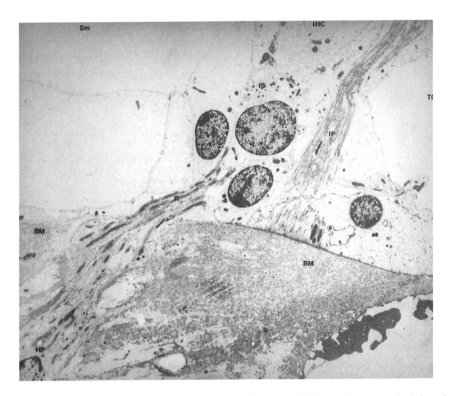

FIGURE 8.10 Transmission electron micrograph showing the radial fibers (*R*) entering the habenula perforata (*HP*) and going toward the base of the inner haircell (*IHC*). The fibers of the inner spiral bundle (*IS*) are also seen in their path below the base of the *IHC*. For orientation, the following are also labeled: *BM* = basilar membrane, *TC* = tunnel of Corti, *IP* = inner pillar cell, *Sm* = scala media. Chinchilla photograph courtesy of Dr. Ivan Hunter-Duvar, Hospital for Sick Children, Toronto.

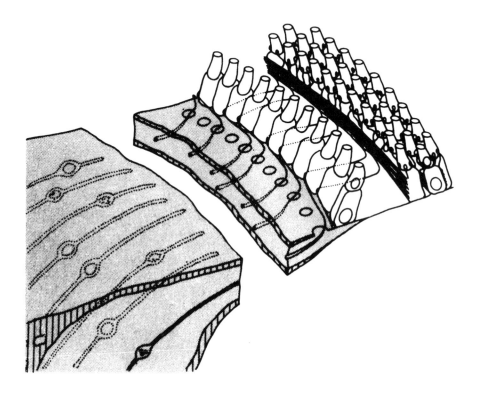

FIGURE 8.11 Schematic diagram of the outer spiral fibers that constitute the afferent innervation of the outer haircells. Adapted with permission from Spoendlin (1974).

thicker than type II fibers. The type I radial fibers may also differ in how they innervate the inner haircells.

The other 5 to 15% of afferent fibers are the outer spiral or type II fibers, and they are said to have a one-to-many connection, as shown in Figures 8.9 and 8.11. In the basal area, the outer spiral fibers innervate the outside row of the outer haircells. As they move in an apical direction, they innervate the middle and finally the innermost row of outer haircells. They then travel further in an apical direction for about 0.6 mm, finally crossing along the floor of the tunnel of Corti between the pillar cells and through the habenula perforata to

the modiolus. The involved course of the outer spiral fibers along the base of the first row of outer haircells is shown in Figure 8.12. Figure 8.13 shows endings of a number of outer spiral fibers at the base of an outer haircell. On average, each outer spiral fiber innervates about 10 outer haircells. Thus, the afferent fibers in the auditory nerve innervate either one or a small number of haircells.

Efferent Fibers

The efferent innervation of the haircells is by the *olivocochlear bundle* (OCB), which is a

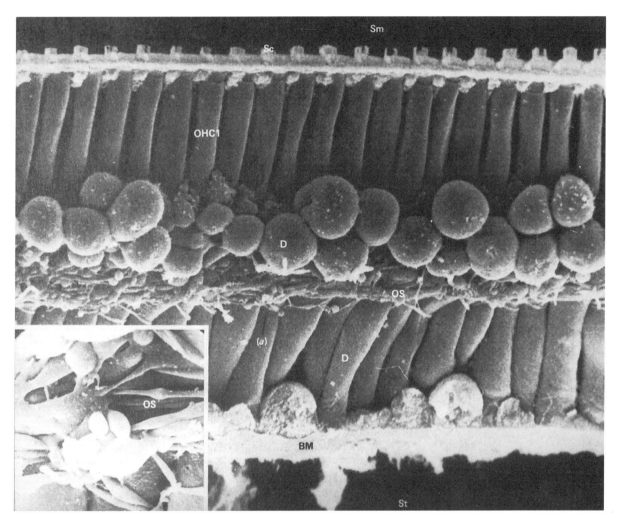

FIGURE 8.12 Scanning electron micrograph of the first row of outer haircells (*OHC1*) and supporting Deiters cells (*D*), as seen from the tunnel of Corti with the outer pillar cells removed. At the base of the outer haircells, the fibers of the outer spiral bundle (*OS*) are seen. The insert shows their interwoven pattern in detail. Note that the Deiters cells at the base of the first row of haircells have bulges, rather than the phalangeal processes shown arising from the Deiters cells beneath the third row of outer haircells as in Figure 7.6. Chinchilla photographs courtesy of Dr. Ivan Hunter-Duvar, Hospital for Sick Children, Toronto.

series of nerves that come from the *olivary complex* (*olive*) in the auditory brainstem to the haircells in the organ of Corti (see Chapter 15 for a discussion of the auditory brainstem). The efferent fibers that come to the inner ear from the olivary region of the brainstem (see Figure 8.14) on the same side are called the *uncrossed olivocochlear bundle*, or UCOCB. That is, UCOCB fibers come from the olivary region on one side of the head to the cochlea on

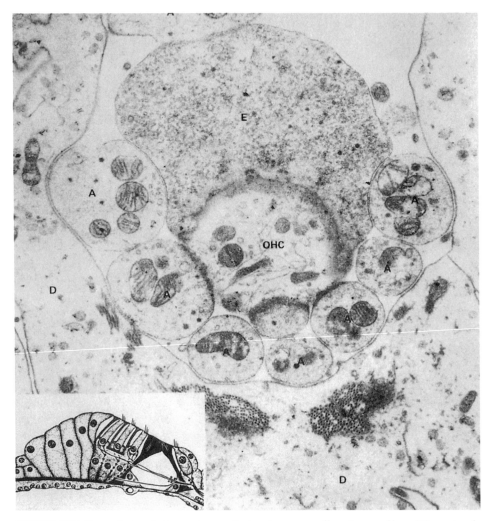

FIGURE 8.13 Transmission electron micrograph showing many afferent nerve fibers (*A*) coming from the outer spiral bundle surrounding the base of an outer haircell. A large efferent nerve ending (*E*) is also seen. (The insert shows the location of the section, in the area of the middle outer hair cell; see Figure 7.5.) Chinchilla photograph courtesy of Dr. Ivan Hunter-Duvar, Hospital for Sick Children, Toronto.

that same side (*ipsilateral side*). The bundle of efferent fibers that arises from the olive on the opposite side is called the *crossed olivocochlear bundle* or COCB. That is, COCB fibers come from the olivary region on one side of the head to the cochlea on the other side of the head (*contralateral side*). The OCB fibers can originate from a location that is lateral within the olivary complex (the *lateral olivary complex*, LOC), or from locations that are medial within the olive (the *medial olivary complex*, MOC). The LOC fibers project predominantly to the inner haircells and the MOC fibers to the outer haircells. Figure 8.14 shows the distribution

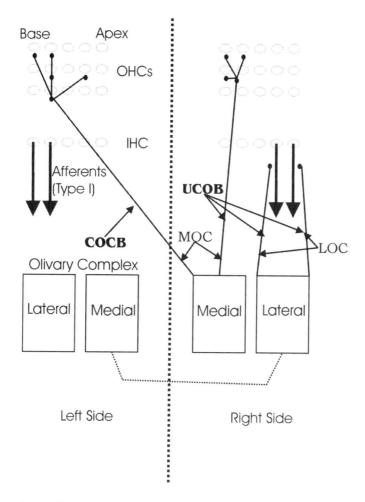

FIGURE 8.14 A simplified schematic diagram showing the OCB connections to the inner ear. The crossed (COCB) MOC fibers (26% of the OCB fibers in the cat) synapse directly on the outer haircells toward the base of the contralateral cochlea, the uncrossed (UCOB) LOC fibers (48% of the fibers in the cat) synapse on the type I afferents coming from the ipsilateral inner haircells located more apically, while the uncrossed MOC (11%) and crossed LOC (15%) fibers make up a much smaller proportion of the fibers in the OCB. Based on data from Warr (1992).

and innervation of OCB fibers for the cat. The LOC fibers synapse on the afferent nerves leaving the inner haircells, while the MOC fibers synapse directly on the outer haircells. Figure 8.15 shows the synapse of an MOC efferent fiber on an outer haircell.

Organization of the Auditory Nerve Bundle

The auditory nerve fiber anatomy may be outlined as follows:

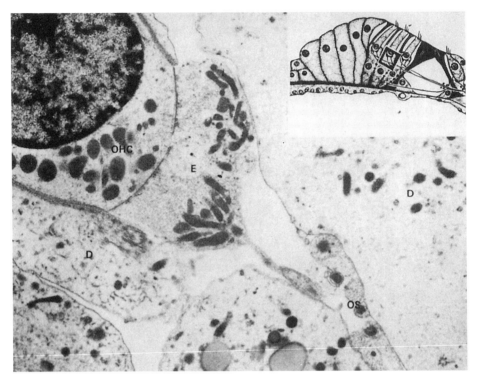

FIGURE 8.15 Transmission electron micrograph showing an efferent nerve ending (*E*) at the base of an outer haircell (*OHC*). Fibers of the outer spiral bundle (*OS*) are also seen. (The insert shows the location of the section, the boxed area at the bottom of the middle outer hair cell; see Figure 7.5.) Chinchilla photograph courtesy of Dr. Ivan Hunter-Duvar, Hospital for Sick Children, Toronto.

1. Afferent fibers

   a. Radial or type I fibers from inner haircells.

   b. Outer spiral or type II fibers from outer haircells.

2. Efferent fibers.

   a. Lateral Olivary Complex fibers coming from the lateral areas of the olivary complex that are either uncrossed olivocochlear bundle fibers coming from the same lateral olivary complex or crossed olivocochlear bundle fibers that come from the opposite lateral olivary complex.

   b. Medial Olivary Complex fibers coming from the medial areas of the olivary complex that are either uncrossed olivocochlear bundle fibers coming from the same medial olivary complex or crossed olivocochlear bundle fibers that come from the opposite medial olivary complex.

Both the efferent and afferent fibers are devoid of their insulating myelin sheaths between the organ of Corti and the habenula perforata. The nerve fibers leave the cochlea through the habenula perforata in an ordered manner and are gathered in a twisted bundle within the modiolus. The nerve fibers that innervate haircells at the apex are in the middle of the nerve bundle. Fibers from cochlear turns toward the base make up the outside fibers of the nerve bundle. Figure 8.16 shows a cross-section of the cochlea with the apex at the top; the middle of the cross-section is the modiolus. The fibers from the apex run down the middle of the modiolus, and those from the other turns of the cochlea form the outside of the nerve bundle. Thus, fibers, carrying low-frequency information (from the apex) are in the middle of the auditory nerve bundle, whereas high-frequency fibers (toward the base) are on the outside of the bundle. The auditory nerve bundle is joined by nerves

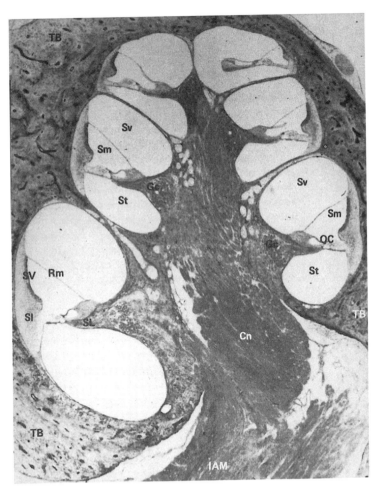

FIGURE 8.16 Light micrograph of a mid-modiolar section of a cat cochlea. The cochlear nerve (*CN*) is seen in the modiolus. Fibers from the apical turns form the center of the cochlear nerve, while the basal turn fibers join with the nerve at its outer edges. The nerve leaves the cochlea via the internal auditory meatus (*IAM*). Photograph courtesy of Dr. Ivan Hunter-Duvar, Hospital for Sick Children, Toronto.

from the vestibular system to form the entire *VIIIth nerve tract*. After the auditory nerve leaves the cochlea, its next junction is the cochlear nucleus. The structure of the nervous system central to the cochlea will be discussed in Chapter 15.

There is, therefore, an orderly connection between the inner haircells and auditory nerve fibers, with the auditory nerve fibers organized from the center to the outside within the nerve bundle. On an anatomical basis, therefore, we might expect that the discharge patterns of the auditory nerve fibers will reflect the frequency, intensity, and timing encoding performed by the mechanics of the inner ear.

SUMMARY

The electric potentials (CM, SP, AP, and single units) of the inner ear are created as a result of hydromechanical disturbances. Resting potentials are also present regardless of stimulation (especially EP). The likely sources of these potentials are the haircells for the CM, the stria vascularis for the EP, and simultaneous discharges of auditory nerve single units for the AP. There are numerous possible sources of the SP. The stereocilia of the haircells are sheared by the differential motions of the basilar and tectorial membranes. This shearing causes neural discharges in the haircells probably initiated in the region of the stereocilia at the top of the inner haircells. The outer haircells are motile and change their length as a result of stimulation. Acoustic echoes and cochlear emissions are possible indications of active processes within the inner ear. There are two types of afferent fibers: radial or type I fibers and outer spiral or type II fibers. The radial fibers make up 85–95% of all afferents. Each

radial fiber innervates only one or two inner haircells, and each inner haircell synapses with many radial fibers. The outer spiral fibers compose the other 5–15% of the afferent fibers. Each innervates about 10 outer haircells in an orderly fashion. The efferent fibers comprise the olivocochlear bundle that arises from either the medial (MOC) or the lateral (LOC) regions of the olive on the ipsilateral (UCOCB) and contralateral (COCB) sides of the head relative to the inner ear, to which the OCB projects. Type I fibers innervating the inner haircells provide the major sensory transduction mechanism for hearing. The outer haircells serve to increase the sensitivity and frequency selectivity of the transduction process, most likely as a result of outer haircell motility. The afferent nerve fibers leave the organ of Corti in an orderly fashion, such that fibers from the base are on the outside of the auditory nerve bundle and fibers from the apex of the cochlea are on the inside.

SUPPLEMENT

The material covered in this chapter can be found in Chapters 3, 4, and 5 of the textbook by Pickles (1988) and in the first two volumes of the series produced by Webster *et al.* (1992), Fay and Popper (1992), Dallos, Popper, and Fay (1996), and Geisler (1998).

The CM was discovered accidentally by Wever and Bray in 1930 when they were recording from the auditory nerve. They considered the CM to be a neural potential, but this misconception was soon clarified by Adrian (1931) and Saul and Davis (1932). The early work of Tasaki and colleagues (Tasaki, 1954; Tasaki *et al.*, 1954) suggested that the origin of the CM was the cilia-bearing end of the outer haircells. Experiments originally performed by Hudspeth (1983) in which recordings were made directly from haircells have significantly improved our understanding of the CM and its origin.

In the 1970s, minor surgical and nonsurgical techniques were developed to allow for recording of the AP, and, to a lesser extent, the CM and SP from humans. This technique is called *electrocochleography*. The recording electrodes are usually placed somewhere between the earlobe and the cochlea, and the results are similar to those of the round window electrode experiments reported on animals. The most sensitive recording location is the promontory, which is the bony protruding wall between the oval and round windows. To make a recording, the tympanic membrane must be penetrated with a fine needle electrode, the point of which rests on the promontory. This approach to recording cochlear potentials is often called the *transtympanic approach*. In a less severe but less sensitive recording procedure, the recording electrode is placed in the external canal (*extratympanic approach*). An external canal electrode may rest against the wall of the canal or the tympanic membrane, or it may penetrate slightly the wall of the canal or the annular ligament of the tympanic membrane. Each approach, transtympanic or extratympanic, has its advantages and disadvantages.

The direction of stereocilia movement (shearing movement toward or away from the outer haircells) that triggers the haircell to neurally discharge appears to depend on the level of stimulation and where along the organ of Corti the haircells are located. It is also possible that the motility of the outer haircells may alter the excitatory direction of stereocilia movement (see Konishi and Nielsen, 1978; Konishi *et al.*, 1988; Zwislocki and Sokolich, 1973; and Pickles *et al.*, 1984). The ability to perform *in vitro* experiments and to record intercellularly from haircells has enabled hearing scientists to explore in fine detail the function of the haircells and their stereocilia.

The motility of the outer haircells was first reported by Brownell *et al.* (1985). The exact function of the slow and fast types of motility and the possible role of efferent control of motility are much researched topics. Almost all of the work on outer haircell motility uses *in vitro* measures of excised other haircells. No complete *in vivo* study has been done to fully explain how outer haircell motility operates in the normal living cochlea (however, see work by Nuttal and Dolan, 1996).

Cochlear emissions and acoustic echoes represent a new class of phenomena that may also help us understand the cochlea. First reported by Kemp (1978), the phenomena have been the subject of a number of subsequent studies. Since the healthy cochlea produces evoked emissions and the damaged cochlea usually does not, cochlear emissions are being tested as a possible means of evaluating hearing in clinical situations, especially in newborns and infants (see Lonsbury-Martin and Martin, 1990). In addition, these phenomena suggest that the emissions may play a role in understanding *tinnitus*, a disorder in which a person suffers from a ringing or other sound in the ear for which no obvious physical source can be found. However, current research suggests that in only 5% of people who suffer from tinnitus can the tinnitus be traced to cochlear emissions (Bilger *et al.*, 1990).

The theoretical issues surrounding outer haircell motility, feedback, and the cochlear amplifier are hotly debated. It does seem clear that there is no longer a need to postulate some form of a *second-filter* mechanism to explain differences between the sensitivity and frequency selectivity measured along the basilar membrane and that measured in the auditory nerve. These measures are now very similar, and differences obtained in the past were probably due to damaged cochleas in

which outer haircell function is greatly compromised.

Spoendlin's work (1970, 1973, 1974) led to
the realization that 90 to 95% of the afferent
fibers innervate the inner haircells. Liberman's investigations (1978, 1980, 1982) have
also contributed significantly to our understanding of the afferent auditory nerve. When
he recorded discharges from several fibers in
the auditory nerve, he was then able to trace
them to their terminations in the cochlea. All
neurons were found to be radial fibers innervating inner haircells. These data support the
notion that essentially none of the single-unit-
discharge data reported so far have come from
fibers that terminate on outer haircells. These
data are consistent with the earlier work of
Spoendlin (1978, 1979), which indicated that
the outer spiral fibers remain unmyelinated
even after they pass through the habenula
perforata on their course centrally. The lack of
a myelin sheath and their small diameter (less
than 0.5 mm) would make it impossible to

record from them by traditional techniques. In
contrast, radial fibers acquire a myelin sheath
as they pass centrally through the habenula
perforata and have diameters between 3 and 5
mm. A number of terms are used interchangeably in describing a neural cell: cell, unit, fiber
(in connection with the central nervous system), neuron, and nerve. All of these terms
will be used in this book to get the student
used to the multiple uses found in the literature.

The peripheral efferent fibers have proven
difficult to investigate. The first investigator
was Rasmussen (1943), who demonstrated
the existence of the efferents. The more recent
work of Warr, Liberman, Guinan, and their
colleagues (see Warr, Guinan, and White,
1986, and Liberman and Brown, 1986) dis-
proved some earlier ideas about the OCB
pathways and have contributed significantly
to our understanding of the synapses of
the efferent system in the inner ear (see also
Geisler, 1998).

CHAPTER 9

The Neural Response and the Auditory Code

*I*n this chapter we will study the function of the afferent and efferent auditory nerves within the VIIIth nerve bundle. We will learn how the frequency, level, and temporal properties of sounds change the neural output of the afferent fibers that innervate the inner haircells, and the way in which the efferent fibers might modify this output. The chapter ends with a description of the code for sound provided by the auditory periphery.

FUNCTION OF THE AFFERENT AUDITORY NERVE

Haircells initiate the neural part of the auditory process. Mechanical deformation of the stereocilia of haircells produces the graded electrical potentials discussed earlier (i.e., potentials whose magnitudes are proportional to the amount of stimulation), which are called *receptor potentials* (namely, the CM and SP). The haircell then liberates a chemical transmitter, which initiates a graded electrical potential in auditory nerve fibers that innervate the base of the haircell. This graded neural

potential is propagated along the nerve fiber to the habenula perforata, where the myelinated portion of the fiber is reached. At that point, neural *spikes* or *discharges* are produced that travel along the auditory nerve to the cochlear nucleus. Because the graded neural response of the unmyelinated portion of the nerve fiber is thought to generate the spikes, it is called a *generator potential*. Neural spike rate is proportional to the velocity of basilar membrane motion.

The electrophysiologist studies the spike discharge patterns of single neurons by amplifying these minute potential changes with amplifiers. The discharge pattern of a neural spike initially exhibits a relatively large potential shift (see Appendix E). It then remains in an exhausted condition for a short period (known as the *absolute refractory period*) during which no stimulus, however strong, can be effective in generating a spike. After the absolute refractory period comes a period of *relative refractoriness*, during which the nerve's ability to respond depends on the level of succeeding stimuli. This total refractory period lasts approximately 1 msec and therefore limits the number of discharges a given nerve may be capable of emitting per second. At this point, we must be careful to make a distinction between the action potential of an individual neuron and the whole nerve action potential, the AP. The individual neuron's action potential (spike) has an all-or-none characteristic. This means that the amplitude of the neural discharge does not vary with the level of the stimulus, as the AP does. The AP is a combination of the discharges of many auditory nerves. See Appendix E for more details concerning neural physiology.

Spontaneous Activity and Neural Thresholds

Electrophysiologists who record the discharges of a single neuron, or *single unit* response, as it is often called, employ several methods for dealing with the data. They investigate the characteristics of the neurons by these different methods in an attempt to understand how the relevant aspects (frequency, intensity, and so on) of the acoustical signal are encoded. One method of studying such discharge patterns consists of establishing what the neuron's discharge rate is without a stimulus and then using that as a base against which to compare various stimulus-evoked discharge rates. *Spontaneous activity* is the neural activity that occurs without a stimulus. For single units of the auditory nerve this would be the neural activity occurring without sound. As establishing a sound-free environment is virtually impossible, there is always the possibility that the neural activity observed without a controlled sound stimulus may be caused by ambient acoustic stimulation. Thus, specifying the exact nature and cause of spontaneous activity is difficult. For our discussion, however, the neural activity without sound introduced by an experimenter will suffice as our definition of spontaneous

activity. The discharge rate of spontaneous activity in single units of the auditory nerve may range from no spikes, or discharges, to more than 100 discharges per second.

One of the most useful indicators of neural activity is the discharge rate of the single neuron (single unit). The discharge rate is simply the number of times a unit discharges or "fires" in a given period. For instance, we know that due to the absolute refractory period the neuron has a 1-msec period after it has discharged before it can discharge again. Thus, the theoretical maximum discharge rate of a single unit is 1000 spikes/sec, although most auditory nerve fibers have a maximum firing rate of less than 500 spikes/sec.

Once the spontaneous discharge rate of a neuron has been determined, then the *single neuron threshold* to a particular stimulus can be calculated. The threshold of a single neuron may be defined as the *minimum* stimulus level that will cause an *increase* in the discharge rate above the spontaneous discharge rate. For instance, if we wish to determine the neural threshold of a neuron with a spontaneous discharge rate of 14 spikes per second to a 1-kHz sinusoid, we increase the level of the 1-kHz sinusoid until the discharge rate is just detectably greater than 14 spikes per second (e.g., 20 spikes/sec); that level will be the threshold. At a lower level, the discharge rate will remain at the spontaneous rate, and a more intense signal may significantly increase the rate of discharge above 14 spikes per second. However, since the spontaneous activity for any given neuron randomly varies, some statistical change in average firing rate is usually used to determine neural thresholds

Figure 9.1 shows the distribution of the single neuron thresholds for different best frequencies of stimulation. The different areas on the graph refer to the range of spontaneous rates that neurons exhibit. That is, for each neuron recorded, its spontaneous rate was de-

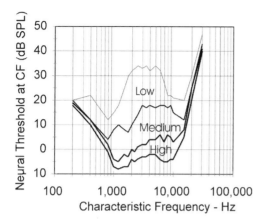

FIGURE 9.1 The neural thresholds (expressed in dB SPL) of auditory nerve fibers of the cat plotted for three different ranges of spontaneous activity (low = <0.5 spikes/sec, medium = 0.5–18 spikes/sec, high = >18 spikes/sec). The x-axis represents the frequency at which the neural threshold for that unit was lowest (i.e., the unit's CF; see text related to Figures 9.3 and 9.5 for further explanation). Adapted with permission from Liberman (1978).

region in which it hears best (see Chapter 10 for a discussion of thresholds of hearing).

Rate-Level Functions

Increasing the level of the acoustic stimulus and measuring changes in the discharge rate of the single neuron above the spontaneous rate is called calculating an *intensity function*, *input–output*, or *rate-level function*. Several such rate-level functions are shown in Figure 9.2. The different curves represent discharge rates from different neurons, each with a different spontaneous rate. Notice that for each neuron the response shows the typical increase in discharge rate for an increase in stimulus level up to 30 to 40 dB above the neuron's threshold, and then the discharge rate remains constant or declines slightly. Essentially all auditory nerve response rates increase over a range of 20 to 50 dB (called the

termined without stimulation and then a tone was presented at different frequencies, and at each frequency the level of the tone that just caused the neuron to fire above its spontaneous rate was determined. This level of the tone was the neural threshold for that frequency. The frequency for which this neural threshold was the lowest was used to determine the best (characteristic) frequency (CF) of the neuron as shown on the x-axis in Figure 9.1. As can be seen, neurons with high spontaneous activity (rates above 18 spikes/sec) have the lowest thresholds; medium spontaneous-rate neurons (between 0.5 and 18 spikes/sec) had the next highest thresholds, and neurons with low rates of spontaneous activity (less than 0.5 spikes/sec) had the highest thresholds. Also notice that the neuronal thresholds are lowest in the frequency region between 1000 and 10,000 Hz, which for the cat is the frequency

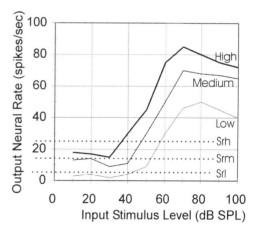

FIGURE 9.2 Intensity or rate-level functions for three neurons: high, medium, and low spontaneous rate fibers. The output of the neurons is plotted in spikes per second as a function of the input level in dB SPL. Sr_h, Sr_m, Sr_l = spontaneous rate of firings for the high (h), medium (m), and low (l) spontaneous rate fibers.

neuronal dynamic range) above their thresholds, after which the discharge rate remains constant or decreases slightly if stimulus level is increased further. The input–output functions for fibers with different amounts of spontaneous activity differ in their intercepts (e.g., in the stimulus level that first causes the fiber to discharge above spontaneous activity).

Response Areas and Tuning Curves

Figure 9.3 shows a family of rate-level functions for a single neuron presented with tones of different frequencies. Any single nerve will respond (i.e., increase its discharge rate above the spontaneous rate) to many frequencies of acoustic stimulation. As can be seen in Figure 9.3a, the neuron does not discharge equally to all frequencies. We can replot the data in Figure 9.3a in one of two ways to display better how the neuron responds as a function of frequency. We can plot an *isolevel* or *isointensity curve* (sometimes called a *response area*) by plotting for each frequency the neuron's response rate for a fixed level of stimulation (i.e., the discharge rate for a 50-dB SPL signal level). The data are replotted in Figure 9.3b as a response area. Notice that this neuron fires most at 2000 Hz, this frequency is often called the *characteristic frequency* (CF) of the neuron (this was the *x*-axis in Figure 9.1). Figure 9.4 shows a family of response areas for a single neuron. Each curve represents a different level at which the tone was presented.

Suppose this neuron has a spontaneous rate of 10 spikes/sec, and, as a result, 15 spikes per second was used as the definition of this neuron's threshold. That is, if the neuron discharges with 15 spikes/sec, then the experimenter is willing to state that the neuron is no longer firing spontaneously and is discharging in response to the stimulus. We can then plot the level of the tone required for the neuron to discharge at this threshold

amount of firing (i.e., 15 spikes/sec) as a function of frequency of stimulation. Such a plot is called an *isorate curve* or a *tuning curve*. Figure 9.3c shows a tuning curve for the data plotted in Figure 9.3a. Notice that the neuron has its lowest threshold at 2000 Hz (the frequency of a neuron's lowest threshold is called its *char-*

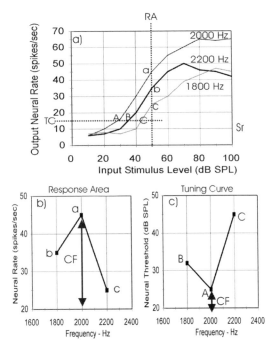

FIGURE 9.3 (a) Three rate-level functions for a single neuron presented with tones with three different frequencies (spontaneous rate, "Sr" is 10 spikes/sec). (b) A response area plotting discharge rate versus tonal frequency (Hz) obtained from Figure 9.3a by determining at a level of 50 dB the discharge rate for each frequency (in Figure 9.3a the vertical dashed line, "RA," passes through points a, b, and c, which are in turn plotted in Figure 9.3b). (c) A tuning curve plotting neural threshold (in dB SPL) as a function of tonal frequency, where the neural threshold is defined as that level required for the neuron to discharge with 15 spikes/sec. The tuning curve is obtained from Figure 9.3a (the horizontal dashed line, "TC," runs through the points A, B, and C, which in turn are plotted in Figure 9.3c).

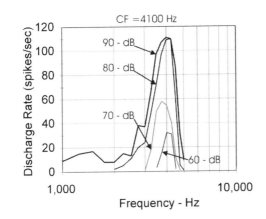

FIGURE 9.4 Response curves for a single unit at six different intensities. The discharge rate is plotted as a function of the frequency of tone stimulation. Each curve is generated by holding the level of the stimulus constant and recording the discharge rate for each frequency. The unit has a characteristic frequency (CF) of 4100 Hz. Adapted with permission from Rose *et al.* (1971).

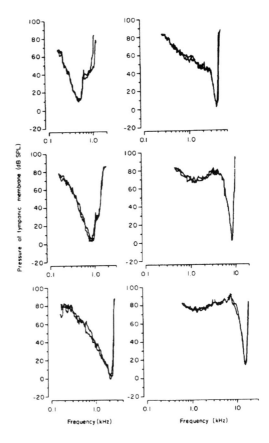

FIGURE 9.5 Tuning curves for six single units with different characteristic frequencies. The stimulus level in dB SPL calibrated at the tympanic membrane needed to reach each unit's threshold is plotted as a function of stimulus frequency. Note the steep slope on the high-frequency side of the tuning curve and the shallow slope on the low-frequency side. The three units on the right have higher characteristic frequencies and show low-frequency "tails," which indicates they are responsive to a wide range of frequencies of stimulation. Adapted with permission from Liberman and Kiang (1978).

acteristic frequency or CF). Figure 9.5 shows a series of tuning curves for different neurons. Notice that each neuron has a different tuning curve with a different CF (also note that the data plotted along the *x*-axis in Figure 9.1 are the CFs of each fiber).

Tuning curves are usually drawn with frequency plotted on a logarithmic scale. The higher-frequency side of the tuning curve appears very steep. The lower-frequency side of the tuning curve is less steep, and for higher CF units, such as the bottom three in Figure 9.5, a long, low-frequency "tail" is obvious. Also notice that the threshold level difference between the tip of the tuning curve at CF and the low-frequency tail of the tuning curve is approximately 40–50 dB for all three units. This *tip-to-tail difference* tends to be 40–50 dB for all auditory neurons. This means that nerves respond best to their CF, are unlikely to respond to many frequencies higher than CF, and will usually respond to frequencies lower

than CF if the stimulation is approximately 40–50 dB above the threshold of CF.

The data in Figures 9.4 and 9.5 show that auditory nerve fibers are very selective to frequency, with each neuron firing best to a limited range of frequencies, and this range is different (the CF is different) for each neuron. These type I nerve fibers synapse with one or a few inner haircells along the cochlear partition (see Chapter 8). It is therefore not surprising that the discharge rate pattern reflects the same type of frequency selectivity obtained at one point along the basilar membrane. That is, *the auditory nerve preserves the frequency selectivity found along the basilar membrane due to the motion of the traveling wave.* Recall that fibers on the outside of the auditory nerve bundle innervate basal haircells, and these fibers have been shown to have high-frequency CFs, whereas those fibers toward the middle of the nerve bundle that innervate the apex of the cochlea have low-frequency CFs (see Chapter 8). This is a further demonstration of the correspondence between the frequency selectivity found in the cochlea and that measured in the auditory nerve.

This frequency selectivity of auditory nerve fibers is often described in terms of bandpass filtering. That is, a single nerve responds best to a limited range of frequencies (near CF) and less to frequencies lower or higher than this range. This is similar to the way in which a bandpass filter passes sinusoids with frequencies in its passband with no change in level and attenuates the amplitudes of sinusoids with frequencies higher and lower than the passband.

Histograms

If we continue to study the auditory nerve fiber, we discover another property of its discharge pattern that might be important for our ability to perceive frequency. Single nerves with CFs below 4 or 5 kHz tend to discharge about once per cycle for low-frequency sinusoid stimulation (below 1 kHz). Figure 9.6 shows the type of discharge pattern we might expect in that situation. This type of response is called a *time-* or *phase-locked response*, or a *following response*, because the response appears locked to, or follows, the peaks in the stimulus. Because of the absolute refractory period of the neuron, the unit cannot dis-

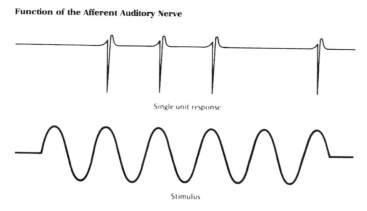

Function of the Afferent Auditory Nerve

Single unit response

Stimulus

FIGURE 9.6 Single neuron discharges to a low-frequency stimulus. Top trace: discharge pattern of the single unit. Bottom trace: the low-frequency stimulus. Note that the neuron responds with a discharge to almost every cycle of the stimulus.

charge to every cycle (peak) of a sinusoid near or above 1 kHz, although the fiber could discharge to every other or every third (and so on) cycle of the stimulus. Note also that the fiber only fires to positive-going peaks in the depiction shown in Figure 9.6. That is, the neuron only discharges when the stereocilia are sheared in one direction (see Chapter 8). The fact that the nerve only discharges to one direction of sound movement is called *rectification*.

To observe phase locking we use a method of data analysis known as *histograms*. Histograms are graphic displays of a single neuron's response to repeated presentations of a given stimulus. The vertical axis of a histogram is usually some measure of the number of responses. It might represent "frequency of responses," "percentage of response," or just "number of discharges." The horizontal axis is always time, but it may represent time after stimulus onset or the time interval between successive neural discharges.

Poststimulus time histograms (PSTs) display the total number of responses at a given time to a repeated stimulus. The beginning point in time is usually some point just before the stimulus onset, such as 5 msec (see Figure 9.7). The data shown in Figure 9.7 illustrate typical PST histograms to tone bursts. The greatest number of discharges occurs at the onset of the tone burst ("on" effect). The number of discharges decreases (*adapts*) rapidly until a nearly constant number of discharges is maintained for the duration of the tone burst. Immediately after the offset of the tone burst, the number of discharges in this neuron is zero or near zero (less than the spontaneous rate). After a short period, the number of discharges returns to the normal spontaneous rate. The PST histograms in Figure 9.7 show the number of discharges at each point in time for 1200 presentations of the tone burst stimulus. It should be clear from comparing Figures 9.6 and 9.7 that the PST histogram can reveal response patterns that would never be seen by simply recording discharge rate or studying the response of a unit to each individual stimulus presentation. Figure 9.7 does not show the time-locked synchrony of the discharges to the lower CF tone because the time scale is too large.

Interval histograms provide a method for studying the time interval between each pair of neural discharges. The horizontal axis of the interval histogram represents the time between successive neural discharges or the *interspike interval*. The vertical axis shows the number of interspike intervals that occurred. Thus, if an interval histogram had its largest peak at 1 msec, then the time between successive discharges was most often 1 msec. Such an interval histogram is shown in Figure 9.8, in which the stimulus was a 1000-Hz sinusoid of 1 sec duration repeated 10 times. Because of the absolute refractory period of the single neuron, we would not expect an interspike interval of less than 1 msec. Although Figure 9.8d shows that a 1-msec interspike interval occurred more than any other specific interval, interspike intervals of 2 to 10 msec were

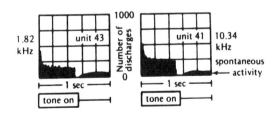

FIGURE 9.7 Typical PST histograms to tone bursts. For each histogram the tone burst is presented 1200 times. The histogram is started 5 msec before the onset of each tone burst. Each time the unit discharges, the histogram increases vertically at the point in time (on the abscissa) the discharge occurred. On average, more responses occur at the beginning of the tone burst than at the end. After the tone burst, there is a decrease in spontaneous activity. Adapted with permission from Kiang *et al.* (1965).

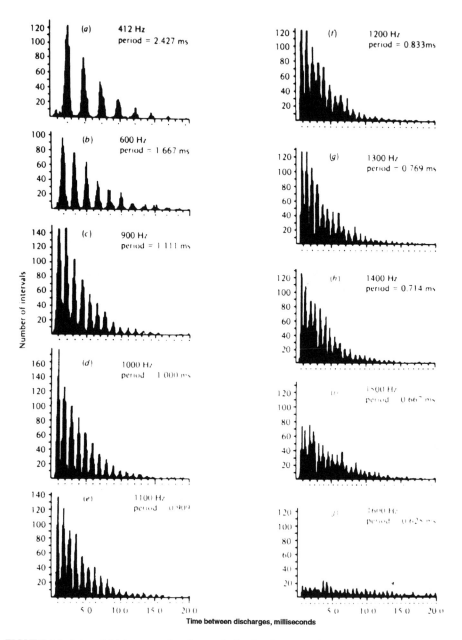

FIGURE 9.8 Interval histograms showing periodic distributions of interspike intervals to pure tones of different frequencies. The stimulus frequency is indicated in each graph. The level of all stimuli is 80 dB SPL, and the tone duration is 1 sec. The responses to 10 stimuli constitute the sample on which each histogram is based. Time in milliseconds between successive neural discharges is plotted on the abscissa. Dots below the time axes indicate integral values of the period of each frequency employed. Reprinted with permission from Rose *et al.* (1967).

also plentiful. Another point of interest in Figure 9.8d is that the neural discharges occurred at integral multiples of 1 msec. This phenomenon is also true of other frequencies of stimulation, as illustrated by the interval histograms to other pure tone stimuli shown in Figure 9.8. The relationship between the period of the stimulus (and integer multiples of that period) and the interval between discharges is shown clearly in Figure 9.8. This supports the idea that single neural discharges are locked to the cycles of the stimulus at low frequencies, as shown in Figures 9.6 and 9.8. Some locking of the responses to the cycles of the stimulating frequency occurs up to about 5 kHz.

Phase-locked PST histograms (sometimes called *period histograms*) provide additional information about timing in the nerve discharge pattern. The time- or phase-locked PST histogram is generated by obtaining a PST histogram while making sure that the counting of neural discharges begins at the same phase of the stimulating waveform each time the stimulus is presented. Quite often, the histogram shows the neural counts over one period of the stimulating signal. An example of this type of histogram to pure tones of different frequencies is shown in Figure 9.9. As can be seen, the neural discharges occur most frequently during only one phase of the tonal waveform (to positive-going displacements, as depicted in the figure). This is consistent with the example shown in Figure 9.6. A measure of *neural synchrony* (*coefficient of synchrony*) can be derived from the phase-locked PST histogram that indicates how well the neural pattern conforms to the stimulus waveform period. To calculate the coefficient of synchrony we measure for one period of stimulation the period histogram. If over one period of stimulation the discharge pattern in the period histogram changes in a sinusoidal fashion (i.e., as if it is following the top half of

a sine wave or the *rectified sine wave*), then the measure of neural synchrony (or *coefficient of synchrony*) is nearly 1.0 (the coefficient of synchrony is almost 1 for the stimuli shown in Figure 9.9). If the neural pattern of the phase-locked PST histogram shows neural discharges evenly distributed throughout the period of stimulation (i.e., flat), then the coefficient of synchrony is near zero. When the coefficient of synchrony is determined as a function of stimulus level, the function is called the *synchrony-level function* to differentiate it from the rate-level function described earlier. Recall from Figures 9.2 and 9.3 that rate-level functions do not increase beyond a 30–50 dB change in input level. Synchrony-level functions show an increase in synchrony over a much larger range of levels.

The response of an auditory nerve fiber to a click stimulus reveals several important points concerning neural transduction. Figure 9.10a,b show period histograms to a rarefaction and a condensation click. The fact that there are periodic peaks in the period histograms is an indication that the clicks undergo processing by the biomechanics of the inner ear that is like the processing produced by a bandpass filter. In Figure 9.10c,d, the time-domain waveforms of a rarefaction and condensation click are shown before and after the click has passed through a bandpass filter. The multiple oscillations in the click waveforms at the output of the filter reflect the bandpass filtering such that the period of oscillations is equal to the reciprocal of the center frequency of the filter. If the nerve discharges to the peaks of the cochlearly filtered click (see Figure 9.9), then the period between the major peaks in the period histogram should be equal to the reciprocal of the CF of the fiber being measured, as is shown in Figure 9.10a,b. This is further evidence that the biomechanics of the inner ear act as if sound is being bandpass filtered. Note also that the time-domain wave-

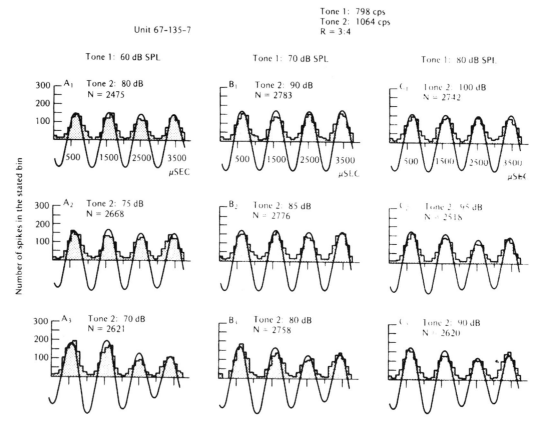

FIGURE 9.9 Time-locked PST histograms to the sum of two pure tones. In the top row, tone 2 is 20 dB more intense than tone 1. In the middle row there is a 15-dB difference between the two tones, and in the bottom row the difference is 10 dB. For all cases the time-domain waveform of the summed sinusoids is superimposed on top of the PST histogram. The nerve discharges only during one phase of the waveform (the positive-going sections of the waveform). The PST histogram displays the ability of the nerve to discharge in synchrony with the period of the input stimulus. Adapted with permission from Hind *et al.* (1967).

forms of the condensation and rarefaction clicks are phase reversed (as they should be since the input clicks are phase reversed). If the nerve discharges when the stereocilia move in only one direction, then the peaks in the period histogram of the condensation click should occur one-half period later than those occurring for the rarefaction clicks, as shown in Figure 9.10a,b.

TWO-TONE SUPPRESSION AND OTHER NONLINEAR NEURAL RESPONSES

The discussion of afferent activity has centered on tonal or simple stimuli. The neural activity to complex stimuli can often be predicted based on what is known about the ac-

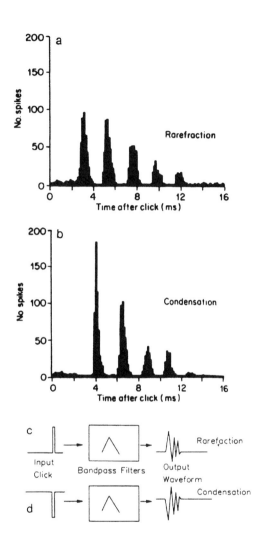

FIGURE 9.10 (a) The period histograms to a rarefaction click. (b) The period histogram to a condensation click. The auditory nerve fiber's CF was 450 Hz. (c) A rarefaction click before and after bandpass filtering. Reprinted with permission from Kiang *et al.* (1965). (d) A condensation click before and after bandpass filtering. The oscillation or ringing of the click in response to the filter occurs with a periodicity equal to the reciprocal of the bandpass filter's center frequency. The multiple peaks in the period histogram that repeat at an interval equal to the reciprocal of the fiber's CF (2.22 ms = 1/450 Hz) indicate that the fiber reflects a process that is like bandpass filtering.

tivity measured for simple stimuli. However, additional phenomena occur when two or more sinusoids are added. One of the more important events arises from presenting two tones together. The presentation of one tone will usually cause a neuron to discharge at a rate well above its spontaneous rate, especially if the tone's frequency is near the fiber's CF. Adding a second tone at a particular frequency and level may cause a decrease in the discharge rate of the fiber to the first tone. This is shown in the top panel of Figure 9.11. When tone A is on alone, the neuron discharges at a high rate. When tone B is added to tone A, the discharge rate of the fiber is greatly reduced. It is as if tone B inhibited or suppressed the neural activity associated with tone A. This phenomenon is, therefore, called *two-tone suppression*. Figure 9.11b indicates the frequencies and intensities of the inhibiting tone (tone B), which causes a decrease in the firing rate to the exciting tone (tone A). The darker line shows the tuning curve for tone A, and the lighter lines indicate the frequencies and intensities of tone B, which reduced the firing rate of the nerve fiber to stimulation caused by tone A. Thus, there are inhibitory regions above and below the best frequency of the fiber's tuning curve. The data shown in Figure 9.11 are based on measuring a decrease in the rate of neural discharge due to the interaction of two tones. A decrease in measures of neural synchrony, such as the coefficient of synchrony, also occurs for proper combination of two tones. Thus, two-tone suppression can be measured neurally as *rate suppression* or as *synchrony suppression*.

Two-tone suppression is a form of a nonlinear response of the cochlea to sound stimulation. The other forms of nonlinearities discussed elsewhere in this book (e.g., Chapters 5, 7, and 14) can also be found in the neural discharge patterns of auditory nerve fibers. That is, at high sound levels, harmonic distor-

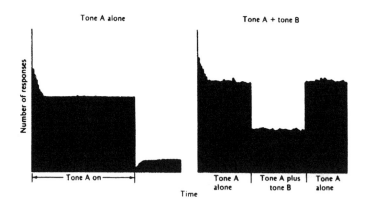

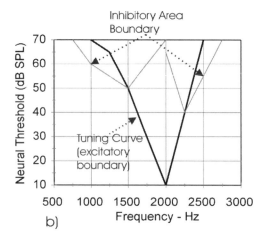

FIGURE 9.11 A schematic diagram showing two-tone suppression. (**a**) In the top figure, the PST histogram to tone A presented alone is shown on the left. If a second tone of the correct frequency and intensity (tone B) is added to the first tone (tone A), the discharge rate of the nerve fiber can be reduced during the time tone B is added to tone A. (**b**) The bottom figure depicts the typical two-tone suppression result in terms of a tuning curve diagram. The dark borders indicate the boundary for the frequencies and intensities of tone A that excite the nerve (i.e., the tuning curve, see Figures 9.3 and 9.5). The light borders indicate the boundary for the frequencies and intensities of tone B that, when added to tone A, decrease (suppress) the discharge rate of the nerve responding to tone A.

tion and difference-tone distortion, especially that associated with the cubic-difference tone (the difference tone existing at $2f1 - f2$, when the primaries are $f1$ and $f2$), can be measured in the neural response of auditory nerve fibers. Although two-tone suppression and the other nonlinear phenomena are easily measured for neural discharges, they can also be measured for biomechanical motion and for the cochlear microphonics (i.e., adding two tones can reduce the magnitude of the CM relative to that which occurred when only one tone was presented). Thus, the origin of these forms of nonlinearities is probably not entirely neural. Other excitatory and inhibitory interactions can occur when two or more tones are presented either simultaneously or in temporal sequence. Care must be taken in predicting the neural processing of a complex stimulus based only on what is known about the processing of simple stimuli.

FUNCTION OF THE EFFERENT SYSTEM

The exact function of the efferent system is not well understood, but electrophysiological evidence indicates that the system is inhibitory in nature. By electrical or sound stimulation of the crossed olivocochlear bundle (COCB) and simultaneous measurement of the cochlear potentials and afferent neural activity, investigators have found several effects measurable in the discharges of single auditory nerve fibers, in the action potential (AP), and in the cochlear microphonic (CM). Most neural studies of the OCB are done by monitoring neural function in the ipsilateral ear and either presenting sounds to the contralateral ear or directly stimulating the COCB electrically with brief electric shocks at the contralateral ear (usually at the floor of the *fourth ventricle*).

Figure 9.12a shows a PST histogram measured in response to a 4.37-kHz tone presented to the ipsilateral ear along with a brief contralateral noise presented in the temporal middle of the ipsilateral tone. When both the contralateral noise and ipsilateral tone were presented, the response rate to the ipsilateral tone was suppressed. Thus, it appears as if contralateral noise stimulates the COCB fibers, which in turn causes inhibition of the neural firing to the ipsilateral tone. Figure 9.12b shows how the rate-level function for ipsilateral stimulation is shifted when the ipsilateral sound is simultaneously presented

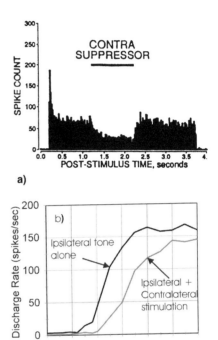

FIGURE 9.12 (a) A PST histogram for a combined ipsilateral 4.37-kHz 3.5-sec tone and a contralateral broadband noise (1 sec long, presented midway through the presentation of the ipsilateral tone). The decrease in discharge rate at 1.5 sec is a result of the contralateral tone exciting the COCB fibers, which in turn inhibit firing of the ipsilateral auditory nerve fiber. (b) A rate-level function measured from an auditory nerve fiber (CF = 1.22 kHz, Sr = 1.6 spikes/sec) when it is excited with a 1.22-kHz ipsilateral tone (solid curve) and then with a combination of the 1.22-kHz ipsilateral tone and a 1.0-kHz contralateral tone (dashed curve). As can be seen, addition of the contralateral tone excites the COCB, which in turn causes the neuron to discharge less to the ipsilateral tone at all levels (i.e., moves the rate-level function to the right). Adapted with permission from Warren and Liberman (1989).

with a contralateral sound (note that for any input stimulus level the neural rate is lower when both tones are present [*light curve*] than when only the ipsilateral tone alone was presented [*dark curve*]). Similar results are obtained when electrical stimulation of the COCB is used rather than acoustical stimulation.

These inhibitory effects may help decrease the neural activity in intense noise situations for protecting the nervous system against noise-induced damage (see Chapter 16). The efferents may also help reduce the effect of background sounds on our ability to detect wanted signals (see Chapter 11 on masking).

ENCODING OF FREQUENCY, INTENSITY, AND TIME

From the information covered in the two preceding chapters we can describe, in a variety of ways, how the auditory system can encode (determine) the frequency, intensity, and temporal/phase characteristics of an acoustic waveform. We have seen that the traveling wave of the basilar membrane vibrates with a maximum amplitude at a place along the cochlea that is dependent on the frequency of stimulation, so that which haircell's cilia are sheared maximally depends on the frequency of the stimulus. The afferent auditory nerve fibers innervate these haircells in a systematic fashion, and each nerve fiber is most sensitive to a particular frequency. The individual fibers within the auditory nerve are also organized according to the frequency at which they are most sensitive and the location along the cochlea at which they innervate haircells; each place or location within the nerve is responding "best" to a particular frequency. The nerve is organized *topographically* (see Chapter 15 for a discussion of tonotopic organization) ac-

cording to the tonal frequency that simulates the auditory system. Thus, the frequency of the input can be determined by noting which nerve fiber (place) within the auditory nerve discharges with the greatest relative discharge rate. This way of determining frequency is called the *place theory*.

It is also apparent from the data shown in Figure 9.6 and from the interval and period histograms that auditory nerve fibers also discharge at rates proportional to the period of the input stimulus when the input frequency is less than approximately 5000 Hz. It is possible, therefore, that information is available to the nervous system about stimulus frequency from an analysis of the periodicity of neural discharge rate. That is, the periodicity of the nerve discharges could be used to determine the frequency of an input stimulus. Thus, for frequencies less than 5000 Hz both the place of maximal discharge and the periodicity of discharge pattern can aid the nervous system in determining the frequency of stimulation. Using discharge periodicity as the basis for frequency processing is called a *temporal theory* of frequency coding.

Intensity is assumed to be encoded by an increase in the discharge rate within the auditory system. This would involve an increase in the discharge rate of a single auditory nerve fiber. Notice, however, in Figure 9.2 that a single neuron's discharge rate will increase only for a relatively small range of level changes (usually less than 35 dB). Thus, a single neuron's response cannot account for the 140-dB dynamic range of level to which we are sensitive (see Chapter 10). This would imply that intensity might be determined by the increase in discharge rate of a large number of fibers rather than just one. For instance, notice that, if we combine the outputs of all three fibers shown in Figure 9.3, we could cover a larger range of intensities than that covered by any one fiber. Thus, combining information from low-, medium-, and high-threshold fibers may serve as the code for sound level. The inclusion of discharges from many fibers whose CFs are other than those of the stimulus (see, e.g., the low-frequency tails of the tuning curves shown in Figure 9.5) may also help to account for the wide dynamic range of the ear. The processing that takes place in the auditory brainstem (see Chapter 15) may also aid in determining a code for stimulus level that encompasses the entire 140-dB range of level to which many animals are sensitive.

Clearly, the fact that nerves discharge in synchrony to the phase of the stimulating sound provides a basis for encoding the phase and dynamic timing aspects of sound. However, such phase and timing information would be limited to periodicities that are greater than 0.2 msec, because 5000 Hz appears to be the approximate upper frequency limit to which auditory nerve fibers can discharge in synchrony with a sound (i.e., 0.2 msec is the reciprocal of 5000 Hz).

Figure 9.13a shows the average rate response of a large number of auditory nerve fibers of a cat to a synthesized speech vowel (the vowel \e\ as pronounced in the word "bet"). This is a *population response* of the auditory nerve to a complex sound with an amplitude spectrum that covers a broad range of frequencies. The data are shown for two levels of stimulation. Each datapoint used to generate the average curve represents the average firing rate of a fiber whose CF is plotted along the horizontal axis (hundreds of auditory nerve fibers were measured in determining the data of Figure 9.13a). The amplitude spectrum of the vowel is shown in Figure 9.13b. The major peaks in the spectrum are called *formant peaks*, resulting from the way in which the mouth, tongue, throat, etc. filter the sound produced by the vocal cords (see Chapter 14). Formant peaks are a key aspect of speech that account for speech recognition. Thus, it is cru-

cial that these formant peaks be well represented in the neural output of the auditory periphery. The data shown in Figure 9.13a indicate that the spectrum of the sound is well represented at low stimulus levels, but a great deal of information is lost when the level is

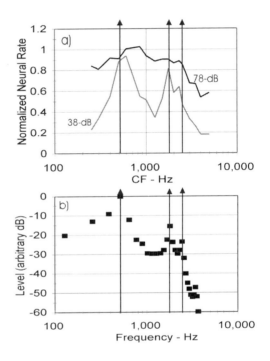

FIGURE 9.13 (**a**) Average normalized rate functions from a population of auditory nerve fibers to the speech vowel, /e/. Normalized rate is the unit's neuronal discharge rate divided by the maximum rate measured. Normalized rate is plotted as a function of the CF of the fibers. The arrows point to the formant frequencies of the vowel, whose spectrum is shown in Figure 9.13b. Two different presentation levels of the vowel are shown. Notice how the spectrum of the vowel is well described by the neural data at the low, but not at the high, stimulus level. (**b**) The amplitude spectrum of the vowel /e/ used in the nerve population studies. The arrows point to the formant peaks. The spectrum was determined near the eardrum of the cat. Adapted with permission from Young and Sachs (1979).

high. That is, the neural rate output from many neurons represents or codes for the levels and frequencies of a complex sound at low, but not at high, stimulus levels. Since the ability to recognize speech does not change much, if at all, when its level is changed, then the information about the spectrum should be preserved at high and low stimulus levels. Figure 9.14 indicates what happens to a neural measure of a population response if rate information is combined with a measure of synchrony (ALSR, *average localized synchronized rate* is a measure that combines the rate at which the nerve discharges and its synchrony). Because nerves continue to respond in synchrony to a sound's periodicities at high

stimulus levels where the rate has saturated (see Figures 8.4, 9.2, and 9.3), using synchrony information along with rate information preserves the neural code for the speech sound's spectrum at relatively high stimulus levels. These population responses indicate that the auditory periphery at the level of the auditory nerve provides a pretty faithful code of the physical attributes of sound, especially at low stimulus levels. Additional sharpening of this information may take place in the neural circuits of the auditory brainstem.

Modeling the Auditory Periphery

Enough is known about the function of the inner ear and the auditory nerve that computer models have been developed to simulate the function of the auditory periphery. Such models have two stages: a stage that simulates the frequency selectivity of the basilar membrane traveling wave and a stage for inner haircell transduction of the vibratory output of the first stage into a neural discharge pattern. Figure 9.15 displays the output of one such computer model to the /e/ vowel sound (shown in Figure 9.13b). Each line is a simulation of the first 50 msec of the PST histogram from auditory nerve fibers. The fibers are organized vertically so that fibers with low CFs are at the bottom of the display and those with high CFs at the top. The vertical axis can also represent distance along the cochlea. Note that CF is roughly logarithmically related to cochlear distance, such that in humans each 3.5 millimeter of basilar membrane is equal to approximately one octave of frequency (CF).

The time delays shown for the low-frequency channels in the model (toward the bottom of the display) correspond to the time it takes the traveling wave to reach the apex (low-frequency end of the cochlea). The

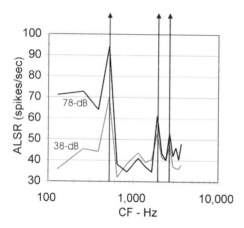

FIGURE 9.14 The Average Localized Synchronized Rate Functions (ALSRs) based on a population of nerve fibers is shown as a function of the fiber's CF. The stimulus is the speech vowel /e/ (its spectrum is shown in Figure 9.13b). The ALSR calculation takes into account both synchronization of the units to the period of stimulation and the rate at which the unit discharges. The arrows point to the frequencies of the formants in the vowel. Two different levels of stimulation are shown. Compare the ability of the ALSR measure to preserve the spectral information of the vowel (see Figure 9.13b) to that provided by just the rate measure (Figure 9.13a), especially at the high presentation level. Adapted with permission from Young and Sachs (1979).

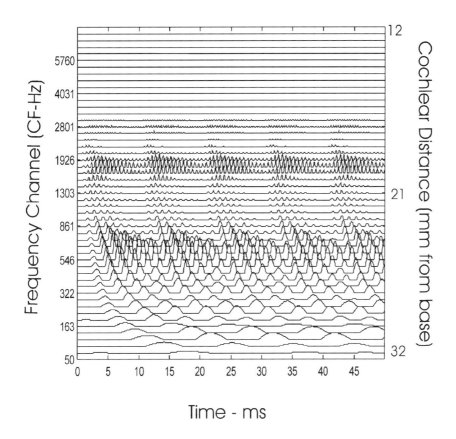

Time - ms

FIGURE 9.15 The output of a computational model of the auditory periphery. The model has two stages, one that represents the traveling wave activity of the cochlea and one that represents transduction of basilar membrane vibration into neural discharges. The stimulus was the vowel /e/, as shown in Figure 9.13b. Each line is a simulation of a PST histogram of a tuned auditory nerve fiber, with low-CF fibers on the bottom and high-CF fibers at the top. The right-hand *y*-axis represents the distance in millimeters from the stapes. The display is a simulation of the pattern of neural activity that is likely to flow up the auditory nerve bundle to the brainstem when the vowel /e/ is spoken. Model from Patterson *et al.* (1995).

higher PST activity in certain channels (e.g., around 550 Hz) represent the fact that the /e/ sound has a spectral peak (e.g., the first formant, see Figure 9.13b) in the region 500–550 Hz, which would cause fibers tuned to 500–550 Hz to discharge at a high rate. The periodic nature of the discharge pattern over time (note that the PST pattern repeats about every 10 msec, or at 100 Hz) represents the periodic nature of the time-domain waveform associated with this /e/ vowel sound (see Chapter 14 for a discussion of speech production). Thus, the display is a stimulation of the neural activity that might flow up the auditory nerve to the cochlear nucleus when the vowel /e/ is spoken.

SUMMARY

The responses of the single neurons of the auditory nerve demonstrate that they have a wide range of spontaneous activity, with low spontaneous fibers having high thresholds and high spontaneous fibers having low thresholds. The unit's discharge rate increases with an increase in stimulus level for a 20–50 dB range above threshold, as revealed in rate-level functions. Single nerves respond to a wide range of frequencies but are most sensitive to one frequency, which is called the characteristic frequency (CF) of a unit, as revealed by response area and tuning curve measures. Neurons with high CFs respond to a wider range of frequencies than do units with low CFs. Interval and period histograms reveal that fibers with CFs below 5 kHz respond to the periodicity of low-frequency stimulation. Two-tone suppression indicates one type of nonlinear interaction that occurs for complex stimuli. Stimulation of the COCB efferent fibers results in changes in a variety of cochlear measures, especially a decrease in single-unit discharge rate. Place and temporal theories are used to account for the neural system's ability to encode frequency. Intensity is thought to be encoded by the number of neural discharges, and the phase and dynamic timing information for rates up to 5.0 kHz may also be preserved in the synchrony with which auditory nerve fibers discharge to the phase of sound stimulation. Computer models can simulate the neural output of the auditory periphery.

SUPPLEMENT

The material in this chapter may be found in Chapters 3, 4, and 5 and part of Chapter 6 in the book by Pickles (1988), Webster et al. (1992), Fay and Popper (1992), Dallos et al. (1996), and Geisler (1998). The data reported in this chapter are based largely on extracellular recordings, where the recording electrode is near but not actually inside the nerve (see Appendices E and F). In 1974, Mulroy et al. and Weiss et al. reported the first recordings of intracellular potentials from the haircells of the organ of Corti of the alligator lizard. Then Russell and Sellick (1977a,b) were able to record intracellular receptor potentials directly from the inner haircells of the mammalian cochlea. In 1980, Sellick and Russell compared the intracellular receptor potentials measured from inner haircells of the guinea pig with the CM recorded from the scala tympani. They concluded that inner haircells respond to the basilar membrane velocity for frequencies below 100–200 Hz. Above that frequency, basilar membrane displacement is the effective stimulus for inner haircell response. They also noted the presence of two-tone suppression at the receptor potential level. Data on intracellular recordings from outer haircells have also been measured (see Dallos et al., 1982; Tanaka et al., 1980). It is now fairly well established that the DC properties of outer haircell function contribute to the discharge rate of auditory nerve fibers, while the AC properties determine the synchronous aspects of neural discharges.

Historically the term CF was defined from the period-histograms to click stimuli, where CF equaled the reciprocal of the time between the major peaks in the click-evoked period histogram (see Figure 9.10). *Best frequency* (BF) was used to define the tip of the tuning curve. Over the years, BF and CF have been used almost interchangeably. We decided to use CF in this book since it appears to be the term most often used to describe the frequency to which a neuron is most sensitive. A number of terms are used almost interchangeably in describing the responses of a nerve: discharges, firings, spikes (e.g., spike rate), and responses. All of these will be used in this book to acquaint the student with their multiple uses in the literature.

The alert reader would have noticed that tuning curves tend to change in their effective bandwidths as their CFs change. Tuning curves for low-frequency CF fibers are about 200 Hz wide (in terms of bandwidth measured at 10 dB up from the tip of the tuning curves) and increase approximately linearly as CF increases, when both bandwidth and CF are plotted on a log axis, such that at a CF of 20,000 Hz fibers have bandwidths that are 2000 Hz wide (the result is that $Q_{10\,dB}$, see Chapter 5, values increase from 1 to 10 as CF increases).

The advent of computer processing of the responses of single auditory nerve fibers brought a mass of new information about the auditory system and a new surge in interest. The responses of single neurons in the auditory nerve can be more complex than we have shown in this chapter. We have presented only the single-neuron response data that are consistent with the cochlear mechanics discussed in the previous chapter, but many exceptions exist in which the neuronal discharges do not reflect basilar membrane motion in a simple fashion.

We have talked about tuning only in biomechanical terms (e.g., the motion of the traveling wave) or in terms of the representation of that motion in the neural discharge patterns of auditory nerve fibers. Fettiplace and Crawford (1980) showed that for many species the haircells themselves are tuned due to bioelectrical properties of the haircell and stereocilia. Thus, additional tuning of the cochlea to the frequency of sound may be provided by the electrical tuning of the haircells (at least in some animals).

The functional role of the efferent system has not been fully established. The OCB clearly has an inhibitory influence on the neural output of the cochlea, and it may play some role in altering the biomechanics of the inner ear, as is discussed in Chapter 8. Since most neural studies of the auditory system have been conducted with the animals under anesthesia and since anesthesia eliminates the normal function of the efferent system, its role may not be fully appreciated, since few data exist that describe how the auditory periphery functions when the efferent system is fully operational. Work by Winslow and Sachs (1988) using electrical shocks of the COCB and Liberman (1991) using contralateral sound stimulation indicate that under the appropriate stimulus conditions OCB stimulation can act to enhance the neural response to signals in quiet or in a background of noise. These results provide additional data for the possibility that efferent stimulation makes the signal-to-noise ratio larger than it would normally be based on the stimulus, enhancing the auditory system's ability to detect signals in noisy backgrounds. Behavioral studies of the significance of the efferent system are few, and their results are generally inconclusive.

The careful reader of Figures 9.11a and 9.12a will notice that the inhibitory change in PST histograms due to simultaneous presentation of two tones is similar in both figures. In Figure 9.11a, both stimuli were presented to the same ear, whereas for Figure 9.12a the stimuli were presented to opposite ears. However, with contralateral stimulation the experimenter must be careful to rule out the possibility that the contralateral stimulus somehow "leaked" across the head and caused direct ipsilateral stimulation. If such a "leak" occurred, then the measured suppression may be more a result of two-tone suppression than COCB activity.

The model's output shown in Figure 9.15 is based on the Auditory Image Model (AIM) of Patterson *et al.* (1995), in which the first stage is a bank of bandpass filters (a Gammatone filter bank, see Chapter 10), and the second stage is based on the Meddis haircell (Meddis, 1986). Greenwood (1990) has documented the relationship between frequency coding and cochlear distance for several species.

PART III

AUDITORY PERCEPTION OF SIMPLE SOUNDS

CHAPTER 10

Auditory Sensitivity

THRESHOLDS OF AUDIBILITY

A first step in understanding auditory perception is to understand the auditory system's sensitivity to frequency, amplitude, and starting phase. When we understand the sensitivity of the auditory system to these variables for sinusoidal stimuli, the results might form a complete picture of the absolute sensitivity of the auditory system to any stimulus because all acoustical stimuli can be defined in terms of sinusoids. However, as we will see in this and following chapters, the solutions are not that simple.

When a pure tone is presented to one ear, the auditory system is actually insensitive to starting phase. This does not mean, as Helmholtz suggested more than 100 years ago, that the auditory system is phase insensitive. For instance, if two sinusoids of different frequencies are added, the perceived quality of the sound may vary as the starting phases of the sinusoids are varied. Also, as mentioned in Chapter 2, if a sinusoid is presented to both ears and there is a change in the interaural (between-ears) phase difference, then observers report a change in the perceived location of the tone (see Chapter 12).

We can measure auditory sensitivity to frequency and level by determining the level required for a listener to detect the presence of a sinusoid at each of many frequencies. There will be very low and very high frequencies to which, no matter how intense the sinusoid, the auditory system is insensitive. These frequency limits define the bounds of the auditory system's sensitivity to frequency. The thresholds relating the smallest level required for detection to the frequency of the tone are called *thresholds of audibility*.

Thresholds of audibility are obtained in the laboratory in the following manner. For each frequency tested, a psychometric function is obtained either directly or indirectly using a psychophysical procedure (see Appendix D). Figure 10.1 demonstrates the results from this part of the experiment. The threshold is determined for each frequency by choosing a performance level, such as 75%, for $P(C)$. The thresholds of audibility are plotted as the threshold in decibels versus frequency. A listener who has a low *threshold* can also be described as being very *sensitive*. Thus, "high sensitivity" and "low threshold" mean the same thing.

Minimum audible field (MAF) thresholds are sound-pressure levels for pure tones at absolute threshold, measured in a free field. A listener listens in a room to tonal sounds presented over loudspeakers. In the absence of

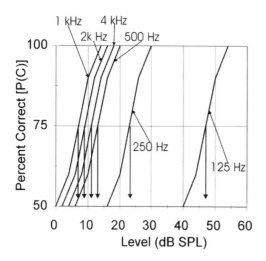

FIGURE 10.1 Psychometric functions obtained for six tones. The thresholds are obtained for these forced-choice psychophysical data at a *P(C)* of 75% (indicated by the arrows). Adapted with permission from Watson *et al.* (1972).

the listener, a microphone is placed at the position of the listener to calibrate the threshold levels. MAF thresholds are usually determined for listeners facing the source, listening with both ears (binaurally), at 1 meter from the sound source.

Minimal audible pressure (MAP) thresholds describe thresholds in terms of the sound pressure level at the observer's tympanic membrane. The listeners listen to sounds presented over earphones, and various procedures are used to determine the sound pressure that occurs at the tympanic membrane. Although the minimum audible pressure at the eardrum is independent of the type of sound source used to deliver the sound, most earphones are not calibrated in terms of the eardrum pressure they produce. To estimate the actual pressure at the tympanic membrane in a standardized manner, the sound-pressure

level is estimated from the sound level in a test *coupler* attached to the earphone during calibration. Such couplers are designed to approximate the average acoustic properties of the ear of a listener with normal hearing and contain a microphone for estimating the sound pressure level that would exist at the tympanic membrane. Since there are different types of earphones and couplers, the threshold sound-pressure levels vary for each earphone–coupler combination. For a particular earphone–coupler combination, the threshold sound-pressure levels measured in the coupler are called the *Reference Equivalent Threshold Sound Pressure Levels* (RETSPLs), and these RETSPLs have been standardized for a variety of earphone–coupler combinations. Each earphone–coupler combination generates a different set of values for RETSPL.

Three classes of earphones, each with its own coupler type, are typically used for most hearing tests. *Supra-aural* phones fit over the pinna and are calibrated using a *"6-cc" coupler* (6 cc represents the average volume of the adult outer ear canal that lies between the earphone and the tympanic membrane). *Circumaural* phones fit completely over and around the pinna and are calibrated using an *artificial ear*. *Insert earphones* fit directly into the outer ear canal and are often calibrated in an *occluded-ear simulator*, sometimes called a *Zwislocki coupler*, after its founder Joseph Zwislocki, or in a 2-cc coupler (since 2 cc is the volume between the tip of the insert earphone and the tympanic membrane). Figure 10.2 and Table 10.1 show the RETSPLs for a few earphone–coupler combination examples.

As can be seen by comparing the MAF with the supra-aural or circumaural earphone MAP thresholds, they do not agree. MAF thresholds are always lower than MAP thresholds. Since the sound impinging upon the tympanic membrane should be the same

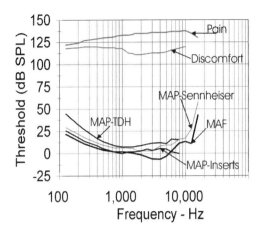

FIGURE 10.2 The thresholds of hearing in decibels of sound pressure level (dB SPL) are shown as a function of frequency for six conditions: RETSPL–Minimal Audible Field (MAF) thresholds, RETSPL–MAP thresholds for a supra-aural phone and 6-cc coupler, RETSPL–MAP thresholds for an circumaural phone and an artificial ear, RETSPL–MAP thresholds for an insert phone and a Zwislocki coupler, thresholds for pain, and thresholds for discomfort. The RESTPL measures are from ANSI 3.6-1996. The thresholds for pain and discomfort represent estimates of the upper limit of level that humans can tolerate.

TABLE 10.1 Thresholds of Audibility for a Number of Testing Conditions (from American National Standard ANSI 3.6-1996 Specifications for Audiometers)

Frequency (Hz)	Thresholds — RETSPL (dB SPL)			
	MAF[a]	Supra aural[b]	Circum-aural[c]	Insert[d]
125	22	45	29.5	26
200	14.5	32.5	–	18
250	11	27	18	14
400	6	17	12	9
500	4	13.5	9.5	5.5
750	2	9	6.5	2
800	2	8.5	–	1.5
1,000	2	7.5	6.5	0
1,250	1.5	7.5	–	2
1,500	0.5	7.5	5.5	2
1,600	0	8	–	2
2,000	–1.5	9	3	3
2,500	–4	10.5	–	5
3,000	–6	11.5	3	3.5
4,000	–6.5	12	8.5	5.5
5,000	–3	11	9.5	5
6,000	2.5	16	9.5	2
8,000	11.5	15.5	16	0
10,000	13.5	–	21.5	–
12,500	11	–	27.5	–
16,000	43.5	–	58	–

[a]Binaural listening, free-field, 0° incidence.

[b]TDH type.

[c]Sennheiser HDA2000, EEC 318 with type 1 adaptor.

[d]HA-2 with rigid tube.

for a threshold response, and since great care is given to calibrating the actual sound-pressure levels for each procedure, all thresholds should be essentially the same. Thus, one might conclude from examining the difference between the MAF and MAP thresholds that a "missing 6 dB" (the average difference between MAF and MAP measures in the mid-frequency range) existed between sound field (MAF) and earphone (MAP). Experimental difficulties led to that erroneous conclusion. When estimates of MAP and MAF data are corrected for head diffraction, outer-ear resonances, and the type of calibration procedure used, the greatest difference between the two types of measurements is maximally 2.5 dB,

indicating that the different reference sound-pressure levels required for calibrating different sound sources reflect differences in the calibration procedure, the fact that the head acts to diffract sound, and that the outer ear has a resonant frequency (see Chapter 6). Therefore, there is no "missing 6 dB."

The shapes of the MAP and MAF absolute threshold functions are approximately the same, showing a loss of sensitivity below approximately 1000 Hz and above approximately 4000 Hz. The loss of sensitivity declines at approximately 6 dB/octave at low frequencies below 1000 Hz and at approximately 24 dB/octave above 4000 Hz (the ability to estimate the slope of the high-frequency side of the thresholds of hearing function is much more difficult than at the low-frequency side due to the difficulty in measuring thresholds above 8000 Hz). Several factors governing the acoustic transfer function of the outer and middle ears (Chapter 6) have been suggested as explanations for the shape of the thresholds of hearing functions. For instance, the loss in middle-ear pressure gain at low and high frequencies (see Figure 6.8) helps explain the loss of sensitivity for these tonal frequencies.

The fact that we can detect tones over a range of frequencies from 20 to 20,000 Hz requires a very good loudspeaker or earphone system. Zero dB SPL corresponds to 20 μPa, which is an extremely small pressure. At this pressure and at a signal frequency of 1000 Hz, the tympanic membrane is vibrating through a total distance approximately equal to the diameter of a hydrogen atom. The horizontal axis in Figure 10.2 was constructed so that the intensity is in decibels of sound-pressure level, or dB SPL, that is, 0 dB SPL is equal to 20 μPa of pressure (see Chapter 3). Therefore, a number such as 50 dB SPL means that the threshold is 50 dB above a pressure of 20 μPa.

The curves in Figure 10.2 represent the thresholds in dB SPL from the ANSI standard on thresholds. These values have been standardized as the recognized thresholds and are the set of threshold values agreed upon when instruments are built and hearing tested. If hearing is tested using such a standard, the thresholds may be expressed in decibels relative to these standardized values. Here, the thresholds are expressed in dB HL (decibel of

hearing level). For instance, a person with a 30-dB HL threshold at 1000 Hz measured with a supra-aural earphone has a threshold 30 dB above that shown in Figure 10.2 and Table 10.1, or the threshold is 37.5 dB SPL (i.e., 30 dB HL plus 7.5 dB SPL).

The thresholds of audibility define the smallest amount of pressure to which the auditory system is sensitive. What about an upper limit in terms of the most pressure the auditory system can tolerate? The upper curves in Figures 10.2 show an estimate of this upper limit. We can ask the listener to say when the sound is "felt," when *pain* is experienced, or when a tickling sensation is felt. The listener can also be asked when the sound level has become uncomfortable (*thresholds of discomfort*). These experiences indicate that the sound pressure is reaching a maximum. This maximum limit is approximately 120–130 dB SPL, and it remains relatively unchanged as a function of the frequency content of the stimulus. Thus, the *dynamic range* (difference between threshold in dB SPL and the maximum limit in dB SPL) of the auditory system is a function of frequency; it is approximately 125–135 dB at 1000 Hz, but 80–90 dB at 100 Hz.

DURATION

One variable that was not specified in the procedures used to obtain tonal thresholds is the duration of the sinusoid. From several points of view, we would expect that, the longer a sound lasts, the easier it is to hear. For the thresholds of audibility shown in Figure 10.2, the tones had a duration of more than 400 msec.

When time (T) is involved in measurement of intensity, we must be careful in noting whether the intensity is expressed in terms of units of power (*P*) or energy (*E*). Remember from Chapter 3 that

$$P = E/T \quad \text{and} \quad E = PT. \quad (10.1)$$

To describe the auditory system's sensitivity to duration, the experimenter can keep either the signal constant in terms of units of power or energy, but not both, while changing its duration. Of course, if energy is maintained constant, then the power of the signal must change as the duration changes as expressed in equation (10.1).

Figure 10.3 shows the thresholds for different frequencies as a function of the duration of signals. In this figure, level is expressed in terms of units of power. Thus, for each dura-

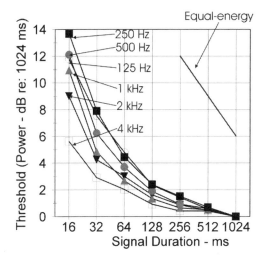

FIGURE 10.3 Thresholds in units of power for six tonal frequencies (ranging from 125 to 4000 Hz) displayed as a function of tonal duration. The thresholds are calculated and plotted in relation to the thresholds obtained at 1024 ms. For instance, the threshold for a 32-ms tone burst with a frequency of 250 Hz (circles) is approximately 8 dB higher than that of a 1024-ms (1-second) tone. The straight line with a slope of 3 dB/duration doubling indicates performance for equal energy detection. Adapted with permission from Watson and Gengel (1969).

tion a psychometric function was obtained in which observer performance was related to signal power. Then for each duration the threshold in units of power was determined from the psychometric functions and plotted in Figure 10.3.

Notice that at durations greater than approximately 250 to 500 msec the threshold in units of power for various tones does not change much; for durations greater then about 250–500 msec the tone does not become easier to detect. However, as the tone's duration is made shorter than 250 msec, the power of the tone must be increased for the observer to detect the tone. This increase is approximately equal to 8 to 10 dB of power increase for each 10-fold decrease in the duration of the tone, although this effect is slightly dependent on frequency. A 10-dB increase in power for each 10-fold decrease in duration means that the signal energy is remaining approximately constant for a constant level of the listener's performance. That is, equation (10.1) states that energy will remain constant if, as duration decreases, power increases. In other words, if $P = E/T$, then $\log P = \log(E/T)$, or 10 $\log P = 10 \log E - 10 \log T$ (see Appendix B, Rule 2). Therefore, $10 \log E = 10 \log P + 10 \log T$ (if T is less than 1 sec). Thus, for a 10-fold change in duration, power must also change tenfold to keep energy constant (or for a doubling of duration signal power must change by 3 dB). For the data shown in Figure 10.3, it is important to realize that duration is expressed in relation to 1 second. Therefore, a change from 1 second to 1/10 of a second (100 msec) is a change of 1/10 or, in decibels, (10 $\log 1/10$) or −10 dB. So as duration becomes shorter (less than 1 second), the power must become greater if energy is to remain constant. For example, if a 500-msec tone has a power of 80 dB, then the energy of this 500-msec tone is 77 dB; that is,

$$10 \log E = 10 \log P + 10 \log T,$$
$$E \text{ in dB} = P \text{ in dB} + 10 \log T$$
$$= 80 \text{ dB} + 10 \log (500 \text{ msec}/1000 \text{ msec})$$
$$(1 \text{ sec} = 1000 \text{ msec})$$
$$= 80 \text{ dB} + 10 \log 0.5$$
$$= 80 - 3 \text{ dB} = 77 \text{ dB}.$$

If the duration is changed from 500 to 50 msec and energy is to remain at 77 dB, then

$$77 \text{ dB} = P \text{ in dB} + 10 \log 50 \text{ msec}/500 \text{ msec}$$
$$= P \text{ in dB} + 10 \log 0.10$$
$$= P \text{ in dB} - 10 \text{ dB}$$

or

$$P \text{ in dB} = 87 \text{ dB} (77 + 10 \text{ dB}).$$

Thus, the power of the 50-msec tone must be 10 dB higher than that of the 500-msec tone, if energy is to remain constant.

Between approximately 10 and 300–500 msec, the energy of the signal must remain approximately constant for constant detection performance by the observer. Note, however, that the constant-energy property of the auditory system is only approximately true and depends on frequency. A complete understanding of auditory processing requires that we explain the exact form of the relation between signal detection and duration.

Once the duration of a sinusoidal signal decreases below 10 msec, much more power is required for detection than is needed to keep energy constant. It appears that the auditory system is not a constant-energy detector below 10 msec. In Chapter 4, we showed that a tone turned on and off spreads its energy over a large frequency region. As the duration of the tone becomes shorter and shorter, this frequency region over which the energy is spread becomes larger and larger. Thus, for very short-duration tones (less than 10 msec) the frequency region over which the energy of the tone is spread becomes so large that not all

of the energy is contributing to its detectability. Because there is some energy at frequencies to which the ear is insensitive, there will be less energy in the region where the ear is sensitive. This in turn means that the total power or energy of the tone must be increased so that enough energy is in the auditory system's frequency region of sensitivity for the tone to be detected.

In other cases (e.g., for a low-frequency tone) the spread of energy associated with the short duration of the tone might produce energy in a frequency region (higher in frequency) where the auditory system is more sensitive (see Figure 10.2). The listener might detect the short-duration tone because of the energy available at these other more detectable frequencies. Therefore, the duration of the signal used to establish tonal threshold is important. If it is longer than approximately 300 msec, the thresholds represent intensity in units of power; if the signal is between 10 and 300 msec, the thresholds reflect approximately constant energy; for signals of durations that are less than 10 msec, the spread of energy makes determination of thresholds difficult.

TEMPORAL INTEGRATION

Notice in Figure 10.3 that further increases in duration beyond 300 msec do not change the detectability of a tone. It is as if the auditory system requires about 300 msec for maximal performance. If the signal is shorter than 300 msec, then its level expressed in units of power must be increased for maximal performance. This property of the detection of signals of different durations is often called *temporal integration*. Figure 10.4 diagrams the basic idea of temporal integration. A signal must have some critical amount of energy (depicted by the hashed area in Figure 10.4a) to

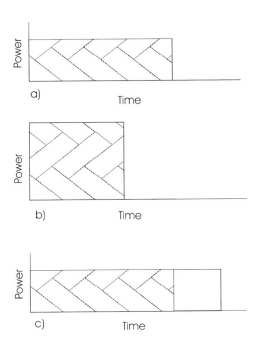

FIGURE 10.4 A schematic diagram depicting the concept of temporal integration. The hashed area at the top Figure 10.4a represents the energy of a signal required to detect the presence of a tone. In Figure 10.4b, the signal is shorter than that in Figure 10.4a; thus, the power of this shorter signal must be increased over that shown in Figure 10.4a to yield a just barely detectable signal. In Figure 10.4c, the signal duration is longer than the time required to achieve the necessary energy for detection; thus, there is no need to change the power of the signal in order for the signal to remain just detectable.

be detected, and once the signal contains that amount of energy (that area), it is detectable. In addition, the process of summing the power (integration) to generate the required energy is completed by 300 msec (the *integration time*). This means that, if the duration is less than 300 msec (the estimate of the integration time from Figure 10.4), then the power of the signal must be increased for the signal to be detected (the height of the rectangle must be increased to achieve the required area, as shown in Figure 10.4b). For durations greater than 300 msec, the threshold expressed in units of power remains constant for a constant level of performance, since the required integration time has been met or exceeded (Figure 10.4c).

The estimate of the integration time for auditory processing can vary a great deal from stimulus condition to stimulus condition. For instance, detection of a tone requires an integration time of about 300 msec; however, detection of a click (impulse) stimulus requires an integration time of only a few milliseconds. Thus, the integration time of auditory processing depends on the type of signal being processed. Estimates range from 1–2 msec to more than 500 msec.

DIFFERENTIAL SENSITIVITY

Although the thresholds of audibility define the frequency and intensity range of the auditory system, it does not describe our sensitivity to *changes* in intensity and frequency. In the early 1800s, Weber observed that it was easy to distinguish between a 1- and a 2-pound weight but not so easy to differentiate a 100-pound weight from a 101-pound weight, although both pairs of weights differ by 1 pound. It was found that the difference between two weights that could just be detected was proportional to the value of the smaller weight. That is, if the *just-noticeable difference* (jnd) for a 1-pound weight was 0.1 pound, then the jnd for the 100-pound weight was 10 pounds. Weber and Fechner stated this relation in equation form:

$$\Delta S / S = \text{constant},$$

where ΔS is the just noticeable physical difference in some stimulus value, and S is the smaller of the two values being discriminated. The ratio $\Delta S/S$ is called the *Weber fraction*. Psychoacousticians have attempted to determine whether the Weber fraction for the level, frequency, and duration of sinusoids is a constant and, if so, over what range of levels, frequencies, and durations.

Extreme care must be taken in studying differential sensitivity. As mentioned previously, not all of the energy of a tone is at the tonal frequency if the tone is very short or if it is turned on and off abruptly. This spread of energy also arises if the frequency, or level, or phase of an ongoing tone is suddenly changed. Thus, we cannot simply study differential sensitivity by changing the level or frequency of an ongoing tone and asking a listener if a change is detected. The listener will probably hear the change because of detecting a click resulting from the spread of energy to other frequencies rather than an intensity, frequency, or duration change per se. In most studies, two tones are presented: one tone is the standard tone and the second is called the comparison tone, which is presented with a slight change in intensity, frequency, or in some cases duration (although, as we will discuss later, there are other methods for measuring duration discrimination).

FREQUENCY DISCRIMINATION

Figure 10.5 shows the value of threshold Δf (frequency difference) required to just discriminate a change in frequency from a given frequency, f. The data are plotted as threshold Δf versus f, with the various curves representing different levels. These data represent the values of the difference threshold (Δf) for frequency. That is, psychometric functions in which the listener's performance was related to various frequency differences could be obtained for many different base frequencies, and then the difference threshold for fre-

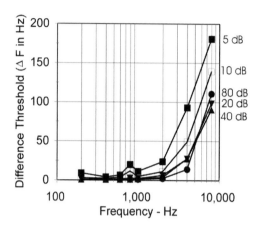

FIGURE 10.5 The value of threshold Δf (Hz) required to just discriminate between two different frequencies is shown as a function of the base frequency for five stimulus levels, expressed in dB SL. Adapted with permission from Weir *et al.* (1977).

quency calculated from the psychometric functions.

As can be seen, the value of threshold Δf increases as f increases above 1000 Hz. In the mid-frequency region, this increase is enough to maintain the Weber fraction for frequency, $\Delta f/f$, approximately constant. This can be seen in Figure 10.6, in which the Weber fraction ($\Delta f/f$) is plotted as a function of f. Notice that over an intermediate range of frequencies this fraction is nearly constant at approximately $\Delta f/f = 0.002$ (or 0.2%). This means that at low frequencies threshold Δf can be as small as 1 Hz. (For instance, if f is 800 Hz and $\Delta f/800 = 0.002$, then $\Delta f = 800 \times 0.002 = 1.6$ Hz.)

INTENSITY DISCRIMINATION

Figure 10.7 demonstrates the value of threshold ΔI in dB required for the observers to detect a difference in level from the initial

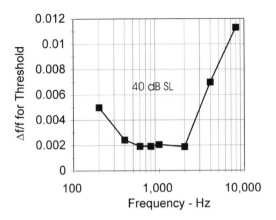

FIGURE 10.6 The value of $\Delta F/F$ (the Weber fraction for frequency) is shown as a function of frequency for the 40 dB SL data plotted in Figure 10.5. The results show that over an intermediate range of frequencies the Weber fraction is approximately constant.

+ $I)/I = K$, which is another form of the Weber fraction. Now by expressing this ratio in decibels, we have: $10 \log [(\Delta I + I)/I] = 10 \log (\Delta I + I) - 10 \log (I) = 10 \log K = C$, another constant. This last equation is the same as the decibel difference between the more intense stimulus, $I + \Delta I$, and the less intense stimulus, I. This is the same as ΔI in dB or ΔI "in decibels." Thus, if ΔI in dB is a constant, then $\Delta I/I$, not in dB, is also a constant. ΔI "in decibels" is plotted along the left-hand axis in Figure 10.7. From the relations shown above, ΔI in decibels is a simple transform of $(\Delta I + I)/I$ or $\Delta I/I$, as shown on the right-hand axis of Figure 10.7. These relationships also show the various forms that the Weber fraction can take when intensity discrimination is studied.

Intensity discrimination data for a variety of tones of different frequencies and a wide-

value I. The value of threshold ΔI in dB is plotted as a function of I, and the different curves represent the results using different tonal frequencies and a wideband noise. Threshold ΔI in dB is the difference threshold for level.

Notice that the auditory system is sensitive to approximately a 0.5- to 1.0-dB change in level across a broad range of frequencies and levels. The fact that threshold ΔI in dB is nearly a constant as a function of I implies that $\Delta I/I$ (not in decibels) for pressure, energy, or power (or the Weber fraction) is a constant. This results from the fact that the logarithm (decibels) of a ratio (such as the Weber fraction) is equal to a difference between the logarithm of the divisor and that of the dividend (see Appendix B).

Let $\Delta I/I = c$, a constant. This is the Weber fraction. By slightly changing this equation, we get: $(\Delta I/I) + 1 = c + 1 = K$, another constant. We can write this last result as: $(\Delta I/I) + 1 = (\Delta I$

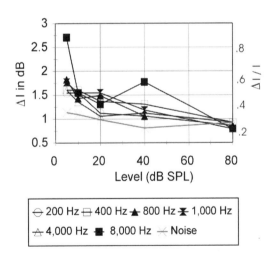

FIGURE 10.7 The value of threshold ΔI in decibels (the difference in decibels between the more and less intense tones) required for threshold discrimination is shown as a function of overall tonal level in dB SL. Data for six tonal frequencies and a wideband noise are shown. On the right-hand axis is $\Delta I/I$ not in decibels. Adapted with permission from Jesteadt *et al.* (1977) and Viemeister (1974).

band noise are shown in Figure 10.7. Notice that the thresholds for threshold ΔI in dB remain very near to 0.5–1 dB for a wide range of levels for the wideband noise, but that the threshold ΔI in dB decreases slightly, but steadily, as the level of the tones increases. The fact that the thresholds for tones are not constant when plotted in terms of ΔI in dB means that they do not strictly follow the Weber fraction. This fact is often referred to as the *near miss to the Weber fraction*. The "near miss" applies to tonal but not to wideband noise stimuli.

TEMPORAL DISCRIMINATION

If we asked a listener to detect the difference between a 50-msec sinusoid and a 60-msec sinusoid, additional variables aside from the tone's duration would be present that might aid the listener in detecting a difference in duration. Recall from Figure 10.3 that, because tones of different durations sometimes have different thresholds, the 50- and 60-msec tones might appear different in detectability. Also, the spectrum of pulsed tones depends on the duration of the tone (Chapter 4), so that the two tones differ in spectra. These two additional variables (spectra and detectability) make the study of temporal discrimination difficult.

To measure sensitivity to duration, some investigators have attempted to avoid these confounding problems by presenting the listener with what they call acoustic markers. To measure duration discrimination the stimuli may consist of the following. A standard stimulus in which a tone of 170-msec duration is followed 10 msec later by an identical 170-msec tone. And, the comparison stimulus in which the two 170-msec tones are separated by 20 msec instead of by 10 msec. The lis-

tener's task is to decide whether these two stimuli are different in time separation between the two 170-msec tonal markers. If the durations are correctly judged to be different, then one assumes that the auditory system can detect this 10-msec difference in duration between the markers.

The data in Figure 10.8 show the results from such an experiment. At various standard separations of T milliseconds between the two tonal markers, the value of the additional amount of time (ΔT) required to detect the increase in duration is shown for the condition when the markers were 1000 Hz and 85 dB SPL. Thus, as the temporal separation (duration) increases, a greater and greater change in the temporal separation is required to make a temporal discrimination. Although threshold ΔT increases as a function of T, it does not increase at a rate such that $\Delta T/T$ (Weber fraction) is equal to a constant. Notice that at T equal to approximately 1 msec threshold ΔT is approximately 2 msec, so the Weber fraction is

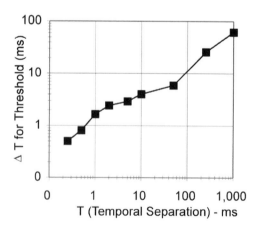

FIGURE 10.8 Value of threshold ΔT (change in temporal separation between two tonal markers in ms) is shown as a function of T (standard separation between the two tonal markers in ms). Adapted with permission from Abel (1971).

2.0. At 300 msec of T, the value of threshold ΔT is 30 msec, so the Weber fraction is 0.1.

Although these results, plus those of other investigations, show that the difference in time required for temporal discrimination increases as the standard time increases, the exact nature of the temporal relationships depends on many different stimulus conditions.

TEMPORAL MODULATION TRANSFER FUNCTIONS

The conditions used to obtain the data of Figure 10.8 are based on a single change in the temporal property of sound. Most real-world sounds have fluctuating temporal changes. Detection of sinusoidally amplitude-modulated (SAM) wideband noise can be used to measure auditory sensitivity to fluctuating temporal changes. Recall from Chapter 4 (see Figure 4.16 and equation (4.6)) that sinusoidal amplitude modulation for a noise can be written as $[1 + m \sin (2\pi F_m t)]n(t)$, where m is modulation depth, F_m is modulation rate, and $n(t)$ is the noise carrier stimulus.

The basic task for the listener, as shown in Figure 10.9, is to detect which wideband noise stimulus (panel **a** or **b**) is sinusoidally amplitude modulated. A psychometric function is obtained for each rate of amplitude modulation (F_m) relating performance, such as $P(C)$, to depth of modulation (m). Threshold modulation depth is determined from these psychometric functions and is plotted as a function of modulation rate such as that shown in Figure 10.10. The value of threshold m (threshold modulation depth, where m ranges from 0 to 1) is often expressed in decibel units as $20 \log m$, where 0 dB means 100% modulation ($m = 1$) and the more negative threshold depth becomes in decibels, the smaller the depth (for $m = 0$ the noise is unmodulated and $20 \log m = -\infty$; if $m = 0.5$, then $20 \log m = -6$ dB).

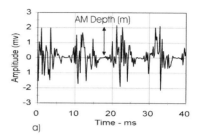

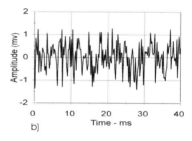

FIGURE 10.9 An unmodulated (**b**) and a sinusoidally amplitude-modulated noise (**a**) used to determine a temporal modulation transfer function. The listener is to determine which sound is amplitude modulated, and the depth of modulation (where AM depth = m) is adjusted to determine threshold.

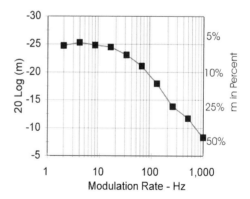

FIGURE 10.10 The temporal modulation transfer function (TMTF) for a wideband noise stimulus showing threshold depth of amplitude modulation (expressed in decibels as $20 \log m$, where m is the threshold depth of modulation) as a function of the rate of sinusoidal modulation. On the right-hand axis is shown modulation depth (m) in percent. Adapted with permission from Bacon and Viemeister (1985) and Viemeister (1979).

As can be seen, the ability to detect sinusoidal amplitude modulation of noise remains fairly constant up to modulation rates of approximately 50 Hz, and as modulation rate increases beyond 50 Hz the thresholds decline, indicating that fluctuations in the amplitude of the noise are more difficult to detect. That is, as modulation rate increases (the time between amplitude peaks in the modulated noise becomes shorter), the depth of modulation must be increased in order for the listener to detect the presence of modulation.

The shape of the data curve in Figure 10.10 is like that of a lowpass filter (see Chapter 5 and Figure 5.2). Thus, imagine that the auditory system acts like a lowpass filter, attenuating the depth of modulation for modulation rates above its cutoff frequency (which according to Figure 10.10 would be approximately 50 Hz), making it difficult to detect amplitude modulation for high modulation rates. Such a lowpass filter describes the way in which amplitude-modulated stimuli might be transformed by the auditory system. These data are, therefore, often described as determining the *temporal modulation transfer function* (TMTF) of the auditory system.

SUMMARY

The threshold of audibility describes the auditory system's sensitivity to level and frequency. The thresholds are measured in either an MAF or MAP paradigm, both of which involve a calibration procedure. The duration of the tone used to determine absolute threshold is crucial. If the duration is between about 10 and 300 msec, its energy must remain approximately constant for a constant level of detection by the observer. For durations of 300 msec or longer, the power must be held constant, whereas for durations shorter than 10 msec much more energy is required for tonal detection because short-duration tones are affected by spread of energy. Temporal integration is often used to describe these duration effects. The data from listeners detecting differences in frequency obey, to a first approximation, the Weber–Fechner law, with $\Delta f/f$ equaling approximately 0.002. The Weber fraction also applies approximately to discriminations of differences in tonal level. In this case, only a 0.5 to 1.0 dB difference is necessary for reliable discrimination performance. Temporal discrimination is difficult to study due to detectability and spectral changes associated with changing duration. The temporal modulation transfer function is used to describe detection of amplitude modulation for different rates of modulation.

SUPPLEMENT

Measures of auditory thresholds have existed since the early part of the twentieth century, and Sivan and White (1933) conducted the first thorough study of them. Parts of Chapters 1, 2, 3, and 4 in the textbook by Moore (1997) cover topics related to auditory sensitivity. Chapter 2 by Green in the book by Yost, Popper, and Fay (1993) also contains information relevant to this chapter. Laming (1988) provides an interesting discussion of differential sensitivity in audition and vision that pertains to some observations made in this chapter.

Because thresholds of hearing are used to define hearing loss for the hearing impaired, they have been standardized both in the United States and internationally. The ANSI standards establish the thresholds of audibility (sometimes called an *audiogram*) for young adults with normal hearing, as outlined in

Table 10.1. Copies of these and other standards may be obtained either from ANSI, 1430 Broadway, New York, New York 10018, or from The Acoustical Society of America, Standards Manager, Suite 1NO1, 2 Huntington Quadrangle, Melville, New York 11747-4502.

Sometimes an individual may have a slightly different threshold in the frequency region near 4000 Hz. Because this is the frequency region near the resonant frequency of the outer ear, it is usually assumed that alterations in normal thresholds in this frequency region are due to some consequences of these resonances.

The difference between the MAF and the MAP thresholds has been of concern to auditory scientists for years. The differences are almost entirely explained if the impedance properties of the middle and inner ear, along with the resonances of the outer ear, are carefully measured and properly combined. Killion (1978) and Yost and Killion (1997) describe these calculations.

The ability to measure thresholds at very high frequencies (above 8 kHz) is difficult because of the resonance and standing-wave acoustics of the outer ear. There are large differences in the acoustics at or near the tympanic membrane depending on the type of sound source (external speaker, supra-aural, or insert) used to present the sound. A major difference between measuring thresholds with a supra-aural and an insert earphone is the effect of internal noise due to the *occlusion effect*. When the outer ear is closed with a supra-aural earphone cushion, the effects of internal noise such as that caused by breathing and the pulsing of blood through the arteries near the ear canal interfere with the ability to detect sounds since the ear is occluded by the earphone. Deeply seated insert phones produce less of an occlusion effect and, therefore, produce lower thresholds than supra-aural phones at low frequencies (see Yost and Killion, 1997).

As mentioned in Chapters 6 and 7, the inner ear can be vibrated directly via bone conduction. A common hearing test is to vibrate the mastoid (bone of the skull behind the ear) or the forehead with a bone vibrator. One can then measure the bone vibration force required to produce a just-detectable sound for a variety of frequencies in much the same manner as the air-conducted thresholds of Figures 10.1 and 10.2 were obtained. The difference between the air-conducted (through earphones) and bone-conducted thresholds may be used to estimate the site (in the middle ear versus the inner ear or nervous system) of a hearing abnormality. The values given in Table 10.2 show the force (relative to 1 dyne) required for normal bone-vibrated thresholds, when a vibrator, meeting the specifications of the standard, is applied to the mastoid (column 1) and the forehead (column 2). These values can then be used to determine if a person has a bone-conducted hearing loss in the same way the values in Table 10.1 are used to determine if someone has an air-conducted hearing loss.

The concepts of temporal integration may be expressed in the following formula:

$$T(I - I_\infty) = C,$$

where T is the tonal duration, I is the tonal level at threshold and I_∞ is the threshold level for a very long (greater than 1 second) tone, and C is a constant. Thus, in order to maintain the value of C constant, I must decrease as T increases, as is consistent with the data shown in Figure 10.3. Recent discussion of temporal integration and alternative methods to account for changes in auditory perception as a function of temporal variables can be found in Viemeister and Plack (1993).

TABLE 10.2 RETFLs for Audiometric Bone Vibrators, from ANSI 3.6-1996 (see Table 10. 1)

Frequency (Hz)	Mastoid locations (decibels) relative to 1 dyne	Forehead location (decibels) relative to 1 dyne
250	67	79
400	61	74.5
500	58	72
750	48.5	61.5
800	47	59
1000	42.5	51
1250	39	49
1500	36.5	47.5
1600	35.5	46
2000	31	42.5
2500	29.5	41.4
3000	30	42
4000	35.5	43.5
5000	40	51
6000	40	51
8000	40	50

To be consistent with the definitions given in Chapters 2 and 3, we should refer to intensity discrimination as level discrimination. However, since the term "intensity discrimination" is used so widely, we have used it to describe sensitivity to a change in sound level. Riesz (1928) made the first accurate estimates of intensity discrimination thresholds by using a beating stimulus. Riesz reasoned that the ability to hear a slowly beating sinusoid must relate to the ability to detect the change in level that is occurring over time. As Jesteadt *et al.* (1977) argue, this method produces somewhat lower intensity discrimination thresholds than the method they used (Figure 10.7). The beating may also have some additional spectral information that aids the listener in

discrimination. Thus, a more accurate estimate of intensity discrimination is probably obtained in a forced-choice procedure, using two tones of slightly different levels.

Table 10.3 outlines the different ways one might calculate these levels in estimating intensity discrimination thresholds. As Grantham and Yost (1982) showed, there are a variety of methods used to calculate the intensity discrimination thresholds. The more intense tone may be generated by adding two tones, S (the signal) and M (the masker). The less intense tone is M. (See Chapter 11 for a discussion of signals and maskers.) When adding two tones, the phase relationship (α) between the tones (see Appendix A) must be used to compute the summed level. Table 10.3 indicates the formulas used to compute a variety of measures of intensity discrimination. The entries marked a and b in Table 10.3 are the two used for the data shown in Figures 10.6–10.8. The work of Viemeister (1974) should be consulted in regard to the near-miss to Weber's law, the differences between tonal and noise intensity discrimination, and issues pertaining to coding of level by the auditory system (see also Green, 1993, and Viemeister and Plack, 1993).

Shower and Biddulph (1931) used a frequency modulation (see Chapter 4) technique to measure frequency discrimination thresholds, for approximately the same reason that Riesz used beats to measure intensity discrimination thresholds. The results of Weir *et al.* (1977), as displayed in Figures 10.4 and 10.5, are in fair agreement with these earlier data. Jesteadt and Bilger (1974) discuss problems in measuring frequency discrimination thresholds.

The potential power of the TMTF is based on treating the obtained TMTF as a lowpass filter transfer function for predicting auditory sensitivity to other forms of amplitude modu-

TABLE 10.3 Expressions for Four Common Measures of Intensity Discrimination in Terms of Power, Amplitude, and the Weber Fraction (the more intense stimulus is the sum of the signal plus masker and the less intense stimulus is the masker)

Measure of intensity discrimination	Power	Amplitude	Weber fraction
Increment power (ΔI)	$P_{m+s} - P_m$	$\frac{1}{2}V_s^2 + V_m V_s \cos\alpha$	$(\Delta I/I) = \Delta I$
Relative increment power (Weber fraction, $\Delta I/I$)	$(P_{m+s} - P_m)/P_m$	$V_s^2/V_m^2 + 2(V_s/V_m)\cos\alpha$	$\Delta I/I^b$
Difference limen ("ΔI in dB")	$10\log P_{m+s}$ $-10\log P_m$	$10\log[V_s^2/V_m^2 +$ $2(V_s/V_m)\cos\alpha + 1]$	$10\log[(\Delta I/I + 1)]^a$ $= 10\log[(\Delta I + I/i]$
Signal-to-masker	P_s/P_m	V_s^2/V_m^2	$[(\Delta I/I + \cos^2\alpha)^{1/2} -$ $\cos\alpha]^2$

[a]Refers to form used on left axis in Figure 10.8.

[b]Refers to form used on right axis in Figure 10.8.

P = stimulus power, V = stimulus amplitude; s = signal stimulus, m = masker stimulus; α = phase angle of addition between signal and masker. All entries not in decibels. All computations assume measurements made into a nominal 1-ohm load, such that $P = 1/T \int_0^T x^2(t)\, dt$, where T (in units of seconds) is the signal duration and $x(t)$ is the time-domain waveform. See Chapter 11 for a discussion of the signal-to-masker ratio.

lation that might be imparted to a sound. The TMTF predicts how we lose our sensitivity to modulation as the rate of modulation increases. One such form of modulation is square-wave (see Chapter 4) modulation, in which the noise is turned on and off abruptly in a repeating fashion that happens if the noise is multiplied by a square wave rather than by a sine wave. The TMTF does a good job of describing a listener's ability to detect square-wave modulation. One extreme version of square-wave modulation is a noise with a temporal gap in the middle. A sound with a temporal gap is also similar to stimuli like those discussed for Figure 10.8 in which two identical sounds mark a temporal interval. A great deal of work has been done on

listener ability to detect temporal gaps of different widths for assessing temporal auditory sensitivity (see Viemeister and Plack, 1993).

One can also measure modulation detection for sinusoidal carriers (see Dau *et al.*, 1997). However, recall that sinusoidally amplitude modulating a sinusoidal carrier produces a stimulus with sideband components (see Chapter 4) that are separated from the carrier frequency by a frequency difference equal to the modulation frequency. Thus, at fast rates of modulation (high modulation frequencies) these sideband components are at frequencies that differ significantly from the carrier frequency. As a result, listeners might detect these sideband components as their cue for discriminating an unmodulated tonal carrier from a modulated one. Such sideband detection would confound the ability to measure modulation processing per se.

The measure of the temporal processing abilities of the auditory system is sometimes referred to as temporal acuity. Green (1971) and Viemeister and Plack (1993) provide excellent reviews of some of these concepts and experiments. Viemeister and Plack (1993) describe the procedures used to estimate temporal processing time from TMTF experiments. Such TMTF estimates of temporal processing time are usually much shorter than most estimates of temporal integration times. Viemeister and Plack (1993) suggest a possible way to reconcile the different estimates of temporal processing time.

CHAPTER 11

Masking

*S*ounds in our environment rarely occur in isolation; often many stimuli occur either simultaneously or close together in time. The study of masking is concerned with the interaction of sounds. The experimenter is interested in the amount of interference one stimulus can cause in the perception of another stimulus. A change in stimulus threshold is a typical measure of the amount of interference produced. Tonal masking, for instance, deals with the change in tonal threshold of one tone associated with the interference, or masking, produced by another tone.

TONAL MASKING

We might guess that there is probably a great deal of interaction or masking between two stimuli with frequencies that do not differ by much. To investigate this interaction, the following experiment is performed. The listener is asked to detect the presence of a weak-intensity sinusoid (perhaps 5 dB SL) at one frequency (perhaps 1000 Hz)—the *signal tone*. Another tone (the *masking tone*), usually with a frequency different from that of the signal, is

presented simultaneously with the signal. The level of the masker that leads to threshold detection of the signal is used as the indicator of the amount of masking the masker has provided for the signal. If the masker is very intense and the listener can still detect the signal, the masker is not very effective in interfering with detection of the signal. On the other hand, if a weak-intensity masker causes the signal to be undetected, then the masker is effective in interfering with signal processing. Figure 11.1 shows the results from just such an experiment, in which the listener was presented with either the signal-plus-masker followed by the masker or the masker followed by the signal-plus-masker. In this two-alternative, forced-choice procedure, the listener had to decide whether the signal appeared the first time (first observation interval) or the second time (second observation interval). At masker frequencies either lower or higher than the signal frequency, the masker level required for threshold detection was higher than when the masker frequency was near the signal frequency. This indicates that frequencies different from the signal frequency are not as effective at masking as those near the signal frequency. Figure 11.1 shows a family of masking

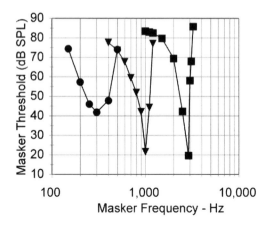

FIGURE 11.1 Three psychophysical tuning curves for simultaneous masking are shown. The different curves are for conditions in which the signal frequency was 300, 1000, and 3000 Hz. Based on data from and adapted with permission from Wightman *et al.*, (1977).

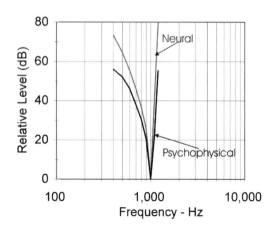

FIGURE 11.2 A comparison of an auditory nerve-tuning curve (from Figure 9.5) and a psychophysical simultaneous tuning curve (from Figure 11.1). The auditory nerve-tuning curve is narrower than the psychophysical tuning curve.

curves obtained with different signal frequencies. In general, their shapes are similar.

Both intuition and knowledge about the neural activity of the auditory periphery should enable you to understand the shape of the curves in Figure 11.1. Consider the "tuning curves" for auditory nerve fibers described in Chapter 8. These curves show that a particular neuron is most sensitive to one frequency of stimulation and that, as the frequency of stimulation differs from the nerve's best frequency, a higher level of stimulation is required to drive the neuron at its threshold value. The similarity between the procedure used to obtain the neural tuning curves and that described above to obtain the masking data shown in Figure 11.1 has led to the name *psychophysical tuning curves* for these masking functions. Not only are the procedures similar, but so are the derived functions. Both the psychophysical and neural tuning curves have a sharp tip at the center or test frequency, and

the high-frequency sides of the curves are steeper than the low-frequency sides. However, the psychophysical tuning curves are not as narrow as the neural tuning curves. These comparisons are shown in Figure 11.2.

A variety of interactions can take place when two tones are presented together. These interactions are important, both for a greater understanding of how the auditory system functions and in terms of recognizing the precautions that must be taken when measuring masking. In Chapter 4, the phenomenon of beats was described. When two tones are close together in frequency, the time-domain waveform has a modulated pattern that is called beating, and we often experience a waxing and waning in loudness or beats. In this situation (i.e., when two tones are close together in frequency, one with frequency $f1$ and the other with frequency $f2$), we hear a tone at a frequency equal to the average of the two frequencies $(f1 + f2)/2$ wax and wane in loudness at a rate equal to the difference between the two frequencies ($f1 - f2$). The "best-beating" sensation for a continuously presented sound is usually heard at a rate of 3–5 Hz (the difference between $f1$ and $f2$ is equal to 3–5 Hz). The beats are strongest when the amplitudes of the two tones are equal. Thus, it is possible in masking experiments that, when the masker frequency is close to the signal frequency, the listener will hear beats. This is especially true because under these conditions the signal and masker are likely to be close together in level (see Figure 11.1). One way to avoid, or at least reduce, the extra sensation of beats during a masking experiment is to present a very short-duration signal. Since the best beats occur with a rate of 3–5 Hz, the signal must be 200 to 333 msec long for just one period of the beat to occur (i.e., the period of 3–5 Hz is 333 to 200 msec). Thus, if the signal is 30 msec long, as it was for the data shown in Figure 11.1, such a short portion of the beat period is presented that the listener usually does not hear any loudness change or beating.

Another independent property of auditory systems that could influence detection of the signal in a masking experiment is the nonlinearity of the ear (see Chapter 5). This nonlinearity can produce audible tones (aural harmonics and combination tones) in addition to the signal and masker, especially at high stimulus levels. Although aural harmonics and combination tones can occur in a masking experiment, they are usually not detected in a psychophysical tuning experiment because of the low signal level and short signal duration.

To indicate how beats, aural harmonics, and combination tones can influence results from a study of masking, consider the following experiment conducted by Wegel and Lane (1924). They asked listeners to detect the pres-ence of signals of different frequencies while an 80 dB SL, 1200-Hz tone served as the masker. For the fixed 1200-Hz masker, the level of the signal was adjusted until the subject could just barely discriminate a difference between the masker and the signal plus masker. The data are shown in Figure 11.3 as the threshold of the signal (expressed in dB SL) versus the frequency of the signal. These data indicate that there is an *upward spread of masking*, in that a masker of a given frequency masks higher-frequency signals more than lower-frequency signals. Because signal frequency was varied in this experiment, the masking data are sometimes referred to as *masking patterns* to differentiate this pattern from the psychophysical tuning curve patterns shown in Figures 11.1 and 11.2, where the signal frequency was kept constant.

The shaded areas in Figure 11.4 indicate that, when the signal was close in frequency to the masker, the observer heard beats that indicated the presence of the signal. Notice that beats were also reported at harmonics (2400 and 3600 Hz) of the 1200-Hz masker. If a nonlinear system is excited with a single frequency, higher harmonics of the input frequency can be detected. When a pure tone of one frequency is presented to the auditory system at a high level, listeners report hearing tones at that frequency and tones at frequencies equal to harmonics (usually the first two or three harmonics) of the frequency presented. These audible higher harmonics are called *aural harmonics*, and their presence indicates that the auditory system is nonlinear. Figure 11.3 shows that at a signal frequency of 2400 Hz (the second aural harmonic of the masker frequency, 1200 Hz) and at 3600 Hz (the third aural harmonic of the masker) the signal frequency is beating with the masker harmonics. In other words, the intense 80-dB SPL, 1200-Hz masker has produced aural har-

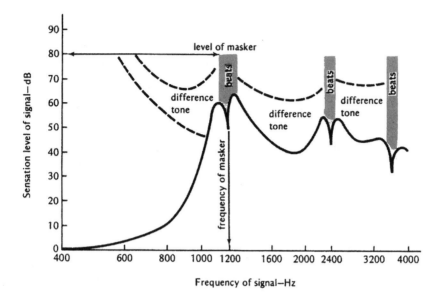

FIGURE 11.3 Masked threshold of signal in dB SL as a function of signal frequency in Hz with a 1200-Hz, 80-dB SL masker. The figure indicates that, in addition to detecting the masker and the signal, the observer can also detect beats and nonlinear difference tones. Adapted with permission from Wegel and Lane (1924).

monics at 2400 and 3600 Hz. When the signal frequency is close to those of these aural harmonics (e.g., 2403 or 3603 Hz), the beating occurs between the signal frequency and that frequency produced by the nonlinear properties of the auditory system (e.g., a 3-Hz beat is generated by a 2403-Hz signal and the 2400-Hz second aural harmonic produced by the 1200-Hz masker).

In addition to the aural harmonics produced by the nonlinearity of the auditory system, combination tones are present when the signal and masker are presented simultaneously. The two types of combination tones (produced by the signal and masker) heard in this experiment were the *primary and secondary difference tones*. The frequency of the primary difference tone is equal to the difference in frequency between the masker and signal. The

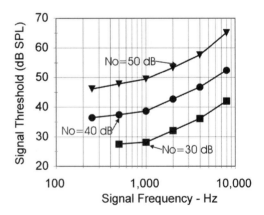

FIGURE 11.4 The signal level (dB SPL) required for noise-masked threshold is shown as a function of signal frequency for three levels (expressed in spectrum level, N_o) of the wideband masking noise. Based on data from and adapted with permission from Reed and Bilger (1973).

frequency of the secondary difference tone (also called the *cubic difference tone*; see also Chapters 8, 9, and 13) is produced by the difference between twice the masker frequency minus the signal frequency, or twice the signal frequency minus the masker frequency (see Appendix A). Wegel and Lane's pure tone masking experiment shows that listeners do detect these difference tones. Although both beats and combination tones can be heard when two tones are presented together, the two phenomena (beats and combination tones) represent very different aspects of hearing.

NOISE MASKING

Since white noise (see Chapter 4) contains a wide range of frequencies, we would expect it to mask tones of many different frequencies. In a noise-masking experiment, a broadband white Gaussian noise, whose spectrum level (N_o, see Chapter 4) was varied is used to mask a tonal signal with different frequencies. The masked threshold of the signal was measured. The data from this type of experiment are shown in Figure 11.4. With no noise present, the thresholds of hearing show that the threshold for a tone depends to a large extent on the tone's frequency. However, as the noise background level is increased, the masked threshold for the pure tone is less dependent on the frequency of the tone. Another important aspect of these data is that, above an N_o of approximately 20 dB, an increase in the spectrum level of the noise means that the signal level must be increased by approximately the same amount for the signal to be detected (i.e., for each decibel increase in the level of the masker, the signal must be increased by the same amount to maintain constant detection). The data indicate that over a wide range of

levels and frequencies the signal energy must be 5–15 dB more intense than the spectrum level of the noise for the signal to be detected, as shown in Figure 11.5.

In these masking experiments, the ratio of the signal energy to the spectrum level of the noise is used to describe the masked threshold. The signal-to-noise ratio (E/N_0) is expressed as the energy of the signal (E) divided by noise power per unit bandwidth (N_0). In decibels, the signal-to-noise ratio is equal to the signal energy (in dB) minus the spectrum level (in dB). The data in Figure 11.5 from the noise-masking experiments suggest that E/N_o must be approximately 5–15 dB for the signal to be detected, with E/N_o being lower for low signal frequencies and increasing as the signal frequency increases.

CRITICAL BAND AND THE INTERNAL FILTER

The data of Figures 11.1–11.3 suggest that those masker frequencies near that of the sig-

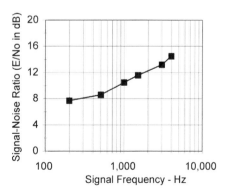

FIGURE 11.5 The signal- (energy) to-noise (spectrum level) ratios required for detection are shown as a function of signal frequency. Adapted with permission from Reed and Bilger (1973).

nal are important in determining masking. Thus, we might expect that, as the noise was passed through a narrower and narrower bandpass filter (i.e., a narrower and narrower band of noise is providing the masking), detection of a tone whose frequency was in the center of the noise's passband would become easier. Conversely, one expects that a signal with a frequency that was not in the passband of the filter would be very easy to detect because there would be no energy in the masker at the same frequency as the signal frequency.

Fletcher performed a band-narrowing experiment in 1940 and made some assumptions about the frequency region of the noise that would be effective in masking the tone. He assumed that some sort of "internal filter" was centered around the frequency of the signal and that the total noise power coming through that internal filter determined the amount of masking for the signal. That is, detection of a tonal signal is determined by the amount of total power present in a narrow range of frequencies. This narrow range of frequencies is determined by the internal filter. Figure 11.6 shows this idealized internal filter schematically for two noise spectra. In Figure 11.6a, the noise spectrum is much broader than the passband of the internal filter, and hence the maximum amount of masking occurs because the maximum amount of total power is coming through the filter. In Figure 11.6b, the spectrum of the masking noise is narrower than the passband of the internal filter, and the signal is easier to detect than that indicated in Figure 11.6a because less than a maximum amount of noise power is coming through the filter. Fletcher called the internal filter the *critical band*, because the frequencies within the passband of the internal filter were *critical* for masking.

A band-reject filtered noise such as that shown in Figure 11.7 offers an excellent

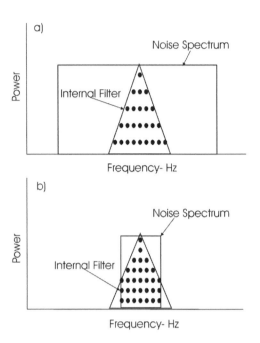

FIGURE 11.6 Schematic diagram of the "internal filter" (the triangle). (**a**) Broadband noise "produces" maximum power (maximum area) at the output of the internal filter. (**b**) The bandwidth of the noise is less than the bandwidth of the internal filter (less area under the filter). There is more masking for a signal whose frequency is at the center of the internal filter for the broadband noise (**a**) than for the narrow-band noise (**b**).

masker for estimating the shape of the critical band. Thresholds for detecting a tonal signal whose frequency is centered in the spectral notch of the band-reject noise are determined as a function of the spectral width of the spectral notch, as indicated in Figure 11.7a. The resulting masking data (Figure 11.7b) can be used to estimate the shape of the critical band. It is assumed that the total power coming through the filter determines the amount of masking. As the spectral notch is widened, there will be less total power coming through the filter and therefore less masking, as indicated in Figure 11.7b.

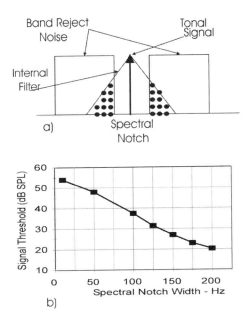

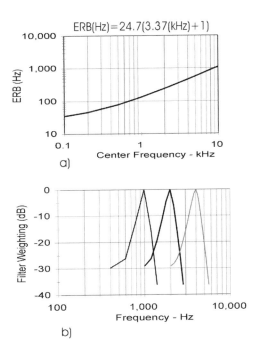

FIGURE 11.7 (**a**) A noise band with a spectral notch or gap is used to mask a signal whose frequency is in the center of the spectral gap. (**b**) The masked thresholds for detecting a 1000-Hz signal are shown as a function of increasing the spectral notch of the band-reject noise. Adapted with permission from Patterson and Moore (1989).

FIGURE 11.8 (**a**) The width (in Hz) of the estimated critical band (ERB or Equivalent Rectangular Bandwidth) is shown as a function of signal frequency. Data were obtained from experiments in which noises with spectral gaps masked tonal signals. The equation in the Figure 11.8a shows a fit to the data that can be used to estimate ERB critical bandwidths for critical bands centered at any frequency. In the equation, "f" is the center frequency of the critical band (i.e., the signal frequency) and is expressed in terms of kHz. Adapted with permission from Moore (1997). (**b**) Estimates of three critical-band internal filters based on the ERB experiments described in Figure 11.7 are shown. (The filters are based on the Rounded Exponential filter model, *roex*, from Patterson and Moore, 1989).

The shape of these masking data can be used to estimate the shape of the internal, critical band filter. The bandwidth of the derived critical band filter (called the *equivalent rectangular bandwidth*, ERB) can be obtained for different signal frequencies. Figure 11.8 shows the ERB as a function of frequency, and the results indicate that the width of the critical band is proportional to its center frequency (i.e., the signal frequency). The bottom part of Figure 11.8 displays estimates of auditory filters centered at different center frequencies (signal frequency). The estimated shape of the filters does not change much with signal frequency, but bandwidth does. The width of the critical band (or ERB) also increases with increasing signal level as indicated in Figure 11.9. This increase in critical bandwidth with level may represent nonlinear properties of cochlear transduction as discussed in Chapters 8 and 16.

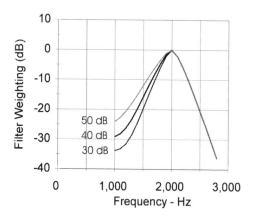

FIGURE 11.9 Estimates of the critical-band internal filters from experiments in which the level of tonal signal was increased in the notched-noise–masking experiment. The width of the filters increase with increasing level. (The filters are based on the Rounded Exponential filter model, *roex*, from Rosen and Baker, 1994).

RELATIONSHIP BETWEEN EXCITATION PATTERNS AND CRITICAL BANDS

We have used the concept of an internal filter (or critical band) to explain masking data when the masker contains frequencies different from the signal. The data of Figures 11.1 and 11.2 involving the psychophysical tuning curve and those of Figures 11.7, 11.8, and 11.9 involving noise maskers with a spectral gap can be used to derive estimates of the shape and bandwidth of the critical band. In these experiments, the signal is kept fixed in frequency, and we assume that the listener detects the signal by monitoring the critical band centered on the signal frequency.

For the masking pattern data shown in Figure 11.3, the frequency of the masker was kept constant and the signal frequency changed. Thus, the listener is assumed to monitor a different critical band for each signal frequency

(each critical band with a center frequency at the signal frequency). In order to explain the masking pattern data, it is assumed that the masking tone stimulates a number of different neurons, one neuron with its best frequency at the masker's frequency and other neurons with best frequencies near that of the masker's frequency. The neuron with its best frequency equal to that of the masker would be stimulated the most, and the other neurons would be stimulated less depending on how close their best frequencies were to that of the masker and on the overall level of the masker. That is, the masking tone sets up a pattern of excitation in the auditory nerve. If we imagine that the detection of a signal tone masked by a masking tone is mediated by the *excitation pattern*, we can explain the results shown in Figure 11.3. That is, the excitation caused by the masker spreads to critical bands located above and below the masker in frequency. When the listener monitors the critical band centered on the signal frequency, the critical band will contain energy due to the spread of excitation from the masker (assuming that the masker and signal are not too far apart in frequency and the level of the masker is sufficiently high for the excitation to spread to that critical band), and this energy will mask the signal whose frequency is at the center of the critical band. Figure 11.10 describes how an excitation pattern can be obtained from critical-band internal filters. Notice the similarity in the shape of the excitation pattern of Figure 11.10b and the masking pattern data of Figure 11.3.

TEMPORAL MASKING

In the masking experiments just described, the masker and signal occurred simultaneously in time. There are many acoustical events in which two stimuli follow one an-

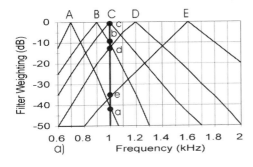

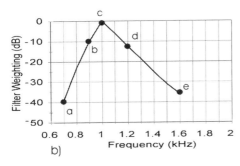

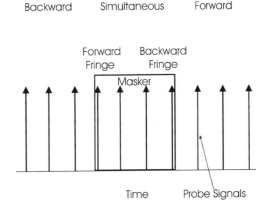

FIGURE 11.11 Schematic diagram of temporal positions of probe signals in relation to a pulsed masker. Signals can occur before the masker (backward masking), during the time of the masker presentation (simultaneous masking), or after the masker (forward masking). Forward and backward fringes also produce masking.

FIGURE 11.10 (**a**) Five critical-band internal filters are shown. (**b**) If a masker tone is placed at 1000 Hz (1 kHz), the masking pattern is obtained by assuming that the excitation produced in any critical band by the masker at 1000 Hz is equal to the amount of energy coming through that critical band at 1000 Hz. Thus, for the filter labeled "A" the amount of excitation for this filter at 1000 Hz is "a" or –40 dB. Thus, in the excitation pattern (Figure 11.10b) at the frequency equal to the center frequency of filter "A" (700 Hz), the amount of excitation is –40 dB (point a). Similar calculations can be made for filters B, C, D, and E yielding the other points b, c, d, and e on the excitation pattern (Figure 10.10b).

other. For instance, in music the notes usually appear sequentially in time, and in speech words appear in sequence. Psychoacousticians have therefore studied the amount of masking provided for a signal that occurs before or after the masker.

Figure 11.11 is a schematic diagram of the stimulus conditions used in studies of *temporal masking*. Signals or probe tones can occur at different times relative to the masker (rectangle in Figure 11.11). The signal probes in the middle of the masker represent the simultaneous-masking conditions we have already studied. When the signal is presented near the beginning or end of the masker, *backward fringe masking* or *forward fringe masking* occurs. When the signal precedes the masker in time, the condition is called *backward masking*; when the signal follows the masker in time, the condition is *forward masking*.

Various stimuli have been used in temporal masking studies (tones, noises, speech, clicks), and the results shown in Figure 11.12 demonstrate the salient data from these experiments. More masking occurs in the fringe conditions than in the simultaneous situation, such that a signal placed in the forward fringe is masked more than one placed in the backward fringe (the masking that occurs when the signal is in the forward fringe is sometimes referred to as

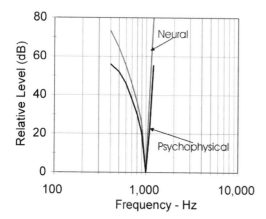

FIGURE 11.12 Schematic diagram of the relative change in signal thresholds as a function of the temporal position of the probe signal in relation to the masker.

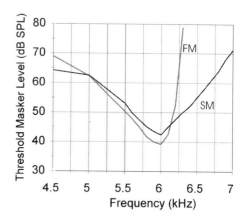

FIGURE 11.13 A comparison of a forward (FM) and a simultaneous masked (SM) psychophysical tuning curve obtained for similar stimulus conditions. The signal was 6000 Hz. Forward-masked tuning curves are very similar to neural tuning curves obtained from the eighth nerve (see Figures 7.17 and 9.5). Based on data from and adapted with permission from Moore (1978).

overshoot, as if the extra masking caused by the masker onset has overshot the masking caused by the steady-state portion of the masker). Forward masking of a stimulus can take place when a temporal difference between the two stimuli is between 75 and 100 msec, and backward masking occurs up to 50 msec. Thus, the amount of backward masking declines more quickly than does the amount of forward masking as a function of increasing the temporal separation between the signal and masker.

TONAL–TEMPORAL MASKING

At the beginning of this chapter, we described the psychophysical tuning curve. If there is forward masking, then we might expect that the effect of the forward masker would be frequency dependent, as it was for simultaneous masking (Figure 11.1). The data shown in Figure 11.13 marked as FM were obtained in the same manner as described for Figure 11.1, except in this case the signal ap-

peared immediately after the masker was turned off (forward masking, FM). These forward masking results show that the masker does influence signal detection when it does not overlap the signal. The effects are about the same as when the signal and masker overlap in time (Figure 11.1). Figure 11.13 shows a direct comparison of psychophysical tuning curves obtained in simultaneous masking (SM) and in forward masking (FM). As can be seen, this comparison indicates that the psychophysical tuning curve is sharper in forward masking than in simultaneous masking. The sharper tuning curve means that the auditory system is better able to detect the presence of the signal in forward masking than in simultaneous masking for the same masker frequency.

These tuning curves are based on one pure tone masking another pure tone. We have already studied the use of noise maskers, but what happens for other complex maskers? Let us consider the case of using a two-tone complex as the masker. In this experiment, the actual masker (M) will be fixed at a particular level (40 dB SPL) and frequency (1000 Hz), and the level of the signal (the signal also has a frequency of 1000 Hz) is varied to determine a threshold. In a test or baseline condition, the signal threshold is determined when the masker is equal in frequency to the signal (i.e., both are at 1000 Hz). In the test conditions, a second tone is added to the masker, so that the masking stimulus consists of two tones: the initial masker M and the second masking tone SU. The level of the second tone will be 20 dB above the level of the masker, or M, tone; the second tone, SU, is therefore 60 dB SPL. We will then vary the frequency of the second tone (SU) and determine the threshold for detecting the signal for each value of SU. Finally, we will perform this experiment for both simultaneous masking and forward masking.

Figure 11.14 shows the two stimulus conditions and the results. The vertical axis in each figure is the change in signal threshold from the baseline condition. Recall that in the baseline condition only the masking tone (M) was presented and the signal and masker were equal in frequency (1000 Hz). The solid horizontal line at 0 decibels represents this masking condition. Thus, if the second tone, SU, provides masking in addition to that caused by M presented alone, the signal thresholds should increase above 0 dB. As can be seen, in simultaneous masking most masking (about 20 dB of masking) occurs when the second tone, SU, is equal in frequency to the masking tone, M (SU, M and the signal are all 1000 Hz). As the second tone (SU) becomes different in frequency from the masking tone (M), the amount of threshold changes above 0 dB decreases until the difference between the two masking tones is so large that only the original masking tone, M, continues to provide masking (i.e., masking is back at the baseline

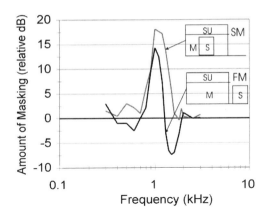

FIGURE 11.14 A comparison of two-tone tonal masking in simultaneous (the SM condition shown with the light curve) and forward masking (the FM condition shown with the dark curve). The frequency of the suppressor tone (SU) is shown on the horizontal axis and the level of the 1000-Hz signal (S) required for threshold detection is shown on the vertical axis. Signal threshold is shown relative to that required to detect the signal when only the masking tone (M) was present. The masking tone is presented at 40 dB SPL and with a frequency of 1000 Hz. The suppressor tone is presented at 60 dB SPL. In the simultaneous-masking condition, the signal is presented in the temporal middle of the masking stimulus (either M alone or M and SU presented together), while in the forward-masking condition the signal is present after the offset of the masking stimulus. Signal thresholds above 0 dB mean that the suppressor tone (SU) increased the amount of masking provided by the masker tone (M), thresholds below 0 dB mean that the suppressor tone (SU) caused less masking than that produced by the masker tone (M) when it was presented alone (e.g., SU suppressed the masking ability of M), and thresholds at 0 dB mean that only the masker tone (M) was providing masking. Adapted with permission from Shannon (1974).

amount of masking, 0 dB). These data are similar to those shown in Figure 11.4.

Let us compare this simultaneous-masking effect to what happens in the forward-masking condition. Notice that in forward masking the values of signal threshold are negative when the second masking tone, SU, is slightly greater in frequency than the masking tone, M. That is, when SU is added to M, signal threshold can be lower than that obtained when just *M* is presented. In this example, the 1000-Hz signal is easier to detect when the 1000-Hz masker (*M*) is present along with a second tone (*SU*) of slightly higher frequency (e.g., 1250 Hz). It is as if the second tone (*SU*) has made the 1000-Hz masking tone (*M*) a less effective masker. The second tone (*SU*) is sometimes referred to as a *suppression tone* because the second tone (*SU*) is suppressing the masking ability of the masking tone (*M*).

Labeling the second tone the *suppression tone* allows us to describe the two-tone masking effect. Care should be used, however, in drawing too many conclusions about the nature of this suppression-like effect. Some investigators refer to the effect as "unmasking" in that the second tone has "unmasked" the effect of the masker (*M*). "Unmasking" is viewed as a more neutral term than suppression because it does not imply that the second tone (*SU*) actually interacts in a direct fashion with *M*. The unmasking or suppression phenomenon is another factor that must be considered in attempting to account for the way in which the auditory system operates when two or more stimuli exist in the environment.

SUMMARY

Masking of a tonal signal by a tonal masker has been used to study the interaction of sounds occurring simultaneously and to probe the frequency selectivity of the auditory system. Psychophysical tuning curves are often used to describe the masking effect of one tone on another tone. The results indicate that low-frequency tones mask high-frequency signals more than high frequencies mask low frequencies. Sometimes, due to the presence of beats and combination tones, the exact relationship between tonal signal threshold and tonal masker frequency is difficult to determine. Beats indicate that the auditory system has a limited frequency-resolving power but that it can follow the amplitude of the input stimulus. Combination tones indicate the extent to which the auditory system is nonlinear. When white Gaussian noise is used as a masker for tonal signals, the ratio of signal energy (E) to masker spectrum level (N_0) required for masked threshold is approximately 5 to 15 dB as the spectrum level of the noise or the frequency of the signal is varied over a considerable range. Critical bands and excitation patterns are used to estimate the bandwidth properties of the frequency-resolving capability of the auditory system. Fringe masking and forward and backward masking are used to determine the interaction of sounds not occurring simultaneously. Tuning curves obtained in forward masking are sharper than those obtained in simultaneous masking. Psychophysical suppression or unmasking can occur when the masker consists of two or more frequencies.

SUPPLEMENT

Throughout the discussion of masking we have stressed the concept of the "internal filter." The psychophysics of hearing suggests that the nervous system operates as if there is a bank of bandpass filters that process sound. We have already learned (in Chapters 7–9) that both the biomechanics and the neural aspects of auditory physiology demonstrate the existence of neural tuning that resembles bandpass filtering. A challenge for auditory scientists is to determine to what extent the neural tuning observed in the cochlea and auditory nerve accounts for the psychophysics of masking. Many of these issues and data are reviewed in the book edited by Moore (1986), *Frequency Selectivity in Hearing*, in Chapter 3 of a textbook by Moore (1997), by Moore and Patterson (1986), and by Moore in a chapter in the book edited by Yost, Popper, and Fay (1993).

Another variable that can affect a listener's ability to detect signals in masking experiments is "off-frequency listening." The basic idea behind off-frequency listening is that the listener may be able to perform the detection task by listening in a frequency region different from that where the signal occurs (i.e., off the signal frequency or off frequency). In many situations, the signal may excite a wide frequency region, and a larger signal-to-masker ratio may exist in a frequency region that is different from that of the signal. Using a noise with a spectral notch (see Figure 11.4) reduces the ability of the listener to "listen off-frequency" to detect the signal in the notch-noise–masking procedure, because this noise energy exists in all frequency regions, except in the narrow region near the signal. An excellent discussion of off-frequency listening can be found in Moore's (1986) book.

Patterson and Moore (1986) used a particular filter shape called the rounded exponential (*roex filter*) to fit masking data like those shown in Figure 11.7. The roex filter function can be written as

$$W(g) = (1 + pg) \exp(-pg),$$

where $W(g)$ is the linear (non-decibel) value of the filter output ($0 \leq W(g) \leq 1$); $g = |f - fo| / f$, where fo is the signal frequency and f is a frequency on the filter function; p is determined by the bandwidth and slope of the filter such that, the higher the value of p, the more sharply tuned the filter, and p is obtained by finding the best fitting function $W(g)$ to the data; "exp" is the exponential argument. Once p is determined from the data, the ERB or the equivalent rectangular bandwidth

(see Supplement to Chapter 5) can be obtained from the roex filter as:

$$ERB = 4fo/p.$$

The *gammatone filter* (see Patterson *et al.*, 1995 and Rosen and Baker, 1994) is another filter function used to describe the shape of the critical band filter. The time domain (the *impulse function*) description of the gammatone function is:

$$t^{n-1} \exp(-2\pi bt) \cos(2\pi fot),$$

where n and b are constants and fo is the CF of the filter. In the frequency domain the filter function is approximately equal to

$$(1 - r)[1 + \{(f - fo)^2\}/b^2]^4,$$

where fo is the filter's CF, f is a frequency on the filter, b controls the filter's sharpness (like p for the roex filter), and r is proportional to the slope of the filter. In these measurements, filter bandwidth is sometimes referred to in terms of ERBs. One can estimate the frequency associated with an ERB via this equation:

$$F = (ERB - 24.7)/107.94,$$

where F is expressed in kHz (Glasberg and Moore, 1990).

Weber (1983) has suggested that there are at least three explanations for the narrower forward-masked tuning curves (see also Lutfi, 1988). Psychophysical suppression appears to be much more effective for high-frequency suppression tones on a lower-frequency masker tone than for low-frequency suppression tones on a high-frequency masking tone. Suppression by low-frequency suppression tones on the masker tone has been observed, but usually only when the suppression tone is fairly high in level (see Shannon, 1974). Although comparisons between psychophysical suppression and two-tone neural suppression (see Chapter 7) are tempting, there are some arguments that the two may not be the same (see Moore and Glasberg, 1982).

The two stimuli in the intensity-discrimination experiment described in Chapter 10 differ only in that one is more intense than the other. The more intense stimulus could have been produced by adding two tones of the same frequency: *Tone A* with a level of *I* in decibels and tone *B* with a level such that when it was added to tone *A* the summed level in decibels would equal *I* + ΔI in dB (recall that when two stimuli of equal frequency are added the sum depends on the phase difference between two tones). Thus, data from the intensity experiment could be plotted as the level of tone *B* required for the subject to determine that it was added to tone *A* versus the level of tone *A*. The fact that the level of tone *B* must be raised above absolute threshold due to the presence of tone *A* is defined as *tone A masking tone B* (see Table 10.3).

Fletcher observed that, when the spectrum of the noise was broad (i.e., the internal filter contained the maximum total noise power required for masking), the power of the just de-tectable masked signal (the signal at masked threshold) was equal to the total power contained within the critical band. Fletcher's observation makes it possible to predict the width of the critical band without performing a band-narrowing or band-reject experiment. According to Fletcher's assumptions, masked signal power (*Ps*) is equal to the power of the noise in the critical band (*Pncb*): $Ps = Pncb$. Since $Pncb = N_o \times CBW$ (see Chapter 4), where *No* is noise spectrum level and CBW is an estimate of the critical bandwidth, $CBW = Ps/N_o$. Expressed in decibels, $10 \log CBW = Ps$ in dB $- N_o$ in dB, and $10 \log CBW$ is referred to as the *critical ratio*. Thus, CBW can be estimated directly from a single masking threshold. However, the signal level (*Ps*) required for detection depends on how efficient the auditory system is at detecting the signal, as well as on the width of the critical band. Thus, if the total power within the critical band does not *equal* the power of the masked signal at threshold (i.e., *Ps* does not equal *Pncb*), then the basic assumption of the critical ratio calculation is violated and the critical ratio will not yield a valid estimate of the width of the critical band.

CHAPTER 12

Sound Localization and Binaural Hearing

LOCALIZATION

The source of sound can be localized in the three spatial dimensions: the horizontal plane (azimuth) or the left–right dimension, the vertical plane or the up–down dimension, and in distance (range) or the near–far dimension (Figure 12.1). Sound itself has no spatial dimensions. Sound localization is therefore the result of the auditory system's ability to process the physical parameters of sound (frequency, level, and time/phase) that correlate with the spatial location of the sound source.

Localization in Azimuth

To understand one set of stimulus cues responsible for our localization abilities, picture a person sitting in a room listening to a sound source without moving his or her head. Figure 12.2 illustrates the temporal and level information arriving at the person's ears that can be used to locate stimuli in the azimuth plane. Notice that the sound travels a shorter distance to the right ear than to the left ear. Hence, it will arrive at the right ear earlier than at the left ear; this yields an *interaural time difference* (ITD) in arrival of the sound. Recall that the speed of sound in air is relatively constant, independent of frequency. Thus, the

interaural temporal difference for any frequency is theoretically the same for all frequencies for a particular stimulus location, whereas the *interaural phase difference* (IPD) will vary according to the frequency of the stimulus. That is, if a tonal sound with a frequency of 1000 Hz (a period of 1 msec) arrives at the right ear 0.5 msec after it has reached the left ear, the tone at the right ear is half a period (or 180°) out of phase with the tone at the left ear. If a 500-Hz sinusoid (a period of 2 msec) arrives at the right ear the same 0.5 msec later than at the left ear, there is only a one-quarter of a period (or 90°) phase difference between the two ears. Thus, two different tones (1000 and 500 Hz) both with a 0.5-msec interaural time difference produce different interaural phase differences.

There is also an *interaural level difference* (ILD) for the condition shown in Figure 12.2 due to two aspects of the physics of sound. First, because the stimulus arrives at the right ear before it reaches the left ear, it is more intense at the right ear (inverse square law relationship, see Chapter 3). However, the difference in level caused by the inverse square law produces extremely small interaural level differences. Like any object, the head can produce a sound shadow (see Chapter 3) if its size is close to the sound's wavelength. Since

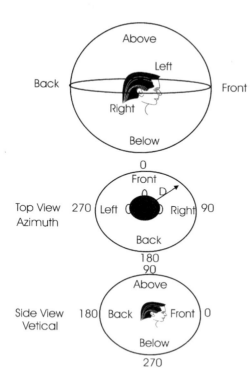

FIGURE 12.1 The three spatial dimensions: azimuth (left–right), vertical (up–down), and distance (D, near–far).

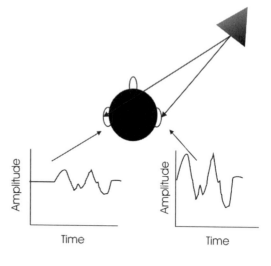

FIGURE 12.2 Schematic diagram of a sound source on an azimuth plane around the head. Distance of the sound from the ears is the range. The sound reaches the right ear first, and thus there is an interaural difference in the arrival time, and the sound at the left ear is less intense than that at the right ear leading to an interaural level difference.

wavelength is directly proportional to frequency, the interaural level difference caused by the sound shadow depends on frequency. The higher the frequency, the shorter the wavelength and the greater the sound shadow (caused by the head) in establishing the interaural level difference. Thus, large interaural level differences exist at high frequencies. Figure 12.3 shows the interaural temporal difference measured at the ears for a stimulus located at different azimuth angles. Figure 12.3 also shows the interaural level difference for different azimuth angles and frequencies. Notice that the interaural temporal difference is a smooth function of frequency and varies from 0 to 0.8 msec as the azimuth of the source

changes. The interaural level difference varies considerably, especially at high frequencies. The binaural auditory system therefore could determine the location of a sound coming from, say, the right side by noting that the right ear received the sound first and that the stimulus was more intense at the right ear. Figure 6.5 also shows how level and interaural time differ between the two ears as a function of frequency when a sound source is placed opposite the left ear.

Figures 12.4 and 12.5 show how well listeners localize sound sources. Figure 12.4 shows the errors in localizing sinusoids of different frequencies. These data indicate that listeners have more errors localizing sounds with frequencies in the mid-frequency region around 2000 Hz than for sounds with lower or higher frequencies. The data in Figure 12.5 indicate

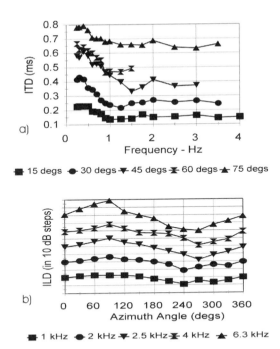

a)

■ 15 degs ● 30 degs ▼ 45 degs ✕ 60 degs ▲ 75 degs

b)

■ 1 kHz ● 2 kHz ▼ 2.5 kHz ✕ 4 kHz ▲ 6.3 kHz

FIGURE 12.3 (**a**) Values of interaural time difference (ITD in ms) measured at different azimuth angles (see Figure 12.1), (**b**) Values of interaural level difference (ILD in dB) measured at different azimuth angles and frequencies (each tic mark on the vertical axis represents 10 dB of ILD). Adapted with permission from Kuhn (1987).

how well a listener locates a broadband noise. Perfect localization would be represented by the data following on the diagonal straight line, since the data are plotted as the judged location of the source versus the actual location of the source. As can be seen, this listener is very good at determining the location of broadband noise in the azimuth plane.

Figures 12.6 and 12.7 represent the data from an experiment in which blindfolded listeners were asked to discriminate between the location of two small loudspeakers, each placed approximately 100 cm from the listener's head. As shown in Figure 12.6, the smallest angular separation between the two loudspeakers that the listener could just detect

is called the *minimal audible angle* (MAA). Thus, the MAA in degrees of angular separation was measured as a function of the frequency of the sinusoid. The various curves of Figure 12.7 represent various azimuth posi-

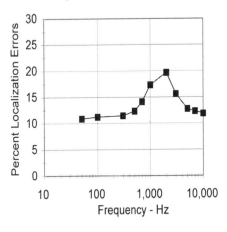

FIGURE 12.4 Errors (in terms of percentage of judgments made) in judging the location of a sinusoidal sound source shown as a function of frequency. Adapted with permission from Stevens and Newman (1936).

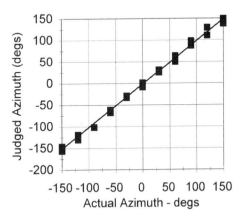

FIGURE 12.5 The judged location in the horizontal or azimuthal direction of a broadband noise source presented at different locations. The diagonal straight line represents perfect judgments (data from one listener, see Figure 12.9). Adapted with permission from Wightman and Kistler (1989b).

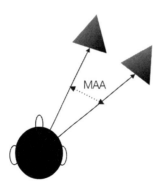

FIGURE 12.6 Schematic diagram of minimum audible angles (MAA). Adapted with permission from Mills (1972).

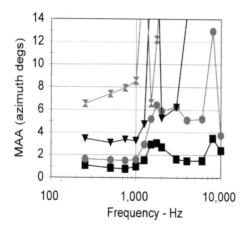

■ 0 degs ● 30 deg s ▼ 60 degs ✕ 75 degs

FIGURE 12.7 Values of MAA are plotted as a function of sinusoidal frequency for different azimuths. The MAA increases as the source moves away from in front of the listener and is large in the mid-frequency region between 2 and 4 kHz.

tions at which the discriminations were made. That is, the curve labeled 0° means that the loudspeakers were directly in front of the listener, whereas the curve labeled 75° means that the loudspeakers were placed 75 degrees

toward one ear (the speakers were to the side of the listener). Notice that the listener required larger and larger angular separation between loudspeakers in order to detect a difference in loudspeaker location as the speakers were moved from directly in front of the listener toward one ear (the listener's head was held stationary). In other words, when the sound is in front of the listener, a change in location can be better detected than when it is to one side. Of course, in the real world this poses no severe limitation because a person can generally move so that the sound source is in the front.

The MAA results also show that listeners made many mistakes in locating sound sources when their frequency content was in the middle-frequency range. Stevens and Newman (1936), and earlier Lord Rayleigh, believed that the middle-frequency region represented those frequencies for which the interaural temporal and level differences were too small to be used as accurate cues for localization. They concluded that there were two cues for determining location: the interaural temporal difference, which provides information for low-frequency stimuli, and the interaural level difference, which provides location information at high frequencies. This idea is referred to as the *duplex theory of localization*.

The sound shadow effect demonstrates why the physical interaural level difference is small at low frequencies and hence why listeners would have trouble using interaural level as a cue at low frequencies. That is, the interaural level difference caused by the head's sound shadow decreases (see Figure 12.2) as the frequency decreases (wavelength increases). One explanation of why the interaural temporal difference might provide location information at only low frequencies is diagramed in Figure 12.8. Assume that the low-frequency tone shown in the top panel was presented so that it arrived first at the

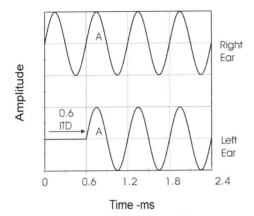

FIGURE 12.8 A 1666-Hz tone presented to the right side of a listener so that it reaches the right ear 0.6 msec before it reaches the left ear. (**Top**) Sinusoid at right ear; (**Bottom**) sinusoid at left ear. After the first peak (A), the waveforms arriving at each ear are in phase, which would indicate that the sound is in front rather than to the right.

right ear. Thus, the sinusoid arriving at the two ears would appear in time as shown in Figures 12.8. The difference in the timing pattern at the ears could be used by the auditory system to determine that the sound source was toward the right ear. In these panels we assume that the period (time between peaks) of the sinusoid was 0.6 msec, nearly the maximum time it takes a sound to travel from one ear to the other (see Figure 12.2 and Figure 6.5b). A 1666-Hz sinusoid would appear at the two ears as in Figure 12.8. Notice that, although the 1666-Hz sinusoid is on the right side and the first period of the left and right waveforms are displaced, the waveforms are identical thereafter (as at point A). Thus, except for the first period there is no difference between the stimulation at the two ears, so the listener might assign the sound to a location in front since stimuli that are directly in front of the listener produce no interaural differences. In this case, the judgment is incorrect because the stimulus was presented opposite the right

ear. For this frequency and this interaural time difference, an ambiguous temporal cue would exist for locating the sound source. This confusion would not exist for frequencies smaller than 1666 Hz, but would occur for those of 1666 Hz and greater. Thus, interaural time produces ambiguous information about spatial location when the frequency is too high (too high relative to the width of the head).

Localization in the Vertical Plane

The cues of interaural time and level have been shown to be of great importance in localizing sounds in the horizontal plane. If a listener's head remains steady, then there are a number of locations that would produce the same interaural differences of time and level; for instance, a sound directly in front would produce the same interaural differences as one directly behind a listener, as well as one directly overhead (see Figure 12.1). Sounds that lie in this plane are in the *mid-sagittal plane*, and this plane forms a *cone of confusion* where all sounds located on the cone produce the same interaural differences. For each sound source location, there is a cone of confusion that describes the location of other sound sources that produce the same interaural differences. While cones of confusion exist for a stationary head, small head movements would enable a listener to accurately localize sound, since the head would be in a different position and the original cone of confusion would no longer exist. Even when we do not move our heads, we can still localize sounds on cones of confusion, such as in the mid-sagittal plane. When sound localization mistakes are made, it is often the case that the mistakes occur along cones of confusion. For instance, in the mid-sagittal plane, listeners often confuse sounds from directly in front with those from directly behind the listener (*front–back confusions*) and vice a versa (*back–front confusions*).

Since the sources of sounds that lie on cones of confusion can be localized, cues in addition to the interaural differences must aid us in determining the location of sound sources in the vertical direction. These cues are usually referred to as *spectral cues*, and are derived from the Head-Related Transfer Functions (HRTFs) as explained in Chapter 6. As explained in Chapter 6 (see Figure 6.5), the many external parts of our head and body, especially the pinna, act as small sound shadows for the path of the sound to the ears. These parts of the body can also delay the sound in reaching the outer ears. These obstacles to sound transmission are most important for high-frequency sounds because the wavelength of high frequencies may be close to the size of these small obstacles (see Chapters 3 and 6).

If the sound is complex, such as a noise, then different frequencies will be attenuated and delayed by different amounts depending on the size of the objects (such as the pinna and various parts of the pinna, the nose, and the torso) the sound encounters before it reaches the ear. The delay will lead to different interaural phases and thus will establish a phase spectrum for the HRTF. Thus, the head and torso provide an HRTF alteration of the sound source. The amount of attenuation and delay (i.e., the spectral characteristics of the HRTF) provided by any obstacle will also depend on the direction from which the sound is coming. For instance, the pinna offers more attenuation from sounds coming from behind than from those coming from the front. Thus, the shape of the HRTF for a complex sound arriving at the outer ear will differ depending on the location of the sound source relative to the body. The amplitude spectral changes can occur at only one ear, making it theoretically possible for us to localize the source of a high-frequency complex sound when we listen with only one ear. Since the major changes in the HRTF occur for high frequencies due to the interaction between wavelength and obstacle size (as explained above), it is not surprising that the major cues for vertical localization occur for the higher frequencies. Figure 6.5a shows that there is a spectral valley in the HRTF near 10,000 Hz. These spectral valleys (and sometimes spectral peaks) are in different spectral locations and have different shapes depending on the vertical location of the sound source. The spectral location of these peaks and valleys are the presumed HRTF cues for vertical localization. Thus, the shape of the HRTF provides information about the location of a sound source, especially vertical location.

Figure 12.9 shows data indicating the ability of a listener to judge the vertical location of a broadband noise source, where the angle is the vertical angle of the source relative to the listener (see Figure 12.1). The fact that the data are scattered about the diagonal line representing perfect vertical localization indicates that the listener is not as good at determining the vertical location of sound as at determining horizontal location (vertical localization is usually poorer than horizontal localization, see Figure 12.5).

Localization as a Function of Distance

Less is known about the cues used to determine the distance or *range* of a sound source. A powerful cue for distance is the sound's loudness or level: loud sounds appear closer than far sounds. Early-arriving reflections from a sound source off nearby surfaces also provide important cues for determining the distance (range) of a sound source.

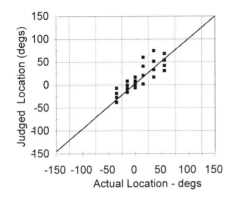

FIGURE 12.9 The judged location in the vertical dimension of a broadband noise source. Data are plotted in the same way as for Figure 12.5 and are for the same listener whose data are shown in Figure 12.5. Adapted with permission from Wightman and Kistler (1989b).

PRECEDENCE-LOCALIZATION IN REVERBERANT SPACES

Everyday experience will tell you that, even in rooms where there are many reflections from walls, floors, and so on, we are still able to localize acoustic events accurately. The reflections cause a complicated pattern of stimulation at the ears, because the reflections come from many different directions. How does the auditory system assimilate these conflicting cues to accurately determine the actual location of a sound source? A number of experiments have shown that it is the first wave arriving at the ears that dominates in establishing the location of the stimulus. That is, in locating sound sources the auditory system appears to process the first wavefront and suppresses the location information in later wavefronts coming from the reflections. Since the first wave will usually arrive at the ears before those coming from any reflections, the first wave contains the information about the sound source. This phenomenon is called the *law of the first wavefront* or the *precedence effect*.

In Figure 12.8, although the later peaks present confusing information, there is no doubt concerning stimulus location if we simply look at the first peak arriving at each ear. By using only the first positive peak in each sinusoid, we can determine that the stimulus arrived at the right ear before it reached the left ear. However, for sinusoidal stimuli the sounds must come on and go off slowly in order to reduce the spread of energy associated with turning sounds on and off abruptly (see Chapters 2 and 4). Thus, these slow onsets and offsets for sinusoidal stimuli eliminate the use of the first wavefront for localization. However, for most stimulus conditions the first-arriving information at the ears will contain reliable information about the source of the sound, and work on the precedence effect suggests that this early-arriving information dominates our ability to localize sound sources.

LATERALIZATION

In studying localization of actual sound sources, it is impossible to separate the variables of interaural time from interaural level because both differences always coexist in localization experiments. In addition, the HRTF-derived differences in spectra cannot be accurately controlled in a free-field study, and the precedence effect will almost always be present. One simple way to more accurately control stimuli than can be done in the free field is to present them over headphones. The experimenter can present only an interaural temporal difference or an interaural level difference, or some particular spectral difference over headphones, thereby controlling the variables. When a tone is presented to a listener by means of headphones, the listener will, under most conditions, perceive an image that lies within his or her head and that has a location that moves as a function of changes in interaural temporal and level differences. To differentiate the perception of the internal (or intercranial) image that usually occurs with headphones from the external image associated with external sound sources, we use the term *lateralization* to describe the former and *localization* for the latter. The image formed from binaural presentations over headphones is sometimes referred to as a *fused image* because the listener reports hearing one image as if the sound sources arriving at both ears were fused. A listener will not perceive a fused image if the interaural temporal difference or the interaural frequency difference is too large. If the interaural temporal difference is larger than approximately 2 msec, the listener will report hearing two images, one at each ear. Also, if the two ears receive independent sinusoidal signals differing greatly in frequency, the listener perceives two images, one at each ear, and both frequencies can be identified.

It has been shown that a fused image in a lateralization experiment appears toward the ear that receives the stimulus first or receives the more intense stimulus in about the same manner that an external image is perceived more toward the ear that receives the sound first (and therefore receives the more intense sound). Thus, a lateralization procedure appears appropriate for studying the effects of interaural temporal and level differences on the ability of the auditory system to locate sound sources.

Figure 12.10 shows the results from experiments in which tones were presented at different frequencies and the interaural temporal phase and level *difference* thresholds were obtained. These and other results show that, as the interaural phase difference increases toward 180°, the fused image was located closer to the ear that received the tone first (leading in time). As the interaural

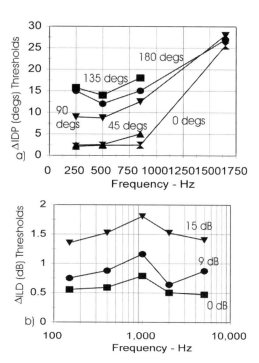

FIGURE 12.10 Value of IPD (interaural phase difference) change threshold (ΔIPD) and ILD (interaural level difference) change threshold (ΔILD) required for a $P(C)$ of 75% discrimination as a function of frequency (Hz). The various curves represent the values of the standard IPD or ILD. (**a**) ΔIPD as IPD changes from 0 to 180°. (**b**) ΔILD is shown as a function of frequency, with each curve representing a different base value of ILD. Adapted with permission from Yost and Dye (1991).

phase difference exceeded 180°, the image was located on the other side of the head (toward the ear lagging in time), and as the interaural time difference approached 360° the image was located toward the middle of the head. A tone presented with no interaural difference was perceived toward the middle of the head or at midline. By introducing an interaural level difference, the tone was perceived closer to the ear receiving the louder tone. Thus, an image was placed in different perceptual positions within the head by introducing an interaural phase or level difference. Assuming that the image was at some location due to a given interaural phase or level difference, the additional amount of interaural phase or level difference the listener required to detect a change in the perceived location of the fused image was determined. The amount of additional phase difference required for threshold detection is called ΔIPD, and the additional interaural level difference is called the ΔILD. The various curves in Figure 12.10 represent the initial interaural phase or level differences introduced such that for phase differences less than 180° the image was located on the left side of the head (Figure 12.10a), and for phase differences greater than 180° the image was on the right side of the head. Data for interaural level discrimination are shown in Figure 12.10b.

A number of aspects of these data are important. First, notice that for any initial phase difference the amount of ΔIPD required for detection remains constant up to frequencies of approximately 900 Hz and then increases. This indicates that at frequencies greater than 900 Hz interaural phase is a poor cue for processing an interaural time difference (for tonal frequencies above 2000 Hz investigators have been unable to move the fused image as a function of changing the interaural phase or time difference). This agrees with Stevens and Newman's (1936) prediction that interaural time is not a usable cue for localization of high-frequency sinusoids. Thus, the binaural system does not use interaural time at high sinusoidal frequencies in order to localize sound.

The second interesting aspect of these data is that, as the image is moved toward one side of the head by introducing phase differences close to 180° or a large interaural level difference, the amount of additional interaural phase or level difference required to discern a

change in perceived location (ΔIPD or ΔILD) also increases. This is consistent with Mills's (1972) finding that listeners are less sensitive to changes in sound source location when the source is located toward one ear than when it is directly in front.

The fact that over headphones the interaural level difference is approximately the same for all frequencies (notice that there is a slight increase in ΔILD for frequencies in the 1000-Hz region) does not mean that in localizing a sound in the free field a listener could use interaural level at low frequencies to locate a sound. That is, even though a listener can detect a 2 dB change in interaural level at 200 Hz with headphones, this 2-dB interaural level difference will not occur in the free field at 200 Hz. The physical interaural level difference at 200 Hz is smaller than 2 dB. Thus, at 200 Hz the listener will not be presented a large enough interaural level difference in the free field to use for localization (see Figure 12.3).

The data described in Figures 12.8 and 12.10a imply that the binaural system can process interaural time for only low-frequency stimuli. This is not the case for complex sounds. If a high-frequency complex sound is presented such that there is a low-frequency repetition in the temporal waveform's envelope, then the binaural system appears almost as sensitive to differences in interaural time as for low-frequency sinusoids. Consider the case for a 300-Hz low-frequency tone, a 3600-Hz high-frequency tone, and a 3900-Hz tone amplitude modulated (AM; see Chapter 5) with a 300-Hz tone. As predicted from Figure 12.10, listeners cannot detect a change in interaural time differences for the 3900-Hz tone and they can for the 300-Hz tone. However, they can also detect an interaural time difference change for the 300-Hz amplitude-modulated 3900-Hz carrier tone almost as well as they could for the 300-Hz tone. Thus, the high-frequency carrier with the slow amplitude modulation (300 Hz) has about the same interaural time difference threshold as the low-frequency tone. Many stimuli that contain only high frequencies but have a low-frequency repetition in the time domain can be discriminated on the basis of interaural time differences. Some examples, in addition to AM stimuli, are beating tones (i.e., adding two high-frequency tones together that differ by a small amount in frequency; see Chapters 5 and 10); narrow bandpassed, filtered noises (i.e., the repetition in the time domain waveform is equal to the bandwidth of the noise, see Chapter 5); and a slowly repeating click stimulus that has been highpass filtered so that only high frequencies are present. The facts of physics along with observations from lateralization experiments lead to the duplex theory for localization being restated as follows: *interaural level is a cue used for locating high-frequency sounds. Interaural time is the cue used to locate any sound with low frequencies or any high-frequency complex sound with a low-frequency repetition in the time-domain waveform.*

The binaural system is remarkably sensitive to changes in interaural time and level. Figure 12.10 shows that the listener could detect a change of 3° of interaural phase. At 1000 Hz this phase difference corresponds to a 0.01-msec change in interaural time. The data from lateralization experiments have also shown that the auditory system is sensitive to temporal differences equal to 10 one-millionths of a second (0.01 msec or 10 microseconds).

In Chapters 7 and 8, we showed that the discharges of auditory nerve fibers were phase locked to the periodicity of low-frequency stimuli. The fact that interaural temporal differences are a cue for localization for low frequencies and for complex stimuli with low-frequency repetitions suggests that phase locking might be crucial for understanding how the nervous system processes interaural

temporal information. If the phase-locked activity of the nervous system is responsible for encoding interaural temporal differences, this would also explain why interaural time is a cue only at low sinusoidal or repetition frequencies (see Chapter 15 for a discussion of a neural mechanism that might code for interaural time differences).

LOCALIZATION VERSUS LATERALIZATION

We have already commented on the fact that sounds presented over headphones are usually perceived as "inside the head" rather than "out in space," where actual sound sources usually appear. Usually, when sounds are presented over headphones, they do not have all the spectral complexity of a real sound source, since the headphone-produced sound does not pass over the torso, head, and pinna of the listener like the sound from an actual sound source would. That is, the headphone-delivered stimuli do not preserve the spectral complexities described by HRTFs. In a real sense, the head and torso filter the sound before it reaches the tympanic membrane, and the HRTF describes the amplitude and phase spectra of this filter function. If a complex sound is actually filtered by a filter made to reflect the amplitude and phase spectra of the HRTF and then presented over headphones, the sound arriving at the tympanic membrane should have all of the spectral complexities of a real sound that has passed over the head and torso. When sounds are presented through headphones after HRTF filtering, most listeners report that the sounds appear much more like those occurring in space than when the sounds are not filtered by the HRTF.

Table 12.1 shows a comparison of a listener locating actual sounds versus "locating" sounds presented over headphones when they have been filtered by the appropriate HRTFs. For each listener and for each sound source location a HRTF was computed for both ears and then used to determine HRTF filters so that over headphones the waveform arriving at the two tympanic membranes for each listener was as close as possible to that which occurs naturally. In listening to both the actual sound sources and the "simulated (*virtual*) headphone sources," the listener indicated where in space they thought the sound occurred. In the actual localization experiment, the listener location judgments are compared to the location of the actual source. For the headphone-delivered sounds, the listener location judgments are compared to the location of the sound source for the HRTF filter used to filter the stimulus for that judgment. As can be seen, there is little difference in that judgments remain excellent in the two listening conditions, which indicates that reproduc-

ing the complex spectrum of real sounds is an important variable in auditory localization.

BINAURAL MASKING

In the preceding section we described the auditory system's sensitivity to changes in interaural time and level, which are principally used to locate sound sources in the azimuth plane. Many experiments have shown that the threshold for detecting a signal masked by noise is lower when the noise and signal are presented in a particular way to both ears. In these experiments, subjects first determined their masked thresholds when both the noise and tonal signal were presented equally to both ears. In one test experiment, the tonal signal was removed from one ear, such that the noise was at both ears and the signal at only one ear. In this case, the signal was easy to detect, and therefore the level of the tone had to be reduced to obtain masked thresholds. Subsequently, many investigators have studied the improvement in detection associated with presenting stimuli to both ears. A certain nomenclature has been developed to describe the various types of binaural configurations of signal and noise.

TABLE 12.1 Comparison of Location Judgments for two Listeners (S1 and S2) Judging the Vertical and Horizontal Locations of Actual Noise Sound Sources and Simulated Sources Using HRTF-Filtered Noises Presented over Headphones[a]

Listener	Actual	Simulated
S1	0.99	0.96
S2	0.97	0.83

The judgments are in terms of correlation coefficients, which means how well the listeners judged the position relative the actual position. A coefficient of 1.0 means the listener indicated for all locations exactly where the source was, whereas a coefficient of 0.0 means that the listener had no idea where the sounds were coming from. Reprinted with permission from Wightman *et al.* (1987).

> **monotic**: stimuli presented to only one ear
>
> **diotic**: identical stimuli presented to both ears
>
> **dichotic**: different stimuli presented to the two ears

Investigators have found that the masked threshold of a signal is the same when the stimuli are presented in a monotic or diotic condition. If the masker and signal are arranged in a dichotic situation, however, the signal has a lower threshold than in either the monotic or diotic conditions. There are several ways to present the signal (S) and masker (M) in a dichotic or diotic manner; again, a set of symbols is used to describe these stimulus conditions:

S_o: signal presented binaurally with no interaural differences (diotic)

M_o: masker presented binaurally with no interaural differences (diotic)

S_m: signal presented to only one ear

M_m: masker presented to only one ear

S_π: signal presented to one ear 180° out of phase with the signal presented to the other ear

M_π: masker presented to one ear 180° out of phase with the signal presented to the other ear

For the binaural conditions described above, the signal or masker is identical in all dimensions except that denoted by a subscript. Thus,

monotic: $M_m S_m$

diotic: $M_o S_o$

dichotic: $M_o S_\pi$, $M_o S_m$, $M_\pi S_\pi$, $M_\pi S_o$, $M_\pi S_m$

To compare detection in one binaural condition with that in another, the data are usually presented as the difference between the signal level required for detection (masked threshold) in a monotic condition and that required in a diotic or dichotic condition. That is, the signal level required for detection in the relevant diotic or dichotic condition is subtracted from the signal level required for detection in the $M_m S_m$ (monotic) condition. Such a difference when expressed in decibels is called a *masking-level difference* (MLD) or a *binaural masking-level difference* (BMLD).

Table 12.2 shows the type of improvement in detection provided by dichotic presentation of maskers and signals (MLD). These data represent approximately the maximum MLD obtained when the masker is a continuous, broadband white Gaussian noise and the signal is a pulsed sinusoid of low frequency (below 1000 Hz) and long duration (greater than 100 msec). Figure 12.11 describes the MLD obtained between the $M_o S_o$ condition and the $M_o S_\pi$ condition as a function of signal fre-

TABLE 12.2 The MLD in dB for a Variety of Stimulus Conditions

Interaural condition compared to $M_m S_m$	MLD (dB)
$M_m S_m$, $M_o S_o$, $M_\pi S_\pi$	0
$M_\pi S_m$	6
$M_o S_m$	9
$M_\pi S_o$	13
$M_o S_\pi$	15

quency. Notice that as the frequency of the signal increases the MLD decreases. However, the MLD never goes to zero; the signal is always easier to detect in the dichotic case than in the diotic case.

The fact that the MLD decreases as a function of frequency has suggested to many hearing scientists that the MLD might be related to the interaural differences of time and level. When the stimulus condition is dichotic, there are differences in the interaural temporal and

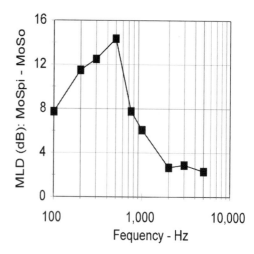

FIGURE 12.11 Difference in masked thresholds (in dB) between $M_o S_\pi$ and $M_o S_o$ conditions (MLD) plotted as a function of signal frequency. Adapted with permission from Webster (1951).

level information between the stimuli arriving at the ears. However, in the diotic condition there are no interaural differences. It is therefore logical to assume that interaural differences associated with dichotic presentations result in improved detection over that obtained in the dichotic condition.

The MLD phenomenon may also be related to another important aspect of binaural hearing. If two or more sound sources are separated in space, it is easier to both locate *and* attend to the individual sounds. For instance, recognizing a particular voice in a choir, distinguishing an instrument in an orchestra, or hearing a particular conversation at a noisy party may be made easier because the sound that is of interest is in a different location than other interfering sounds. This ability to discriminate sounds in complex acoustic environments based on the sources' spatial separation is often called the *cocktail party effect*. The MLD is a result of the signal-plus-masker having a different interaural configuration than the masker alone. Thus, like in the cocktail party effect, this interaural separation between the masker and the masker-plus-signal may make it easier to detect the signal. Recall that the main MLD effect is the ability to detect a signal, whereas the cocktail party effect refers to recognition or identification of a signal. Thus, although the two effects (MLD and cocktail party effect) are probably related, they may not be measuring exactly the same aspect of binaural processing. The ability to locate a sound in space is important, therefore, not only as an aid for determining the position of sounds, but also for attending to a particular sound in an environment with many sound sources.

SUMMARY

For localizing sound sources in the azimuthal plane, interaural time is the relevant cue for stimulus location at low frequencies and for complex stimuli with low-frequency repetition; and interaural level is the cue at high frequencies. Spectral differences provided by the Head-Related Transfer Function (HRTF) are the cues used for vertical localization. Loudness and early reflections from reflective surfaces are the probable cues for localization as a function of distance. The precedence effect stresses the importance of the first wave in determining stimulus location. In the lateralization procedure, stimuli are presented over headphones, and the location of the fused image is dependent on interaural time for low-frequency tones and on interaural level at all frequencies. By using the spectral differences of the HRTF, sounds presented over headphones may be judged similarly to naturally occurring sound sources. Signals and maskers presented in dichotic stimulus configurations have lower thresholds (MLDs) than signals and maskers presented in either diotic or monotic configurations. The MLD and the cocktail party effect demonstrate how binaural cues can be used to detect and recognize signals in noisy environments.

SUPPLEMENT

A number of books describe in more detail binaural hearing, localization, and lateralization: *Spatial Hearing* by Blauert (1997), *Directional Hearing* by Yost and Gourevitch (1987), and *Localization and Spatial Hearing in Real and Virtual Environments* by Gilkey and Anderson (1997). Chapter 6 in the textbook by Moore (1997) and a similar chapter by Yost and Dye (1991) cover many aspects of binaural hearing. Also see Wightman and Kistler (1993). The article by Litovsky *et al.* (1999) provides a review of the precedence effect.

All of the data described in this chapter relate to stationary sound sources and not sensitivity to moving sound sources. Chandler and Grantham (1992) and Saberi and Perrott (1990) have published a number of articles on auditory motion. These studies involve the study of sound sources that actually move and of simulated moving sound sources. If a sound is presented to one loudspeaker at one location at one instant in time and then is presented to a different loudspeaker at another location at the next instant in time, listeners almost always report that the sound appears to move from one location to the other even though nothing actually moved. By changing the time interval between exciting one loudspeaker and then the other, the velocity of the apparent motion can be varied.

The fact that we can move our heads when localizing sounds was briefly mentioned in this chapter. Head motion (see Mills, 1972, for a brief review) may be an important variable for many aspects of localization, especially when one tries to present sounds over headphones in order to produce the illusion that there are sound sources in actual space. These experiments in which headphone-delivered sounds are judged to be "externalized" are often referred to as experiments in *auditory virtual environments*. That is, it is possible to present sounds over headphones such that listeners report that actual sound sources exist, and as such the headphone-delivered stimuli have produced a "virtual" rather than a "real" acoustic environment.

Binaural beats are another phenomenon not described in the chapter. Binaural beats occur when two tones of slightly different frequency are presented, one to each ear. In this case, listeners report hearing a movement of the sound image between the two ears that occurs at a rate equal to the frequency difference. For instance, if a 500-Hz tone is delivered to one ear and a 505-Hz tone to the other ear (5-Hz difference), a movement of an image between the two ears will occur at a rate of five times per second (see McFadden and Passenan, 1975). Henning (1977) studied interaural time discrimination for low-rates of amplitude modulation.

Cherry (1953) first coined the term "cocktail party effect" when he discussed how we perceive one world with two ears. Another way to explain the MLD (see Green and Yost, 1975) has been provided by Durlach (1972) in his equalization–cancellation (EC) model of binaural hearing. In the EC model, the waveforms at the two ears in an MLD procedure are first assumed to be equalized and then the neural information at one ear is subtracted from that at the other ear. Consider what would happen in the M_oS_o and M_oS_π conditions if the waveforms at both ears were subtracted. In the M_oS_o condition there would be complete cancellation since both ears receive the same waveform. However, in the M_oS_π condition only the masker would completely cancel since the masker but not the signal is identical at both ears. The signal would actually add (i.e., subtracting one waveform that is 180° out of phase with another waveform is the same as adding the two waveforms, see Appendix A). Thus, the cancellation process provides a large signal to detect in the M_oS_π condition and nothing to detect in the M_oS_o condition, predicting a lower signal threshold in the M_oS_π condition. The EC model makes additional assumptions about the equalization and cancellation processes so that exact predictions for MLD data can be obtained.

Although this book is largely about human hearing, a variety of different neural methods appear to enable different animals to localize sound sources (see Yost and Gourevitch, 1987, and Fay and Popper, 1998). Studies of animals that echo-locate (e.g., bats and dolphins) present a particularly interesting and well-studied area of sound localization (see Yost and Gourevitch, 1987, and Au, 1993).

CHAPTER 13

Loudness and Pitch

*I*n Chapters 10 through 12, we discussed measures of the auditory system's sensitivity to the frequency, intensity, and temporal properties of acoustic events. This chapter deals with listeners' subjective descriptions and evaluations of frequency and intensity. As we stated briefly in Chapter 2, we most often refer to intensity as loudness and to frequency as pitch. Chapter 12 described those changes in phase and time that can result in the perception of stimulus location in space. Since this subjective aspect of phase has already been discussed, this chapter will be concerned with loudness and pitch, the subjective aspects of intensity and frequency.

LOUDNESS

In order to study loudness, psychophysical procedures such as scaling and matching (see Appendix D) are used. Figure 13.1 displays the results of a loudness-matching experiment. The data are plotted as the level in dB SPL of a comparison tone, required by the listener to match (perceptually equate) the *loudness* of the comparison tone to that of the standard tone, as a function of the frequency of the comparison tone. The standard tone was a 1000-Hz sinusoid presented at various levels expressed in dB SPL. The curves in Figure 13.1 represent different levels of the standard. For instance, for the curve labeled 40 phons the standard was a 40–dB SPL 1000-Hz tone, and the listener varied the level of the comparison tones presented at other frequencies until for each comparison tone the comparison and standard tones were perceived as equally loud. Each contour (curve) is called an *equal loudness contour* since for every frequency presented at the level described by the curve all sinusoids appear the same in loudness and all tones appear equally loud. For instance, tones presented at 100 Hz, 52 dB SPL; 1000 Hz, 40 dB SPL; and 4000 Hz, 37 dB SPL are all judged to be equal in loudness, although they are different in physical level (these tones form the 40-dB equal-loudness contour).

Two terms are used to describe or measure the loudness of a stimulus. *Loudness level* is measured in *phons*; a phon is the level in dB SPL of an equally loud 1000-Hz tone. All tones judged equal in loudness to a 40-dB SPL, 1000-Hz tone have a loudness level of 40 phons. The tones presented at levels such that they are equal in loudness to a 70-dB SPL, 1000-Hz tone all have a loudness level of 70 phons, and so on. The equal loudness contours in Figure 13.1 form the phon scale. The *sone* scale is

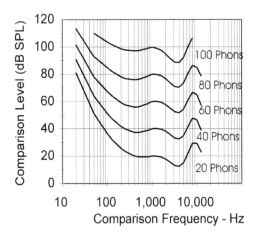

FIGURE 13.1 Equal loudness contours showing the level of a comparison tone required to match the perceived loudness of a 1,000-Hz standard tone presented at different levels (20, 40, 60, 80, and 100 dB SPL). Each curve is an equal loudness contour. Based on ISO standard (1981).

another way to measure loudness, where one sone is the loudness of a 1000-Hz tone presented at 40 dB SPL. One sone is equal to 40 phons. A stimulus that is n sones loud is judged to be n times as loud as 1 sone; n times as loud as the 1000-Hz, 40-dB SPL standard. Figure 13.2a is a plot of the loudness in sones of a 1000-Hz tone versus its level in dB SPL (at 1000 Hz, dB SPL is the same as dB in phons, and thus the horizontal axis could also be phons). The sone scale in this figure, like the level scale, is logarithmic. The data are fit by approximately a straight line above about 30 phons. From the slope of this line, there is a doubling of loudness for approximately every 10 dB increase in level. A listener, therefore, must increase a tone's level by 10 dB before judging its loudness to have doubled. The sone scale was obtained by a ratio-scaling technique in which the listener is asked to adjust the level of a comparison tone so that it appears half as loud or twice as loud as the

1-sone standard (1000-Hz, 40 dB SPL tone). This adjusted comparison tone becomes a new standard and a new comparison tone is presented; this procedure is repeated many times to obtain the sone scale.

The 40-dB equal loudness contour is also used to calculate a measure of level called the decibel weighted by the A scale, or dBA. The dBA measure is often used in noisy environments. The dBA measure is the total amount of sound power, measured in decibels, that is passed through a filter with cutoffs and attenuation rates that match the 40-dB equal-loudness contour. That is, a sound is filtered by a filter whose amplitude spectrum matches

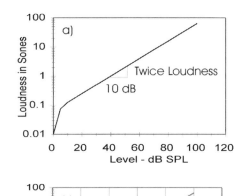

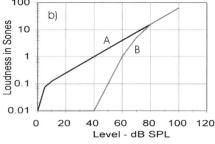

FIGURE 13.2 (a) Loudness in sones of a 1000-Hz tone as a function of the level of the tone (dB SPL). The slope of the function indicates that approximately a 10-dB change in level is required to double loudness. (b) Curve A is the loudness curve from Figure 13.2a; curve B represents the perceived loudness of a 1000-Hz tone masked by a wideband noise. Curve B shows loudness recruitment. Based on diagrams by Steinberg and Gardner (1937).

that of the 40-phon equal-loudness contour before its total power is determined. As a result, those frequency components in the sound that are in the 500- to 5000-Hz region receive the most weight when the total noise power is computed in dBA. This means that the contributions of very-high- and very-low-frequency components to the dBA measure are small. Thus, only those frequency components to which we are most sensitive (see Chapter 10) contribute most to the dBA measure.

Figure 13.2b is also a plot of loudness versus level, but in this case curve B represents the loudness of a tone that was masked by a 30-dB spectrum level (N_o) wideband noise. Curve A is the same as the one in Figure 13.2a. Notice that the threshold for the masked tone is 40 dB above the unmasked tone's threshold. Because both tones are at threshold (one at absolute threshold, the other at masked threshold), the two tones are of equal loudness when their actual levels are 40 dB apart. When the actual level of both tones is 80 dB SPL, they are also judged equally loud. This, in turn, means that the loudness of the masked tone (curve B) increased faster (steeper slope) than the unmasked tone. This increase in loudness (or the steep loudness slope) is sometimes called *loudness recruitment*. The loudness of the masked tone changes much more for each 10-dB increase in level than does the loudness of the unmasked tone. Curve B might also represent the data from someone whose threshold of hearing was 40 dB above normal threshold (i.e., the person had a 40-dB hearing loss) but who judges an 80-dB SPL tone the same in loudness as a normal subject would. The fact that the individual's loudness function shows loudness recruitment is important in treating certain hearing abnormalities. Note, for instance, that amplifying sound (e.g., by use of a hearing aid) will lead to a more rapid increase in loudness for the person with a hearing loss than for the person with normal hearing.

Most of the interactions that have been studied using threshold as a psychoacoustical measure can also be investigated using loudness. For instance, the duration of a tone can be varied, and the listener can adjust the tone's power so that it remains equally loud. In so doing, loudness measures are used to determine temporal integration. A short sound will not be as loud as a long sound if their powers are equal and their durations are less than approximately 250 msec. The loudness of a sound can also be maintained at a constant phon level as the bandwidth of a stimulus is narrowed in order to measure a critical band. In this case, narrowing the bandwidth of a stimulus will decrease the loudness once the bandwidth is less than a critical band. In these and other cases, the data from threshold experiments are not substantially different from those obtained in loudness studies.

Loudness adaptation (or *perstimulatory fatigue*) occurs during exposure to a long-duration (on the order of seconds or minutes) adapting stimulus. The change in loudness adaptation takes place while the adapting stimulus is being presented. That is, a stimulus appears softer if it is kept on for a very long time. The loudness of the stimulus is usually measured by having the listener match a comparison stimulus to the adapting stimulus in terms of loudness. As the adapting stimulus remains on for longer and longer periods of time, listeners decrease the level of the matching stimulus, indicating that the loudness of the adapting stimulus has decreased. Thus, the context in which a sound occurs can affect its perceived loudness.

As we mentioned in Chapter 2, although changes in level are highly correlated with loudness changes, the relationship is not perfect. That is, changes in frequency, duration, intensity, and bandwidth all affect the perceived loudness of a stimulus even though the level of the stimulus remains fixed. Remember that loudness is a subjective evaluation of sound, whereas intensity is an objective measure of vibratory magnitude.

PITCH

The subjective label of pitch has come to represent different aspects of an auditory stimulus. Generally, pitch is highly correlated with frequency. However, in many instances listeners report perceiving a pitch in the absence of any energy at the frequency that corresponds to the reported pitch. The often-discussed relationship of frequency to pitch is predictable on the basis of the encoding of frequency performed by the cochlea and auditory nerve fibers or, more generally, by the tonotopic or place organization of the auditory system. The fact that listeners can report pitches for stimuli that contain no energy at the frequency of the reported pitch means that pitch is probably encoded by mechanisms in addition to those involved only with the place theory of hearing, most likely on the basis of the temporal structure of neural firing.

Experiments on pitch are usually performed with the matching procedure in which a standard stimulus (sometimes a sinusoid) is used as the basis for pitch matches of comparison stimuli. Scaling procedures have also been used in the measurement of pitch, but listeners often respond in a different manner when they are asked to judge pitch than when they are asked to judge loudness. This difference stems from the qualitative aspects of pitch. That is, as the pitch of a stimulus changes, it does not appear to vary along a single dimension of greater to smaller or more to less. The change appears to occur along

many dimensions at once, whereas the changes in loudness are quantitative, in that they can be scaled along a single dimension. One tone can be said to be greater in loudness than another tone, but one pitch may be greater in magnitude than another pitch.

Many attempts have been made to scale pitch, such as the seven-tone musical scale (B C D E F G A). In general, the relationships among musical notes are arranged in an octave manner with 12 intervals per octave. In the *equal temperament scale* the octave is divided into 12 equal logarithmic intervals called *semitones* and each interval is divided into 100 equal logarithmic steps called *cents*. A semitone is 100 cents. Thus, an octave has 1200 cents, and each interval is 100 cents or a semitone. These relationships are shown in Table 13.1. For this table we are assuming that the note A has a fundamental period of vibration that is the reciprocal of 440 Hz. Thus, the interval of cent values forms a type of *pitch scale*. Other schemes are often used to express the physical relationship among intervals. The *just intonation* and the *Pythagorean schemes* have a slightly different number of cents between intervals than the equal temperament scale. These scales are preferred by certain musicians.

Other nonmusically based scales of pitch have also been developed. The *mel scale*, for instance, is a scale derived for pitch in the same way the sone scale was obtained for loudness, by using direct scaling techniques. However, the mel scale is difficult to obtain, probably due to the qualitative nature of the subjective dimension of pitch. Just as loudness is associated with intensity but is considered a subjective measure, pitch may be highly associated with frequency; but many other variables such as level, waveform repetition, and bandwidth can change the pitch of a waveform. Thus, pitch is also a subjective measure.

TABLE 13.1 The relationship among Musical Note, Cents, and Frequency (Hz) for the Three Major Scales of Musical Pitch

Musical note	Just intonation		Equal temperament		Pythagorean tuning	
	Cents	Frequency	Cents	Frequency	Cents	Frequency
C	0	264	0	264	0	264
D	204	297	200	296	204	297
E	386	329	400	333	408	334
F	498	352	500	352	498	352
G	702	396	700	395	702	396
A	884	440	900	443	906	445
B	1088	495	1100	498	1100	501
A	1200	528	1200	528	1200	528

In discussing the threshold of audibility, we mentioned that listeners might be asked to detect the presence of a tonal-sounding stimulus instead of just detecting any sound. Tonality implies that the observer is detecting the presence of pitch. Von Bekesy found that a tone with a frequency less than 1000 Hz must have a duration equal to 3 to 9 periods if the tone was to have a definite pitch. Above 1000 Hz, this critical duration for the perception of tonality or pitch was 10 msec regardless of the frequency of the tone.

These experiments demonstrate the relationship between sinusoidal frequency and pitch; the frequency of the tone in Hz is also its pitch. If a comparison stimulus is judged to have the same pitch as a 440-Hz standard tone, then the comparison stimulus has a 440-Hz pitch. Broadband stimuli such as noises will also have a pitch-like quality if their energies are concentrated in bands of frequency. The perceived pitch as indicated by the frequency of the matching tone is associated with the frequency region where the most energy is concentrated.

Thus, there is a strong correlation between the pitch of a stimulus and the spectral location of its frequencies. However, observers re-port a 100-Hz pitch associated with a stimulus consisting of a sum of the frequencies of 700, 800, 900, and 1000 Hz. Although there is absolutely no energy at 100 Hz, listeners judge the sound to have a 100-Hz pitch. These four tones are all harmonics of 100 Hz (in fact, 100 Hz is the highest frequency for which the tones could be harmonically related). The observation that a pitch could be associated with the fundamental of a complex stimulus even when the fundamental was absent in the spectrum of the complex stimulus is called the *case of the missing fundamental.*

The amplitude spectrum and the time waveform (assuming all the tones had the same phase) associated with the tonal complex (300, 400, 500, 600, and 700 Hz) are shown in Figure 13.3. Notice that there is a 10-msec spacing between the major peaks in the complex time waveform. Since the frequency associated with a 10-msec period is 100 Hz, it may be possible that the auditory system perceived the 100-Hz pitch because of the 10-msec periodicity of the time waveform. In fact, most stimuli that have a periodic time waveform will have a perceived pitch equal to the reciprocal of the period, called a *periodicity pitch.* For instance, since the time waveform of the

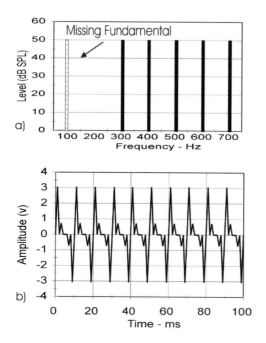

FIGURE 13.3 The amplitude spectrum (**a**) and the time domain waveform (**b**) of a missing fundamental-pitch stimulus consisting of a complex sound with frequency components of 300, 400, 500, 600, and 700 Hz. The pitch is 100 Hz, which is the missing-fundamental (dotted) component in Figure 13.3a. The time domain has an envelope with a 10-ms period, the reciprocal of 100 Hz.

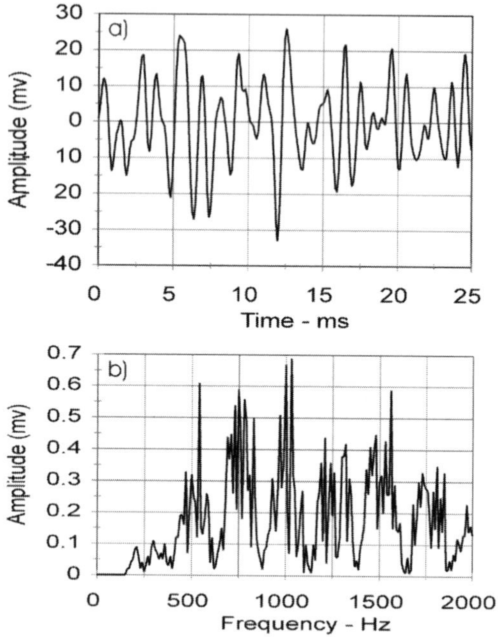

FIGURE 13.4   The amplitude spectrum (**a**) and the time domain waveform (**b**) of a complex sound that leads to a perceived pitch of 250 Hz. The amplitude spectrum has noisy peaks at integer multiplies of 250 Hz, but not a spectral peak at 250 Hz. There is no periodicity in the time-domain waveform envelope at 4 ms, the reciprocal of the pitch (1000/250 = 4 ms).

square wave shown in Figure 4.8 consists of a periodically occurring square wave whose period is 2 msec, this waveform will have 500-Hz periodicity pitch. In this case, as can be seen from Figure 4.8b, there is also energy at 500 Hz in the amplitude spectrum. Thus, we can perceive a periodicity pitch when the complex stimulus has energy at the frequency of the pitch (the square wave) or when there is not energy at the frequency of the pitch (the missing-fundamental tonal complex).

Although the periodicity concept appears to account for the data of the missing fundamental, there is a counterexample. The sound

with the spectrum and time-domain waveform shown in Figure 13.4 produces a 250-Hz pitch (i.e., listeners indicate that a 250-Hz tone has the same pitch as the stimulus described in Figure 13.4). The spectrum of the sound is continuous and there are noisy spectral peaks at 750, 1000, 1250, 1500, 1750, and 2000 Hz. Thus, the spectral peaks are like the tonal components for a missing-fundamental pitch stimulus. However, there is neither any energy at the reported pitch (250 Hz) nor any periodicity in the envelope of the noisy waveform. Thus, neither a concentration of energy in a particular frequency region nor the periodic nature of the envelope can be used to determine the sound's 250-Hz pitch. These data involving other complex waveforms indicate that, although the envelope-periodicity theory might be correct in some cases, other processing is taking place in addition to envelope–period analysis. The pitches of these complex sounds are referred to in a number of ways: *pitch of the missing fundamental*, *complex pitch*, and *virtual pitch* (as opposed to the *spectral pitch* of a sound, whose pitch is determined entirely by the spectral location with the most energy).

Figure 13.5 indicates one more aspect of these complex pitches. In this case the tones of the complex are at 425, 525, 625, 725, 825, and 925 Hz. This sound is generated by shifting the spectrum of a 100-Hz missing-fundamental pitch stimulus consisting of components at 400, 500, 600, 700, 800, and 900 Hz up in frequency by 25 Hz. Notice that, although all frequencies are equally spaced at 100-Hz intervals, 100 Hz is not the "missing fundamental"; 25 Hz would be the highest fundamental of this complex sound. Listeners match the pitch of this complex to that of a sinusoid with a frequency of 104 Hz, indicating that this complex has a perceived complex pitch of 104 Hz, which is neither the frequency of the missing fundamental nor the frequency spacing of the tones. This pitch is referred to as the *pitch shift of the residue*. Again there is no energy in the spectrum at 104 Hz, nor is there a clear periodicity in the time-domain envelope at the reciprocal of 104 Hz.

When the pitch of a stimulus can be determined on the basis of periodicity in the time-domain waveform, hearing scientists believe that some aspect of the auditory nerve's ability to be phase locked to the periodicity of low-frequency stimuli accounts for the way in which the auditory system processes these pitches. The area of pitch perception still poses a real challenge for auditory theorists.

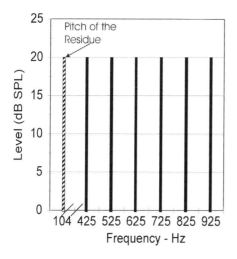

FIGURE 13.5 The amplitude spectrum of a pitch-shift of the residue stimulus consisting of a complex sound with frequency components of 425, 525, 625, and 725 Hz. The pitch is 104-Hz, which is neither the missing fundamental (25-Hz is the fundamental) nor the frequency spacing among the frequency components.

NONLINEAR TONES

In Chapter 11, we described the aural harmonics and difference tones produced by the nonlinearity of the auditory system. The second and third *aural harmonics* ($2f1$ and $3f1$), the *difference tone* ($f1 - f2$), and the *cubic difference tone* ($2f1 - f2$) are those nonlinear tones most often perceived (see Chapter 5 and Appendix B for a discussion of nonlinearity). That is, when a complex sound consisting of a number of sinusoids is presented to a listener, especially at loud levels, listeners hear pitches in addition to those corresponding to the frequency of the sinusoids in the stimulus. These additional pitches are associated with the aural harmonics and difference tones produced by the nonlinear properties of the auditory periphery (see Chapters 8 and 9). The cubic difference tone is of particular interest because in many conditions it is the most perceptible nonlinearly produced tone. For instance, if 1400- and 1680-Hz primary tones are summed, a cubic difference tone of 1120 Hz is perceived; that is, (2 × 1400 Hz) – 1680 Hz = 1120 Hz. This cubic difference tone can be heard when the levels of the 1400- and 1680-Hz primary tones are less than 40 dB SL, whereas the difference tone of 280 Hz (1680 Hz – 1400 Hz = 280 Hz) cannot be detected at these low primary tone levels. Remember that the perception of the 1120- or 280-Hz pitch is not due to the presence of these frequencies in the stimulus. The pitches result from the nonlinear distortion caused by the peripheral auditory system.

In order to describe the nonlinear tones, we must specify their levels and phases as well as their frequencies. The *cancellation method* is often used to obtain estimates of the level and phase of nonlinear tones. In the cancellation method, one complex stimulus is used to elicit the nonlinear tone, for instance, an 840-Hz and a 1000-Hz primary tone pair that produces a 680-Hz cubic difference tone; that is, (2 × 840 Hz) – 1000 Hz = 680 Hz. Another stimulus is presented along with the primaries and is used to cancel the pitch of the nonlinear tone, in this case a 680-Hz *cancellation tone*. That is, a 680-Hz cancellation tone is added to the 840- and 1000-Hz primary tones. Without the addition of the 680-Hz cancellation tone, listeners can detect sound with pitches of 840 and 1000 Hz (the primary tones) and 680 Hz (the nonlinearly produced cubic-difference tone). If the 680-Hz cancellation tone is presented 180° out of phase with that of the nonlinear 680-Hz cubic difference tone, then the 680-Hz pitch might be canceled, and the listener would detect only the two primary tones. This cancellation should only occur when the two tones (cancellation tone and

cubic difference tone) are at equal levels and 180° out of phase (i.e., the addition of two tones of the same frequency and level but 180° out of phase will lead to complete cancellation). In the cancellation procedure, the listener is presented the 840- and 1000-Hz primary tones (tones used to elicit the 680-Hz cubic difference tone) and the 680-Hz cancellation tone. The listener is instructed to adjust the level and phase of the 680-Hz cancellation tone until the listener no longer hears a pitch of 680 Hz (the cubic difference tone). The level of the cancellation tone and 180° minus the phase of the cancellation tone that the listener picked that eliminated the pitch of the cubic difference tone is used to estimate the level and phase of the cubic difference tone (or other nonlinear tones).

Figure 13.6 displays the results from a cancellation experiment. The upper curve shows the level of the 680-Hz tone required to cancel the cubic difference tone as a function of the overall level of the two primary tones (840 and 1000 Hz) used to elicit the cubic difference tone. The lower curve shows the phase (minus 180°) of the 680-Hz tone required to cancel the cubic difference tone.

The intensities and phases of the nonlinearities of the auditory system are crucial values to be determined if we are to describe how the system produces these nonlinearities. The cubic difference tone has been of special interest because it appears at low levels, and the level and phase of the cubic difference tone changes in a complex way as a function of its pitch (i.e., as a function of separation in frequency between $f1$ and $f2$). Because of these relations, most scientists believe that the source of the cubic difference tone is in the inner ear; the perceived pitch of the cubic difference tone probably results either from the nonlinear motion of the basilar membrane or

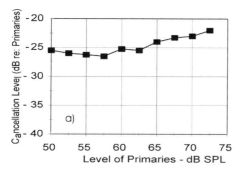

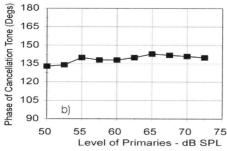

FIGURE 13.6 (**a**) The level of a cancellation tone (relative to the levels of the primaries) required to cancel a 680-Hz cubic-difference tone produced by 840- and 1000-Hz primaries presented at different levels (dB SPL). The cancellation-tone level is about 25–28 dB below that of the primaries for primary levels ranging from 45 to 75 dB SPL. (**b**) The phase of the cancellation tone required to cancel the 680-Hz cubic difference tone shown as a function of the level of the primaries. The estimated phase of the cubic difference tone is 180° minus the values shown in Figure 13.6b. Based on data from Hall (1975), with permission.

from some nonlinearities that exist when the haircells stimulate the auditory nerve fibers.

OTHER SUBJECTIVE ATTRIBUTES OF SOUND

Complex stimuli have subjective attributes in addition to pitch and loudness, one of which is *timbre*. Timbre is often defined as that subjective attribute of a sound which differentiates two or more sounds that have the same pitch, loudness, and duration. For instance, the quality difference between a violin and a cello playing the same musical note, at the same loudness, and for the same duration, would be defined as a difference in timbre between the two instruments. Timbre appears to be related to the bandwidth of the complex stimulus, especially for complex waveforms consisting of harmonically related sinusoids. A stimulus with a larger number of harmonics is usually perceived as having a fuller or richer timbre than more narrowband stimuli, such as a pure tone.

In addition to timbre, musicians often refer to the *consonance* and *dissonance* of complex stimuli, such as notes of music. Consonant pairs are those notes played at intervals that result in a "pleasant" sound. Dissonant pairs are played at intervals that sound "unpleasant" to many musicians. We have already discussed the beats (Chapter 10) associated with mixing two sinusoids whose frequencies differ by a few hertz. The sensation of beats gives way to *flutter* and then to *roughness* as the frequency difference between the two tones is increased.

S. S. Stevens determined that sounds also have attributes of *density and volume*. The density of a tone increases as its frequency or intensity increases. Increases in volume are generally associated with decreases in frequency and level. Thus, volume and density are approximate opposites. The fact that different sounds elicit different subjective descriptions is not surprising. To what extent these subjective labels are an integral part of auditory processing is not known at this time. Many scientists believe that labeling may also depend on culture and experience. For instance, many modern composers are writing music with dissonant intervals. Because some of this music is becoming popular, perhaps which musical intervals are labeled as consonant and dissonant might change over time.

SUMMARY

From equal-loudness contours and loudness-scaling experiments we can construct the phon and sone scales of loudness. These scales enable us to relate the subjective description of loudness to the physical descriptions of frequency and intensity. The musical scale of pitch contains octaves, intervals, semitones, and cents. The mel scale of pitch is constructed from a pitch-scaling experiment. Scales of pitch are more qualitative in nature than are loudness scales. Pitch and loudness are not perfectly correlated with their physical counterparts frequency and intensity. The concept of the missing fundamental illustrates that pitch processing is sometimes dependent on neither spectral information at the frequency of the pitch nor envelope periodicity information associated with the period of the pitch. The levels and phases of nonlinear tones (especially the cubic difference tone) are measured by the cancellation technique. Complex stimuli have additional subjective attributes such as timbre, consonance, dissonance, beats, flutter, roughness, density, and volume.

SUPPLEMENT

Loudness and pitch as subjective attributes of sound have been studied extensively by S. S. Stevens. His book *Psychophysics* (1975) and his article in *Science* (1970), "Neural Events and the Psychophysical Law," provide an insight into his work. Fletcher and Munson (1933) used the loudness-matching technique to obtain the equal-loudness contours. The book by Hartmann (1998) should also be consulted for more details about pitch, loudness, and nonlinearities. The book by Moore (1997) also covers topics on loudness and pitch.

Loudness is often measured using an alternating binaural loudness balance (ABLB) technique. In this procedure, the standard tone is presented to one ear and the comparison tone to the other ear. The tones are alternated in time, and the listener adjusts the comparison tone until it appears as loud as the standard. Steinberg and Gardner (1937) provided insights about the relationship between masking and loudness that led to the concept of recruitment.

S. S. Stevens devised a method that has become standard for determining loudness of complex, nontonal sounds (ISO standard R532-Method A; see the supplements to Chapter 10 of this book for a discussion of standards). This method involves combining the sone measurements for various frequency bands. Another method devised by Zwicker (ISO standard R532-Method B) has also been used to measure the subjective magnitude of a complex stimulus. In general, critical bandwidths estimated using loudness methods are three to four times wider than those obtained from masking studies (see Chapter 11).

Fletcher observed the missing fundamental at Bell Laboratories in 1924. This area of research is important in indicating that neither the place theory nor the temporally based theory alone is sufficient to explain our perceptions of pitch. The student might have noted for the missing-fundamental pitch (in Figure 13.3 where the frequencies added together were 700, 800, 900, and 1000 Hz) that a nonlinear difference tone exists at 100 Hz. Thus, the 100-Hz pitch might be due to the nonlinearity of the ear. This, however, does not seem to be the case. Licklider (1954), for instance, showed that the difference tone due to nonlinearity could be masked by a noise with frequencies near 100 Hz. Licklider then showed, however, that the pitch of the missing-fundamental stimulus was unaffected by a masking noise with a frequency of 100 Hz. Since the nonlinear tone at 100 Hz was masked and the pitch was still perceived, nonlinearity did not yield the missing-fundamental pitch.

There is considerable evidence that the pitch of a tone changes (usually increases) as its level increases. Although S. S. Stevens (1975) studied this effect extensively, the pitch changes are highly variable and usually fairly small. Jesteadt (1980) provides an interesting procedure for determining pitch shifts associated with intensity changes.

There is evidence that we hear primarily the following aural harmonics and combination tones: $2f$, $3f$, and $4f$, $f1 - f2$, $2f1 - f2$, $2f1 - 2f2$, $3f1 - 2f2$. Summation tones have generally not been reported as audible. This is presumably because low-frequency tones mask high-frequency tones very well (see Chapter 11). Because the summation tones are higher in frequency than the primary tones but also usually close to the frequency of one of the primaries, the summation tones are masked by the primary tones.

The cancellation method is usually used with one additional tone added to the input stimulus. Since the pitch of the nonlinear tone is often difficult to detect, it can be made easier to hear if a tone is added that is slightly different in frequency than the cancellation tone (thus, the input stimulus consists of the primaries, the cancellation tone, and a tone that is slightly different in frequency, e.g., 3

Hz, from the cancellation tone). This additional tone and the cancellation tone (and the nonlinear tone) will produce a beating sensation. If the cancellation tone is now added out-of-phase and at the same level as the nonlinear tone so that the nonlinear and cancellation tones are eliminated, then the beating stops and the listener hears the pitches of two primary tones and the tone used to beat with the nonlinear tone. Most listeners find it easier to make the cancellation procedure measurements when they are asked to eliminate the beating rather then to eliminate the pitch of the nonlinear tone.

A review of the cubic difference tone can be found in Zwicker and Fastl's book (1991). Appendix A shows how one could obtain difference tones and summation tones from a nonlinear equation ($y = x + x_2$). The cubic difference tone ($2f1 - f2$) is obtained if the nonlinear equation is in the following form:

$$y = x + x^2 + x^3.$$

Thus, the name cubic difference tone occurs because the tone is obtained by including the cubic (x^3) term in the nonlinear equation.

The stimulus shown in Figure 13.6 is a Regular Interval Stimulus (RIS), called iterated rippled noise (IRN). It can be generated by delaying a noise and adding the delayed noise back to the undelayed noise. The perceived pitch of IRN stimuli is equal to the reciprocal of the delay (Yost, 1996). Thus, in Figure 13.6 the delay was 4 msec, yielding the 250-Hz pitch (1/4 msec = 250 Hz) and 250-Hz spacing between the noisy spectral peaks.

PART IV

COMPLEX SOUNDS PROCESSING, THE CNS, AND AUDITORY DISORDERS

CHAPTER 14

Auditory Perception and Sound Source Determination

*I*n Chapter 1, we describe hearing as an ability to determine the sources of sounds in our world. It was pointed out that the sounds from various sources are combined into one complex sound field before they arrive at our auditory system (see Figure 1.1). The rest of the chapters in this book have described the physical aspects of the sound field, how those physical attributes are processed by the peripheral auditory nervous system, and the behavioral consequences of this neural processing. In this chapter, we will briefly describe some of what is known concerning how the information in the peripheral code for sound is further processed to allow us to determine the sources of sound. We will also briefly discuss speech production and perception. In Chapter 15, we will review the anatomy and physiology of the central auditory nervous system, the neural site of most of the processing that we will describe in Chapter 14.

Figure 14.1 shows the output of a model (the same one used for Figure 9.15) used to simulate the coding of sound by the auditory periphery. The sound in this case is a complex consisting of four tonal components. Let's imagine that this complex sound is the sound field generated by two sound sources, where the spectrum of each sound source consists of two tones. Let's further assume that the tonal components that make up each source are intermingled in the complex sound field as indicated in the sound field spectrum shown in Figure 14.1. Recall from Chapter 9 that the display in Figure 14.1 is not unlike the spectro-temporal pattern of neural information that the auditory periphery provides to the central nervous system about the complex sound field.

Figure 14.1 captures the major aspects of the frequency, intensity, and timing characteristics of the complex sound field. Thus, it represents the way in which these physical attributes of sound are coded in the auditory periphery. However, there is little in this display or code that indicates that there are two different sound sources. Somehow the information in the neural codes for the two tones that make up sound source A need to be *fused* together and *segregated* from those associated with sound source B. The physical characteristics of the sound from source A must differ in some way from those from source B if the nervous system is to determine that the complex field consists of two sources. In this

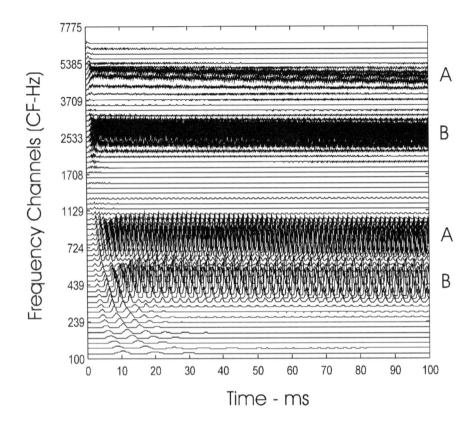

FIGURE 14.1 The outputs of the critical-band filter/basilar membrane stage of the Patterson *et al.* (1995) model, which simulates peripheral auditory processing. The outputs of 64 filter/basilar membrane simulations are plotted for 100 ms of the complex sound. Each line represents the output of one filter channel or one place along the basilar membrane. The filter center frequencies begin at 100 Hz, and the highest center frequency is 7114 Hz. This display provides an estimate of the spectral and temporal information available to the auditory periphery concerning this complex sound of four tones. The tones from one source (Source A) were 450 and 2500 Hz, while those from the other source (Source B) were 725 and 5000 Hz. Thus, the spectral components from the two sources overlap. The plots show a spread of activity because a single tone will excite a number of locations along the basilar membrane near the site of maximal displacement (i.e., near the filter's center frequency). The delay at onset for the low frequencies as compared to the high frequencies represents the time it takes the traveling wave to reach the apex where low frequencies are coded (see Chapters 8 and 9). Although the temporal and spectral information regarding the four tonal components that make up the complex sound is well preserved at the auditory periphery, there is little information displayed in this figure that indicates that two components are from one source and two from another source. Either the physical characteristics of the sounds from each source need to differ in some way or additional processing of the complex sound is required in order for the auditory system to determine that there were two sound sources.

chapter, we will consider seven physical characteristics that may be used as a basis for sound source determination. Some of these have already been discussed and others will be described in this chapter.

1. Spectral separation

2. Spectral profile

3. Harmonicity

4. Spatial separation

5. Temporal separation

6. Temporal onsets and offsets

7. Temporal modulations

SPECTRAL SEPARATION

We have already described the frequency-resolving capability of the auditory periphery and its many consequences (e.g., masking) for hearing. Figure 14.1 (see also Figure 9.15) clearly indicates that the different spectral components of a complex sound may be coded as separate neural events. Thus, for a relatively simple complex sound field the frequency-resolving abilities of the nervous system may aid the system in distinguishing one sound (i.e., one frequency) from another sound (i.e., another frequency). However, for more complex sound fields such as the one shown in Figure 14.1, where the spectral components from the sources are interlaced, frequency separation alone is not sufficient to allow the auditory system to differentiate one sound source from another.

SPECTRAL PROFILE

If one sound is more intense than another, the central nervous system might be able to determine the sources of the sounds based on their relative level differences. In order to ac-

complish this task, the nervous system must be able to differentiate the spectral profile of the loud sound from that of the soft sound as diagrammed in Figure 14.2. Most sound sources produce a particular amplitude spectrum that remains relatively constant in terms of its spectral profile as the overall level of the sound is changed. That is, the characteristics of a sound source are largely determined by the shape of the sound's amplitude spectrum. Making the sound louder or softer changes the overall level of the sound but not the shape of its spectral profile.

The auditory system is remarkably sensitive to small changes in the spectral profile of sounds. Consider an experiment in which listeners are asked to detect a small-level increment of a signal tone when a number of other tones are simultaneously added to the signal tone. From Chapter 11, we know that the ability to detect a signal tone of one frequency is interfered with when tones with frequencies close to that of the signal are simultaneously presented. The other tones mask the signal if the signal and maskers are close together in frequency, and there is little interference when the signal and masker are very different in frequency. Figure 14.3 shows the experimental context: listeners are asked to determine which stimulus in a two-alternative forced-choice task (see Appendix D and Chapter 11) contains the more intense signal tone. The key ingredient in the experiment is that the overall level of both complexes (the signal complex with the signal level increment and the non-signal complex consisting of all tones presented at the same level) is randomly varied over a 40-dB range from observation interval to observation interval. Thus, regardless of the relative levels of the two sounds, the listener is to decide which presentation contains the signal stimulus with the middle signal tone's level incremented. There are at least three ways the auditory system might process these stimuli in order for the listener to perform this

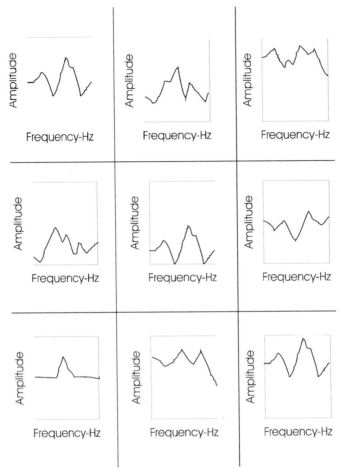

FIGURE 14.2 A schematic diagram of the spectrum of a complex sounds coming from different sources. Which spectra belong to the same sound source? The spectra along the diagonal from upper left to lower right are from the same sound source, but with the sound source producing different overall levels. The other spectra are from different sound sources. As the overall level of the sound increases, that is, as the sound from the source gets louder, the spectral profile of the source does not change. That is, the relative amplitudes of the frequency components remain the same as the overall amplitudes increase. Thus, a key to processing such stimuli is for the auditory system to monitor the relative differences of the amplitudes of the spectral components as the overall level changes.

task: (1) The system could determine which stimulus is louder overall, since the signal stimulus could be more intense due to the signal tone level increment. However, given the 40-dB randomization of overall level, the signal tone would have to be incremented by a great deal (nearly 40 dB) to overcome this 40-dB random variation. (2) The auditory system might process the stimuli in only the critical band tuned to the frequency of the signal, as was assumed for much of the discussion of masking in Chapter 11. However, even if only

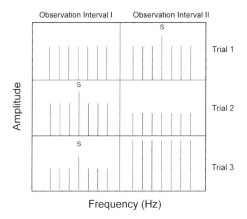

FIGURE 14.3 In a typical "profile analysis" experiment, two stimuli are presented per trial and the listener's task is to determine in which observation interval (I or II) the stimulus contains a level increment of the signal (S, the center frequency component). From trial to trial and from observation interval to observation interval, the overall level of the stimulus is randomly varied over a 40-dB range. Thus, the listener would be correct if he or she indicated that the signal occurred in observation interval II for trial 1 and observation interval I for trials 2 and 3. Because of the overall level variation, a very large signal level increment would be required if the listener used either the overall level of the sound or the level at only the signal (center) frequency as the basis for his/her decision.

the signal tone is processed by this critical-band filter, its level is still varying randomly by 40 dB, which in turn would require an extremely large signal tone level increment for a listener to determine that the signal was incremented in level versus simply changed because of the 40-dB random variation in overall level. (3) The auditory system might be able to compare the level of the signal tone to that of the nonsignal masking tones. That is, the levels of the nonsignal tones are all the same, while when the signal level is incremented it is an increase in level above this constant background. Since this relative level difference remains constant as the overall level of the complexes is randomly varied over the 40-dB range, only a small-level increment

might be required to detect the tonal signal increment.

Figure 14.4 shows the results from an experiment like that outlined above for four conditions: four maskers flanking the 1000-Hz signal, 10 maskers flanking the signal, 20 maskers flanking the signal, and 42 maskers flanking the signal. In addition, the threshold for detecting a level increment for a 1000-Hz signal presented alone without any maskers is also shown (data taken from Figure 10.8). As

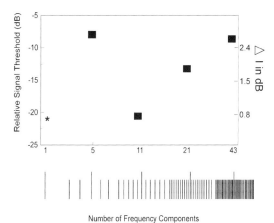

FIGURE 14.4 The results from a profile analysis experiment in which the number of masker frequency components surrounding a 1000-Hz signal component increased from 4 to 42. The thresholds for detecting an increment in the 1000-Hz signal component (the center component) are shown in decibels relative to that of the rest of the masker component intensity. The asterisk on the far left indicates the typical threshold for detecting a level increment of a single 1000-Hz tone. Thresholds for the 10-masker condition are almost as low as those for the single-tone condition, and the thresholds first decrease and then increase as the number of masker components increases from 4 to 42. The level of the signal is expressed in terms of the signal-to-background level (see Table 10.3 and the signal-to-masker ratio calculations). Figure 14.3 should be consulted for a description of the profile stimuli and the listener's task. Based on data from Green (1989) with permission.

can be seen, the thresholds for detecting the signal level increment when the maskers are present and the overall level is randomly varying is almost as low as that obtained for a single tone presented without maskers and without any level randomization. Also note that the thresholds are high for the 5-component stimulus, decrease for the 11-component stimulus, and then increase again for the 21- and 43-component stimulus.

Both the fact that the thresholds are low and the way in which they change as a function of the number and spacing of the components is consistent with the assumption that the auditory system uses the relative difference in level between the signal component and the masker components as cues for detection. That is, the auditory system detects the peak in the spectral profile of the signal stimulus independent of the overall level of the spectrum. When there are a few widely spaced masker components, it is difficult to discern the difference in level across these widely spaced spectral components (see the spectral profiles along the bottom of Figure 14.4). When the components get closer together in frequency, the difference in the level of the signal component as compared to the constant level of the maskers becomes easier to discern; thus, thresholds decrease as the number of masker components increases from 4 to 10. However, as the number of components increases further, the masker components begin to interfere directly with the signal and with each other because they are close enough together in frequency to excite the same critical band, and, thus, the maskers begin to provide direct masking of the signal (see Figure 11.1). That is, as the number of masker components increases beyond 10, they mask the signal, making it more difficult to detect a signal level increment. Data like these suggest that the auditory system is performing a *profile analysis* in order to make compari-

sons across the spectrum of a sound so that subtle spectral intensity changes can be detected. Such processing is advantageous for a system design to determine the sources of sound.

HARMONICITY

In Chapter 13 the phenomenon of complex pitch was described in which complex sounds with particular spectral structures are perceived as having a pitch, even when there are no apparent periodicities in the time-domain waveform envelope or energy in the amplitude spectrum that could be used to predict the perceived pitch (e.g., missing fundamental pitch). Such complex or virtual pitches are often obtained when the complex sounds consist of harmonics of some fundamental. Many sound sources in our world consist of spectral components that are harmonically related. Most musical instruments and voiced speech are some typical examples of such harmonic sound sources. Thus, the ability to perceive these sound sources as having a unitary pitch is an excellent example of the auditory system fusing the information from many spectral regions to produce a single auditory image, in this case described as a pitch. As such, the pitch of a complex sound might be used as a cue for sound source determination.

However, we are rarely able to perceive two or more pitches when a complex sound consists of two or more different harmonic series. For instance, a sound consisting of the harmonics of 200 and 333 Hz (e.g., 200, 333, 400, 600, 666, 800, 999 Hz) would not usually be perceived as having two complex pitches (and therefore, perhaps, come from two sources) of 200 Hz (i.e., 200, 400, 600 and 800 Hz) and 333 Hz (i.e., 333, 666, and 999 Hz). The auditory system would in most cases synthesize (the auditory system is behaving *syn-*

thetically) the entire sound into a single percept rather than analyze (auditory system behaving *analytically*) it into its two harmonic parts. In the example given above, the two complex pitches may be perceived if something else is done to help separate the sound sources. For instance, if the sound with the 333-Hz fundamental were turned on before that with the 200-Hz fundamental, then the two complex pitches of 200 and 333 Hz might be detected (see section below on Temporal Onsets and Offsets). Thus, although phenomena like the "missing fundamental pitch" indicate one means by which the auditory system may synthesize spectral information to form a single percept, such pitch processing does not always appear to aid the auditory system in segregating different harmonic series into a multisource sound.

If one of the tonal components of a complex harmonically related tonal complex is changed in frequency it might be perceived as different in pitch from the fundamental pitch of the complex. Consider a stimulus consisting of the 12 harmonics of a 155-Hz fundamental (e.g., 155, 310, 466, 620) producing a 155-Hz complex pitch. If the 4th harmonic (620 Hz: 4×155 Hz) is changed by about 8% to 670 Hz (($(0.08 \times 620$ Hz) + 620 Hz = 670 Hz)), then the listener can perceive two pitches; the 670-Hz pitch (or something close to 670 Hz) of the mistuned 4th harmonic tone and the 155-Hz complex pitch associated with the complex 12-tone pattern. This example of a *mistuned harmonic* demonstrates that harmonic stimuli must contain spectral components within about 8% of their harmonic relationship in order for the complex sound to be perceived as a single sound source with a single pitch percept. It is also an example of segregating a complex sound into two sources: one a tone with a 670 Hz frequency and one a complex sound consisting of harmonics of 155 Hz.

SPATIAL SEPARATION

In Chapter 12, we described the concept of the "cocktail party effect." The cocktail party effect refers to the auditory system's ability to determine the sources of sounds when they are located at different points in space. A number of different experimental findings (e.g., the MLD results) indicate that the auditory system's ability to localize the sources of sounds assists in sound source determination. Thus, interaural differences of time and level and aspects of the head-related transfer functions that we know are crucial for sound localization probably also aid the auditory system in determining the sources of sound. However, our ability to localize sound sources cannot be the only way in which we determine sound sources. Consider a case where a symphony played by an orchestra was recorded through a single microphone and played back through a single loudspeaker (refer to Figure 1.1). There is very little, if any, information in this monaural recording about where the instruments of the orchestra are located, and yet we have little difficulty in determining a large number of the instruments when we listen to the symphony (e.g., we can determine that there were violins, drum, horns).

There is also growing evidence that spatially separating sound sources as the *only* means to achieve sound source segregation may provide only a weak ability to determine the actual sound sources. Some have characterized this work as suggesting that cues other than spatial separation must be used to determine *what* a sound source is and then spatial cues help determine *where* that source is located. Thus, although spatial separation in one means by which we can determine the source of a sound, it is not the only means.

TEMPORAL SEPARATION

Clearly, if two sources produce sounds at very different times, the separation in time would allow us to determine the sources. In Chapter 11, we described aspects of temporal masking (forward and backward masking) which indicate that when short target sounds precede or follow other sounds they may be masked. Thus, temporal masking will play a role in one's ability to determine the sources of sounds that appear close together, but separated, in time.

Many sounds that occur more or less concurrently are also intermittent sounds that go on and off. In many situations, the sound from one source alternates with that from another source, yet we are more likely to perceive the sound as two sources occurring together at the same time rather than one source with an alternating percept. This is diagrammed in Figure 14.5. For instance, source A might be a tone of one frequency and source B a tone of a different frequency. Under the proper stimulus circumstances, listeners describe the stimulus as if there are two concurrent sounds, each with its own distinct pulsating pitch, rather than one sound that alternates in pitch. That is, the perception is not of one source that has an alternating pitch, but of two different sources, each with its own pitch that is going on and off. The perception of the two alternating tones is two "streams of sound" running concurrently as if they were two sound sources. Understanding the conditions that lead to *stream fusion* (the stimulus conditions that allow nonsimultaneous sounds to appear as one source) and *stream segregation* (the stimulus conditions that lead to the perception of different sources when there is more than one alternating sound) is important for understanding how we determine the sources of sounds in multisource acoustic environments. Under the appropriate condi-

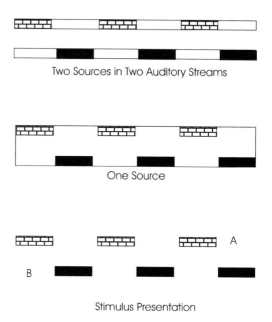

Two Sources in Two Auditory Streams

One Source

Stimulus Presentation

FIGURE 14.5 A schematic diagram indicating the type of procedure used in many streaming experiments. Two alternating sounds (e.g., two different frequencies alternating in time) are presented. Under the appropriate conditions, the listeners do not report hearing a single sound (i.e., from one sound source) that alternates in pitch, but rather they report hearing two sound sources (A and B as if there were two streams) each with its own pulsating pitch.

tions, sounds differing in spectral content, temporal modulation pattern, level, and/or spatial dimensions and presented in an alternating pattern will be perceived as streams of simultaneously occurring sound. However, it is usually the case that sounds that differ in spectral content in one way or another produce the strongest stream segregation.

When complex sounds are generated from simple sounds, it is not always possible to predict how the complex sounds will be perceived based on what is known about how the

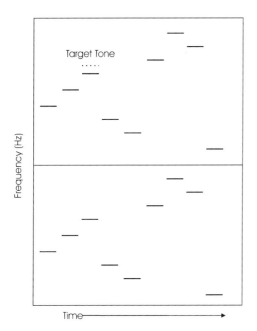

FIGURE 14.6 A 10-tone pattern (each short line represents one tone presented at a particular time in the pattern and with a particular frequency) typically used to study informational masking. The listener's task is to determine if the frequency of the target component was changed in a same–different task (i.e., the listener was presented either the same 10-tone pattern twice or a slight change in target frequency is introduced for the second presentation of the pattern; shown in the bottom pattern in which the shift in the target frequency is shown by the dotted line). In a high-uncertainty task, the spectral components would vary from trial to trial. In a minimum-uncertainty task, the same 10-tone pattern would be presented on each trial, with the same target tone having its frequency altered in the same-different task.

simple sounds are processed. Consider the example depicted in Figure 14.6. A temporal sequence of 10 tones is presented in a same-different psychophysical procedure such that the 10-tone pattern is presented twice per trial. On the second presentation, the frequency of 1 of the 10 tones is changed on half of the trials on a random basis. The listener is asked to determine for each trial if the two patterns are the same or different, and the responses are used to determine a frequency-difference threshold for tones presented at each of the 10 positions in the pattern. From trial to trial, a different 10-tone pattern is generated, but for any one test condition the test tone whose frequency might be changed remains at the same temporal location (e.g., the 3rd tone is the one that might be changed in frequency) for the duration of the condition. When the frequencies in the 10-tone pattern are varied randomly, listeners have very high frequency-discrimination thresholds, with those tones coming at the beginning of the pattern having the highest threshold. In this case, there is a lot of uncertainty about the stimulus structure from trial to trial and it is difficult to determine if one of the tone's frequency changed, especially if that tone occurred at the beginning of the pattern. If a *minimal uncertainty task* is used in which the same 10-tone pattern is presented on every trial (there is no randomization of the frequencies and thus little uncertainty about the frequency content of the 10-tone pattern), the frequency-difference thresholds are much smaller and the tones appearing at the beginning of the pattern no longer have the highest thresholds.

The elevation in frequency-discrimination thresholds described above is due to the amount of uncertainty in the stimulus not to a physical aspect of the stimulus. Thus, this increase in threshold is referred to as *informational masking* to differentiate it from the type of masking we discussed in Chapter 11. Another example of informational masking is a signal-detection experiment with a single tone to be detected in the presence of a masker complex that can be made up of many tones. In one example, the masker may consist of up to 100 tones added together. These 100 tones are chosen from a frequency range that brackets the signal frequency such that some tones in the masker may have frequencies higher and some lower than the signal frequency. The number of tones used for the masker may range from 2 to 100, and the choice of tones from the range of possible tones is selected at random from trial to trial. Thus, in some conditions only two tones are added to make up the masker, while in other conditions 100 tones may be chosen. There is considerably more masking (as much as 50 dB more) when the masker contains only two tones as opposed to when it contains 100 tones. That is, signal threshold is as much as 50 dB greater for a two-tone masker than for a 100-tone masker, when the tones in the masker are chosen at random from trial to trial. This should appear surprising based on what you learned in Chapter 11. Many times when only two tones are chosen (at random) for the masker, the frequencies of these two tones will be very different from that of the signal, and these masking tones should provide very little masking. Even when the two masking tones are close in frequency to the signal frequency, why should two maskers produce more masking than when there are 100 tones making up the masker? The explanation deals with informational masking caused by the uncertainty about the masking spectrum that exists from trial to trial in this type of experiment. That is, the listeners need more information about the stimulus context. For instance, information might be needed about the exact master frequencies, as would occur if master frequencies were not randomized (as in Chapter 11). This additional information appears to depend on the uncertainty the auditory system has about the stimulus context. The more uncertainty there is, the more informational masking there is. With only two tones, the variability in the masker from trial to trial is much larger than when 100 tones are used. In this sense, there is more uncertainty about the masker spectrum with 2 rather than with 100 tones. These and other data indicate that the nature of the listening task, especially those tasks involving complex sounds and uncertainty, may determine a great deal about the sensitivity of a listener in processing many complex sounds. Thus, in understanding sound source determination of complex sounds in real-world situations we need to understand both the stimulus context and the nature of the listening task.

TEMPORAL ONSETS AND OFFSETS

If the sound from one source starts after that from another source has already begun, then the differences in the onsets and/or offsets of the two sounds may be a cue for sound source determination. We have already studied situations in which the onsets of two sounds are different. In Chapter 11, we mentioned the fact that thresholds for detecting signals during the forward or backward temporal fringe of a masker may be higher than those measured more toward the temporal middle of the masker. In Chapter 12, we introduced the precedence effect, in which the ability to accurately locate a sound source depends to a large extent on the first wave reaching the ears, in that interaural information arriving after the first wavefront appears to be suppressed.

In many situations, the characteristic quality of a sound (its timbre) may depend on how the sound is turned on or off. The *attack* and *decay* of the sound produced by many musical instruments provides most of the information that allows us to differentiate among different instruments. A major aspect of music synthesis and music synthesizers is to accurately simulate the temporal and spectral aspects of the attack and decay of a note played by the instrument being synthesized.

A powerful method to enhance detection or discrimination of many complex sounds is to delay the onset of the stimulus of interest relative to that of background or competing stimuli. The immediate change from the background to the target sound of interest often provides a cue to highlight the target sound. For instance, the ability to hear out the pitch of a mistuned harmonic (see section of Harmonicity) can be significantly influenced if the mistuned harmonic comes on before the rest of the harmonic sound.

TEMPORAL MODULATIONS

Many naturally occurring sound sources produce complex sounds that have a slow amplitude or frequency modulation (see Chapter 4). Thus, all the spectral components that constitute such a complex sound will have either their amplitudes or their frequencies slowly modulated in time in a coherent manner. Thus, the auditory system might be able to fuse spectral components that are modulated with the same temporal pattern as one means of grouping together spectral components of a source to aid sound source determination.

Stimuli that are located in different regions of the spectrum but share a common pattern of temporal amplitude modulation produce some interesting auditory phenomena. Consider a masking situation in which a narrow band of noise (called the target band) is used to mask a tonal signal centered in the noise. We know from Chapter 11 that such a noise will elevate the threshold of the signal and hence mask the signal. If another band of noise (called a cue band) with the same bandwidth as the target band is presented in a different frequency region simultaneously with the target band, we would not expect much additional masking unless the cue noise band was close in frequency to the target noise

band. It turns out, however, that the amount of masking generated by the target band may depend on the cue band, even when the cue band is located in a spectral region far away from that of the target band. To understand the results you must first recall the fact that a narrow band of noise has a strong slowly amplitude-modulated envelope, with the frequency at which the envelope amplitudes fluctuate being proportional to the bandwidth of the narrowband noise (see Chapter 4 and Figure 4.16).

It is possible to generate two narrow bands of noise in different frequency regions that have the same (coherent) amplitude-modulated envelopes or different (incoherent) amplitude-modulated envelopes. If the cue and target bands have incoherent envelopes, then the cue band does not alter the amount of masking of the signal provided by the target band. However, if the cue and target bands have the same (coherent) envelopes, then the masked threshold of the signal drops by 10–12 dB. That is, the signal is 10–12 dB easier to detect when the target and cue bands are modulated coherently (i.e., comodulated) than when they are modulated incoherently. The release in masking caused by the addition of the coherently modulated cue band is called *comodulation masking release* (CMR). Figure 14.7 describes both the basic stimulus paradigm (with schematic diagrams of the noise bands) and the results from a typical CMR experiment.

Although there is not a complete understanding of how the CMR effect occurs, CMR does show that the auditory system uses information from a wide region of the spectrum to aid in the processing of some complex sounds. Notice that, because the amplitude of the noise is modulated, brief periods of time occur when the noise level is low, making the tone easy to detect at that time. One possible explanation of the CMR effect is that the co-

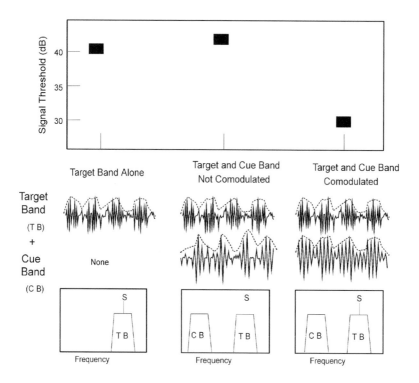

FIGURE 14.7 Both the basic CMR task and results are shown. At the bottom, the time-domain waveforms for the narrowband maskers (Target and Cue Bands) and the amplitude spectra for the maskers and signal are shown in schematic form. The dotted line above each time-domain waveform depicts the amplitude envelope of the narrowband noises. The listener is asked to detect a signal (S) that is always added to the target band. In the target-band-alone condition, the signal is difficult to detect. When a cue band is added to the target band such that it is located in a different frequency region than the target band and has an amplitude envelope that is different from (not comodulated with) the target band, there is little change in threshold from the target-band-alone condition. However, when the target and cue bands are comodulated, the threshold is lowered by approximately 12 dB, indicating that the comodulated condition makes it easier for the listener to detect the signal. The waveforms are not drawn to scale. Based on data from Hall *et al.* (1984) with permission.

modulated cue band provides additional information about when the noise band's level will be low, making it easier for the auditory system to detect the signal during these brief periods of low masker level.

The procedure and data shown in Figure 14.8 describe another form of interaction between stimuli that occur in different spectral regions but have common patterns of ampli-

tude modulation. In this procedure, listeners are asked to detect a change in the depth of amplitude modulation provided for a probe tone (with a carrier frequency of 4000 Hz), using the same procedures that were used to obtain the TMTF described in Chapter 10. This condition is referred to as the Probe Alone condition. Next, another tone with a different carrier frequency (the masker tone that for the

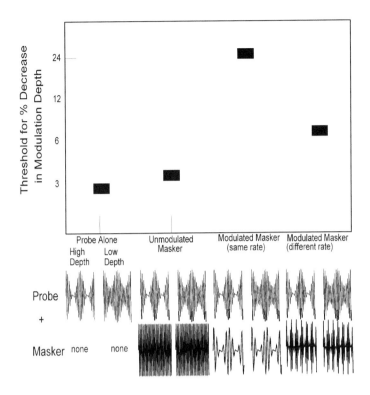

FIGURE 14.8 Both the basic MDI task and results are shown. The basic task for the listener is depicted along the bottom of the figure. The listener is to detect a decrement in the depth of probe amplitude modulation (difference between low and high depth). When just the probes are presented, the task is relatively easy. When an unmodulated masker tone with a frequency different from that of the probe is added to the probes, thresholds for detecting a decrease in probe modulation depth are not changed much from the probe-alone condition. However, when the masker is modulated with the same rate pattern as the probe, the threshold for detecting a decrement in probe modulation depth increases greatly, indicating that modulation depth is difficult to detect when both the probe and masker are comodulated. When the masker is modulated, but with a different rate (shown as a faster rate in the figure) than the probe, then the threshold for detecting a modulation-depth decrement is lowered. The waveforms are not drawn to scale. Based on data from Yost (1992b) with permission.

conditions of Figure 14.8 has a carrier frequency of 1000 Hz) is simultaneously added to the modulated probe tone in the Unmodulated Masker condition. Since the masker tone is far removed in frequency from the probe tone, there is little change in the threshold decrease in modulation depth for the probe tone. However, when the masker is modulated with the same pattern as the probe (Modulated Masker condition), the threshold decrease in modulation depth is much higher, indicating that it is difficult to detect the change in modulation depth when both the masker and probe are comodulated. That is, the comodulated masker interfered with detection of the modulation of the probe, and hence the effect is referred to as *modulation detection interference* (MDI).

It has been argued that MDI reflects a process by which the common modulation pattern applied to the masker and the probe fuses them into one auditory image as if they had originated from the same modulated sound source. As such, it is difficult for the auditory system to process information about amplitude modulation for the individual components (the probe and the masker) of this image. If such an argument is valid, then one might predict that, if the masker and probe were modulated with different patterns (e.g., the probe and masker were amplitude modulated at different rates), they would no longer be fused into a single auditory image since a single sound source would not produce components with different modulation patterns. Thus, if the masker and probe were amplitude modulated at different rates, they would be treated as separate sound sources, and there would be less MDI. The datapoint on the far right in Figure 14.8 shows just such a result; that is, when the masker was modulated at a higher rate than the probe, the threshold for detecting a decrease in modulation depth returned closer to those obtained in the Probe Alone and Unmodulated Masker conditions (i.e., there was less MDI).

While amplitude modulation has been shown to affect sound source determination and segregation, frequency modulation per se does not appear to provide a potent cue for sound source determination. The pitch shifts, and other cues that covary when a sound contains frequency modulations appear to be the cues used in sound source determination and not the frequency modulation per se. Thus, coherent amplitude modulation, but probably not frequency modulation, can aid in sound source determination.

Thus, CMR and MDI results suggest that stimuli that share a common pattern of amplitude modulation can be processed across a wide range of the auditory spectrum. The work on profile analysis also shows that the auditory system is capable of using information from different regions of the spectrum to aid in performing a complex sound detection task. Such across-frequency processing would appear to be a critical aspect of sound source determination, in that the complex sound field will usually contain spectral components from a number of sound sources that are likely to cover a wide region of the spectrum. Similarly, research such as that on streaming show how the auditory system integrates information across time when processing many types of complex sounds. If the auditory system is to use this widely spaced spectral and temporal information in sound source determination, then it must be able to make comparisons across wide spectral regions and across time. Thus, experiments such as those on profile analysis, CMR, MDI, streaming, and complex tonal pattern processing provide some insight into how these aspects of cross-spectral and cross-temporal processing might operate.

SPEECH

Speech represents the major complex acoustic stimulus for most humans. Understanding speech perception requires some knowledge of speech production and language, as well as information about how the auditory system functions. Much work has been devoted to understanding our ability to perceive speech; however, the speech waveform is extremely complex, and hence our knowledge about speech perception is incomplete. An inability to use the auditory system may result in poor speech. Thus, the *speech pathologist* often attempts to correct a patient's speaking problems. Because humans rely to such a great extent on speech for communication, there are large areas of mutual interest among scientists working in audition and those investigating speech.

A variety of methods are used to describe and analyze the speech waveform. We can divide spoken language into small units. Words and letters are small units of speech, but they do not relate directly to our auditory perception of speech (i.e., the "sounds" of speech). From the auditory point of view, we are interested in the "sound" of utterances such as words or letters; hence, the basic auditory unit of speech is the *phoneme*. Phonemes allow us to describe the differences among, for instance, had, head, heed, and hid. These four words sound different because of the middle sound or middle phoneme. Another means of describing the speech waveform is in terms of *speech production mechanisms*. English consonants can be described in terms of *place of articulation* and *manner of articulation*. Places of articulation are the *glottis*, the *palate*, the *gums*, the *teeth*, and the *lips*. The major manners of articulation—which are *plosive*, *fricative*, and *nasal*—refer to the ways in which the air pressure coming from the voice mechanisms is changed to produce certain sounds. Plosives are created by a blockage followed by a sudden release of air pressure; fricatives are created by a turbulent flow of air from the mouth, and nasals by allowing the nasal cavities to participate as a place of articulation. We can relate the type of phoneme to the place and manner of articulation.

We also know that the speech waveform is a complex stimulus, one which may be viewed as a sum of sinusoids. The speech waveform has temporal and spectral units, which are crucial for the auditory processing of speech. We shall refer to the spectral and temporal properties of the speech waveform as the *acoustic properties of speech*. To describe the acoustic properties of speech it is useful to consider the speech signal as coming from a resonator or a set of filters (*vocal tract*) being driven by a pulsating sound pressure wave (*vocal folds*). That is, the vocal folds vibrate air forced across them by the diaphragm; the vibrating air pulses are then fed through the vocal tract, which resonates or filters the sound caused by the vocal folds. Thus, the speech waveform has acoustic properties associated with both the vocal folds and the vocal tract. The frequency at which the vocal folds vibrate is referred to as the *voicing fundamental frequency*. The frequencies at which the vocal tract resonates are called *formant frequencies*. Most speech sounds (phonemes) differ in the frequencies of the various formants, that is, the way in which the vocal tract resonates. Figure 14.9 shows the time-domain waveform (Figure 14.9a) and frequency-domain spectrum (Figure 14.9b) of a speech sound. The spectrum of the repeated pulses (the vocal fold vibration) is a line spectrum. This spectrum is then changed by the vocal tract, creating peaks that are the formant frequencies (labeled F_1, F_2, and F_3 in Figure 14.9b; also see Figure 9.13). The formant peaks are a result of different parts of the vocal tract (e.g., the lips) acting as a filter (or resonator) for the pulses provided by the focal folds. These spectral properties are acoustic properties of speech.

Figure 14.9b may represent the acoustic spectrum of a vowel, because for most vowel sounds the vocal tract and its various parts (e.g., mouth, tongue, and lips) do not change during the production of a speech sound. For most other phonemes, the spectrum of the sound does change because the vocal tract properties change over the period of the time the phoneme or part of speech is uttered. This results in formants that change over time. To describe this type of stimulus, we need a three-dimensional plot of amplitude versus frequency versus time. In Chapter 4 (see Figure 4.12) we introduced the spectrographic plot. This type of plot can be used to display the changes in formant struc-

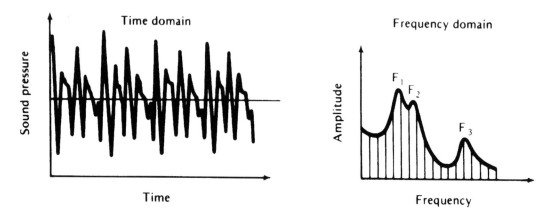

FIGURE 14.9 The time domain (**a**) and frequency domain (**b**) amplitude spectra of a steady-state speech waveform such as that produced by a vowel. The vocal fold vibration is a repetition in the time domain that also yields the spectral lines in the amplitude spectrum (the frequency spacing of the spectral line-components are spaced at the reciprocal of the period of the vocal cord vibrations). The vocal tract produces resonant peaks in the spectrum at the formant frequencies F_1, F_2, and $_F3$. These figures indicate some of the acoustic properties of speech.

ture that take place over the time of a speech utterance. Figure 14.10 shows a spectrograph for a sentence. For this spectrograph, frequency is plotted along the vertical axis, time along the horizontal axis, and the amplitude is proportional to the darkness of the plot. Notice that sometimes there is a shift upward in frequency of a dark band (frequencies with high amplitudes). This type of shift is called a *formant transition* or *shift*, because the center of the formant band is shifted in frequency (note the second formant transition sweeping up in frequency between 1.2 and 1.4 sec in Figure 14.10). These formant transitions represent changes (movement) of some part of the vocal tract during the production of the sound. A careful study of the spectrographs of speech in relationship to the speech production mechanisms can reveal important physical characteristics of speech sounds.

Speech is often used as a complex sound source in studying both *speech perception* and auditory processing. An often used method for studying sound source segregation is the

two-vowel procedure. In the two-vowel procedure, two vowels are produced (usually by a computer program) and played simultaneously. Then one of the stimulus parameters we have discussed in the chapter (e.g., spatial separation) is used to determine if it will aid listeners in segregating the two sounds so that they perceive the two vowels. The logic of the experiments is that each vowel represents a different sound source. For instance, if two artificial vowels are computer generated such that they each have the same and completely periodic fundamental frequency representing perfectly periodic vocal cord openings and closings, the vowels presented in isolation will be perceived as the speech vowels, even though for real voices vocal cord function is not periodic (i.e., the fundamental frequency varies slightly from moment to moment, and this variation in vocal cord frequency is *vibrato*). When these computer-generated periodic speech sounds are played together, it is difficult to segregate the sound into the two vowel sources. Thus, even though the two

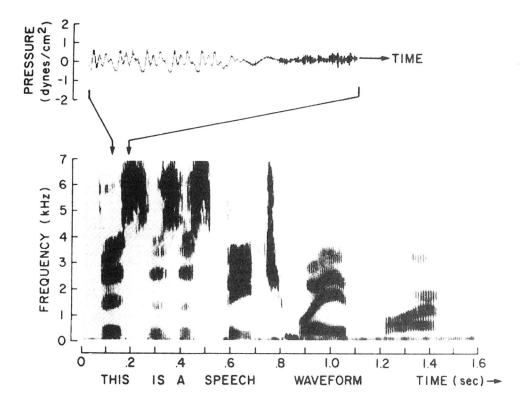

FIGURE 14.10 A spectrograph of the sentence "This is a speech waveform." A segment of the time-domain waveform, which occurs during the end of "this," is shown above. Time is plotted along the *x*-axis, frequency along the *y*-axis, and the darker the lines become, the more intense the sound is at that time and frequency. The bands of energy that exist (e.g., at 1.5, 2.5, and between 3 and 4 kHz for the "th" sound at the beginning of the first word) are the formants. A formant transition can be seen starting at about 0.9 sec at the beginning of the word "waveform." The formant begins at around 500 Hz and ends near 2 kHz, where it remains for the rest of the word. Reprinted with permission from Klatt (1982).

vowel sounds have different formant structures, they cannot be perceptually segregated when played together, probably because they have the same periodic fundamental frequency. If the fundamental frequency of one vowel is made different from that of the other vowel, then it can be easier to identify the two vowel sounds when they are played simultaneously. This is an example of a change in harmonicity promoting sound source segregation, since changing the

fundamental frequency changes the harmonic structure of each artificial vowel (see Figure 14.9a). Other stimulus changes also assist in segregation using this two-vowel procedure. For instance, amplitude modulating one vowel with a different modulation pattern than that used for the other vowel aids in sound source segregation. Thus, the two-vowel procedure takes advantage of speech perception to study complex auditory processing.

SUMMARY

After the basic acoustic attributes (frequency, intensity, and time/phase) of a complex sound field are coded by the auditory periphery, the neural information is further processed in order to determine sound sources. Seven possible physical variables may aid in sound source determination: spectral separation, spectral profile, harmonicity, spatial separation, temporal separation, onsets and offsets, and temporal modulation. Experiments studying profile analysis, CMR, MDI, and streaming indicate that the auditory system processes information across a wide frequency and time range. In conditions of stimulus uncertainty, informational masking can be involved in complex sound processing. The study of complex nonspeech sounds helps us in understanding speech. The speech utterance can be described by its phonetic units, speech production mechanisms (manner and place of articulation), or acoustic properties (fundamental frequency, formant frequency, formant shifts). The two-vowel procedure uses speech sounds as possible sound sources for the study of sound source segregation.

SUPPLEMENT

The books by Yost and Watson (1987), Handel (1989), Bregman (1990), and Warren (1999); the articles by Yost (1992a,b); and Chapters 7 and 8 in the book by Moore (1997), chapters 11 and 13 in the book by Rosen and Howell (1991), and the chapter by Yost and Sheft (1993) all contain additional information on the topics of this chapter. Bregman (1990) describes sound source segregation as *auditory scene analysis*, in which auditory perception is viewed as processing an auditory scene of sound images representing different sound sources. Bregman (1990) is most responsible for the work on auditory stream fusion and segregation.

Dave Green's book on *Profile Analysis* (1989) describes a range of topics related to profile analysis and intensity perception. The topic of auditory perception has often been relegated to a discussion of speech and music perception. There is a growing recognition that sound source determination or auditory image perception is a global topic of auditory perception that can encompass special acoustic waveforms such as speech and music. Yost and Sheft (1993) provide a review of the seven variables that might support sound source determination.

A phenomenon similar to streaming occurs in which two alternating sounds are presented, one soft and one loud (the sounds usually also differ spectrally); under the appropriate conditions, the softer sound appears to be steady even though it is pulsating. One can adjust the level of the sound until it is just perceived to be pulsating (or, conversely, steady). In this case, one is estimating the *pulsation threshold* of the sound. Many phenomena that have been studied using masked thresholds have also been studied using pulsation thresholds (or *auditory induction* as it has also been called, see Warren, 1999).

Informational masking has been studied by Watson and colleagues (see Watson, 1976) in their work on 10-tone pattern recognition. Neff and Green (1987) have also investigated properties of informational masking in a manner similar to that described in this chapter. Carlyon (1991) has demonstrated that frequency modulation per se probably does not support sound source segregation. Work on mistuning harmonics in an otherwise harmonic series has been studied by Moore *et al.* (1985) and Hartmann *et al.* (1990). Darwin and colleagues have studied the role of interaural and spatial differences in sound source segregation as well as the use of onset cues (Darwin and Ciocca, 1992). Several investigators—including Darwin (1981) and Summerfield and Assmann (1991) have used the two-vowel procedure to study sound source segregation.

Textbooks on speech such as those by Borden and Harris (1980) and Zemlin (1981) cover speech at a level similar to that of this book. The book by Rabiner and Juang (1993) cover speech at a more advanced level. The *Sensimetrics Series in Human Communications: Speech Production and Perception* (1997) provides an informative CD-ROM for understanding the fundamentals of speech production and perception.

CHAPTER 15

The Central Auditory Nervous System

*A*s explained in Chapter 1, the majority of this book (Chapters 2–14) is devoted to understanding how the basic attributes of sound are coded within the auditory periphery and the perceptual consequences of this coding. Chapters 1 and 14 also explained that hearing involves more than neural coding of frequency, intensity, and time. A great deal of this additional processing takes place in neural centers that lie in the auditory brainstem and cortex. In addition, because localization and other binaural perceptions depend on the interaction of information arriving at the two ears, we need to study the central auditory centers, since auditory nerves from the two cochleas interact only in the brainstem and cortex. This chapter deals briefly with the structure and function of the central auditory nervous system (CANS) as it receives information from the cochlea. Once again, a review of Appendices E and F will be helpful to the reader who lacks general knowledge of anatomy and physiology.

ANATOMY OF THE CENTRAL AUDITORY PATHWAYS

Figure 6.1 depicted the general anatomy and physiology of the auditory system but not the anatomy of the *central auditory system*. Figure 15.1 illustrates in schematic form the principal connections of the *ascending* or afferent (from the cochlea toward the cortex) auditory system. A glance at this figure will demonstrate the intricacies of the system. The afferent fibers of the auditory nerve leave the cochlea and travel to the *cochlear nucleus*. Those fibers *higher* in the system (closer to the cortex) have many pathways, called *tracts*, some traveling *contralaterally* (to the opposite side of the brain) and others remaining *ipsilateral* (on the same side). Furthermore, some fibers may leave one point and go directly to the next, but others may bypass the obvious next point and travel to a higher location. Still others will send *collaterals* or branches to one point as the main tract travels past that point to terminate at a higher location. Because of the complexity of this intricate system of pathways connecting one point to another, it is often convenient to label the fibers according to the number of connections (synapses) that occur earlier than the region under discussion. The fibers of the auditory nerve that leave the cochlea are the *primary* or *first-order fibers*, all of which make connections, or synapses, in the cochlear nucleus. The fibers that leave the cochlear nucleus after one synapse are called *second-order fibers*. The fibers that originate after the next

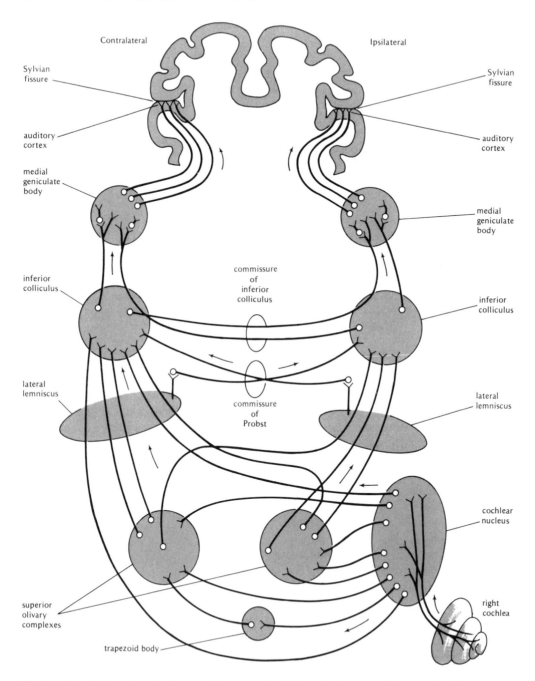

FIGURE 15.1 Highly schematic diagram of the ascending (afferent) pathways of the central auditory system from the right cochlea to the auditory cortex. No attempt is made to show the subdivisions and connections within the various regions, cerebellar connections, or connections with the reticular formation. Based on similar diagrams by Ades (1959), Whitfield (1967), Diamond (1973), and Harrison and Howe (1974a,b).

synapse are called *third-order fibers*, and so on. Within the central nervous system are *nuclei* at which a group of nerve cell bodies are located. These nuclei should not be confused with the nucleus of a single nerve cell body. Within these nuclei lie *interneurons*, which interconnect various nerves within the nuclei but do not carry information between the nuclei. Sometimes a nerve fiber will pass through a nucleus without making a synapse with any units in the nucleus; these fibers are called *fibers of passage*. Figure 15.1 represents only the most important afferent connections that are thought to serve the function of hearing per se. Regions that receive an auditory input in addition to inputs from other sensory systems have been excluded. Other information not elaborated in Figure 15.1 involves cerebellar connections, the innervation pattern within the various auditory nuclei, and the various types of cell bodies and their interconnections within auditory nuclei.

Figure 15.2 is another simplified schematic illustration showing that the main tracts and nuclei above the cochlear nucleus are stimulated *binaurally* (that is, by both ears). By studying Figure 15.2 we can trace the general pathway of the neural signal from the cochlea to the cortex. After the neural impulses leave the cochlea, they travel to the *cochlear nucleus* where the first synapse is made. From the cochlear nucleus, tracts lead to both the ipsilateral and contralateral *olivary complex*, so most bilateral representation occurs at this point and above. From the superior olive, neural impulses are transmitted to the *inferior colliculus* through and/or around the *lateral lemniscus*, from there to the *medial geniculate body*, and finally to the *auditory cortex*. These are the major nuclei of the central auditory nervous system, although other nuclei exist. Figures 15.1 and 15.2 are tremendous oversimplifications. It is not our purpose, however, to present a detailed account of the anatomy of the

central auditory system, but rather to stress the importance of the major nuclei shown in Figure 15.2 and to bring an awareness of the complexity of the auditory nervous system shown in Figure 15.1.

We have seen that the ascending (afferent) auditory system is anatomically complex. Descending (efferent) fiber tracts, shown in Figure 15.3, may arise in the auditory cortex or in a variety of nuclei and terminate at other nuclei, especially in the cochlear nucleus and the olivary complex. Chapter 8 has already discussed part of this system, which consists of an olivocochlear bundle arising in the olivary complex and terminating in the cochlea. Those fibers appeared to have an inhibitory action on electrophysiological responses of the cochlea, although their exact function is as yet unknown. All descending pathways cannot be considered as simply inhibitory neural networks. For instance, electrical stimulation of a particular segment of the superior olive can cause an increase in discharges in certain cochlear nucleus neurons. Thus, both excitatory and inhibitory connections might exist within this descending system. Therefore, rather than considering this system as only one that limits the passage of information from "lower" to "higher" levels of the ascending system, a more general view is that the efferent (descending) auditory pathways represent a control system that varies the routing and hence helps shape sensory input.

The anatomy (*morphology*) of individual neural units within the CANS can vary considerably. Figure 15.4 shows some of the various forms of neural cell types found within the cochlear nucleus and inferior colliculus. Similar diversity exists throughout the other neural centers of the CANS. These anatomical difference probably have significant physiological consequences. *Bushy cells*, with their large dendritic trees with multiple synapses, may process neural information from many

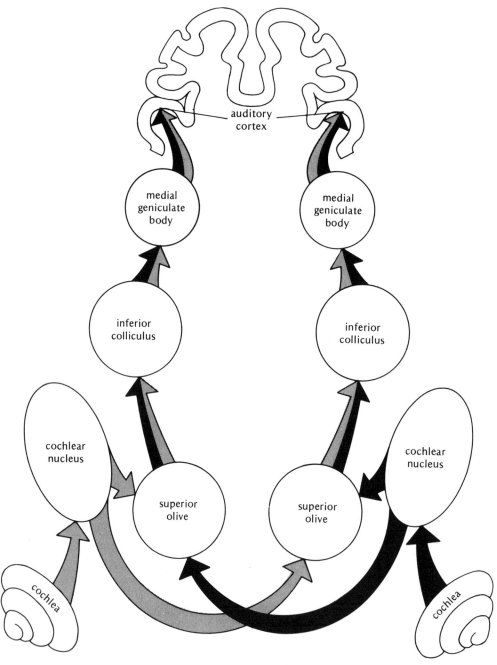

FIGURE 15.2 Highly schematic diagram of the bilateral central auditory system; the main pathways and nuclei are shown for both cochleas. Bilateral representation from binaural stimulation occurs at the superior olive and in all regions above. Based on a similar diagram by Lindsay and Norman (1972). Adapted with permission.

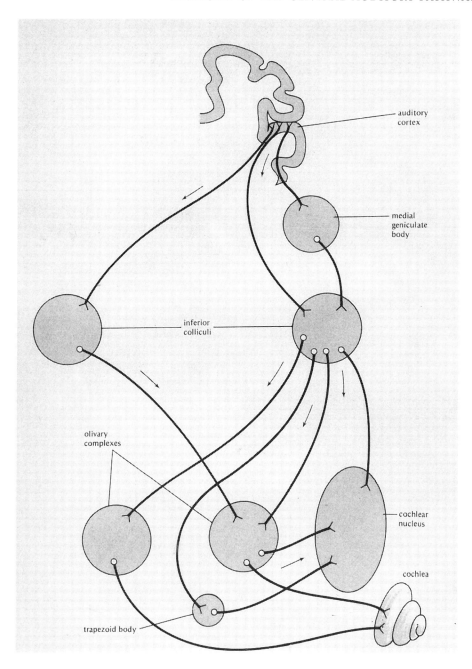

FIGURE 15.3 Highly schematic diagram of the descending pathways of the central auditory system from one side of the auditory cortex to the right cochlea. No attempt is made to show the subdivisions and connections within the various regions. The complex crossing and bilateral innervation shown for the ascending system in Figure 8.1 is also present in this system. Based on a diagram by Harrison and Howe (1974b). Adapted with permission.

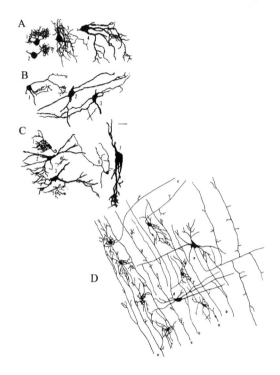

FIGURE 15.4 (**A–C**) Cell types in the cochlear nucleus. (**A**) 1 and 2 = *spherical bushy cells*, 3 = *globular bush cells*, (**B**) 1–3 = *multipolar cells*, (**C**) 1 = *cartwheel cells*, 2, 3 = *fusiform cells*, 4, 5 = *granule cells*, 6 = *stellate cells*, 7 = *giant cells* (scale bar = 1 μm). Reprinted with permission from Cant (1992). (**D**) Part of a neural circuit in the inferior colliculus (central nucleus of IC). a and b = *banded cells*, d, e = *stellate cells*. Reprinted with permission from Oliver and Heurta (1992).

different input axons. If these axons come from nerves with different frequency tuning (i.e., different CFs), bushy cells may play a role in processing changes in the spectral profile of complex stimuli. When the cells are seen in a neural circuit such as that shown in Figure 15.4D, one sees an organization of cells with axons going from top to bottom in layers (i.e., the *banded cells*—a and b) and cells that cut across the circuit like the *stellate cells* (d). The stellate cells may help integrate neural infor-

mation in the layers of cells and axons flowing from top to bottom. Exact structure–function relationships have been established for some, but not most cell types. However, different type of neural units often display different physiological properties. Some of these properties will be described in the section on Single-Fiber Responses.

STRATEGIES FOR STUDYING THE CENTRAL AUDITORY SYSTEM

Topographical Organization

In Chapters 7, 8, and 9, we discussed the concept of frequency being represented by a particular place along the basilar membrane, the place being determined by the location of maximum displacement of the basilar membrane in response to a particular frequency of stimulation. Fibers with high characteristic frequencies innervate the base of the cochlea and those with low CFs innervate the apex. The maintenance of this neural spatial representation of frequency throughout the nuclei of the central auditory pathways is referred to as *tonotopic organization*. The test for tonotopic organization is to determine where there is an orderly neural spatial representation of fibers with various CFs. The most common organization that is studied within the CANS is tonotopic organization. In order to determine how a nucleus is tonotopically organized, the tuning curves for nerve fibers in the nucleus are measured and their characteristic frequencies determined. The organization of CFs in terms of the fiber's anatomical location within the nucleus usually determines the tonotopic organization of the various parts of the CANS. However, the enormous variation in the ways in which CANS nerves respond to sound often makes the determination of tuning curves and CF a more difficult task in the CANS than at the periphery. All neural centers have a tonotopic organization, although the neural map for frequency in some nuclei is quite complex.

In most other sensory systems, the response of neural units in the central nervous system is systematically organized in terms of a number of different features of the stimulus that drives these units. In the CANS, investigators search for similar topographical organizations. For instance, are there fibers that respond best when sound is presented from a particular spatial location? If so, then these cells that are "tuned" to the spatial location of sound might provide a *neural map for auditory space*. There is also evidence that parts of the inferior colliculus and auditory cortex may be topographically organized to process amplitude modulation, in that neural units are selective for different rates of amplitude modulation. Based on what happens in other sensory systems, spatial maps or modulation rate processing might occur in conjunction with a tonotopic organization. Topographical organization for features of sound other than frequency, and perhaps space and modulation, have been found in a few species (e.g., owls and bats), but a number of emerging data hint that such neural organization exists in the CANS of many other species.

Excitation and Inhibition

Processing neural information can take place at the level of individual neurons or within circuits of many neurons. The tuning of an auditory nerve fiber is an example of how an individual neuron may be selectively responsive to a particular stimulus parameter. The tonotopic organization of neurons with different CFs may represent the basis of a neural circuit designed to process spectral information. Both the ability of a single neuron or a circuit of neurons to process information important for auditory perception depends to a large extent on how inputs to the neurons interact. The interactions can be reinforcing (additive) or canceling (subtractive). Thus, the excitatory and inhibitory properties of neural transmission across a synapse is a valuable piece of information for understanding neural processing.

The transfer of a neural event across a synapse is crucial for the integration of information within the central nervous system. The basic circuitry that exists in the CANS is caused by nerves that interact with each other in either an excitatory or inhibitory manner. That is, two nerves may converge on another cell, and the combined excitation of the input nerves may produce a firing pattern in the target nerve as if it were adding the information from the two input nerves. If one input nerve is excitatory and one inhibitory, the target nerve may serve as if it were differencing the information between the inputs. In some circumstances, these cells are referred to as *E–E* or *E–I*, in terms of how the inputs to the cells interact in an excitatory (*E*) or inhibitory (*I*) manner.

Chemical structures like *acetylcholine* (ACh) are believed to act as excitatory neurotransmitters for information from neuron to neuron, while *amino acids* such as *gamma-aminobutyric acid* (GABA) and *glycine* may act as inhibitory neurotransmitters. By studying the concentrations of these and other neural chemicals, the excitatory or inhibitory function of a neural site or part of a neural site can be inferred (see Appendix F). Quite often, different chemicals are used to *block* the generation or flow of these types of neurotransmitters and biochemical agents. Sometimes the interaction of these chemical with others within the nervous system may be altered. These attempts to change the normal chemical activity in a neural structure can help determine how information is transmitted from one neuron or group of neurons to other neurons.

Single-Fiber Responses

Measures of single-fiber responses used to characterize the discharge patterns of auditory nerve fibers are also useful for studying the central pathways. In auditory nerve fibers, spontaneous activity varies greatly in discharge rate from fiber to fiber. The interval histograms of spontaneous activity of various neurons, however, are essentially unchanged regardless of the CF of the unit and its discharge rate. In the central auditory system, this regularity of interval histograms of spontaneous activity is not maintained. Interval histograms of spontaneous activity at these levels are more variable, changing as a response to a variety of nonauditory events, such as the state of alertness.

Single auditory nerve fibers were shown to have rate-level functions that produced an increased discharge rate with an increase in level over a 20- to 50-dB range. Single fibers from the central pathways generally have a smaller dynamic range than those of the auditory nerve. Furthermore, central fibers may show a decrease in discharge rate with an increase in stimulation at high levels. Thus, the rate-level function of a central fiber may be similar to that of an auditory nerve fiber or it may be shaped like an inverted ∪, showing an increase in discharge rate with an initial increase in stimulus level and then decreasing at higher levels. Such rate-level functions can be further complicated by simultaneous acoustic stimulation of the opposite ear.

We pointed out in Chapter 9 that single-unit responses of the auditory nerve show phase-locking discharges to the stimulating sinusoid. Phase locking to individual cycles of the tone also occurs in many units of the cochlear nucleus for the low-frequency stimuli. In addition, synchronous responses to low-frequency stimuli have been observed in the trapezoid body, the inferior colliculus, and the superior olive. As we investigate higher in the system, however, we find that the relationship between neural discharges and the period of low-frequency stimuli is less clear.

In Chapter 9, we showed that PST histograms to tone bursts were essentially the same for all fibers of the auditory nerve. In the central auditory system, however, PST histograms to tone bursts may illustrate any number of patterns, as seen in Figure 15.5. The higher-order fibers may respond in the same manner as the primary fibers, or they may produce *"on" responses, "off" responses*, or *"on–off" responses*, or they may exhibit more complex responses such as those called *pausers* or *choppers*. Not all of these patterns can be recorded from every nucleus within the system. On the other hand, one pattern may be recorded from one region within a nucleus and another from a different region within the same nucleus. In addition, one neuron may have different response patterns depending on the type of stimulation (e.g., the PST histogram may appear different for a CF than for a non-CF tone). The tuning curve and the interval histogram structures also vary from neural unit to neural unit. Some neural units demonstrate two-tone inhibition and some do not. Usually, neural units with the same morphology exhibit about the same physiological properties, while units with different anatomical structures often have different physiological properties. That is, the different cell types shown in Figure 15.4 often generate one of the characteristic PST histograms shown in

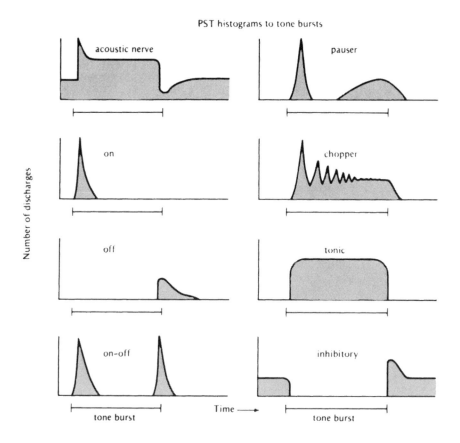

FIGURE 15.5 Idealized PST histograms to a variety of tone bursts recorded from the central auditory system. The time patterns of responses of the central system are thought to reflect the wide variety of tasks that it must accomplish.

Figure 15.5. Examples of this structure–function relationship will be seen later in Figure 15.14.

Presumably, these different neural response patterns help process information about sound. For instance, the chopping seen in the PST histograms of copper cells probably results from some intrinsic property of the cell that causes it to discharge at a particular rate. If the rate of envelope fluctuations of an amplitude-modulated stimulus drives a chopper cell at a rate that is similar to the intrinsic chopping rate, then this chopper cell may respond best for a particular modulation rate. As such, chopper cells may encode the rate of amplitude modulation.

As the neural impulse ascends the auditory pathways, it is delayed relative to stimulation at the cochlea. A measure of the delay time can be used as an indicator of a variety of physiological or functional events. For instance, if two cells in the same region of the central nervous system receive information at different times, one can assume that the neuron with the later-arriving information is fed by pathways that involve more synapses than the other neuron. A convenient way to investigate the timing information of a neuron is to obtain a PST histogram to a click. The time between the onset of the click and a neural discharge is an indicator of the travel time from the cochlea to the neuron under study.

Poststimulus time histograms to click stimulation for auditory nerve fibers demonstrate either a single peak or multiple peaks, depending on this CF (recall Figure 9.10). For higher-order fibers, PST histograms of click stimuli may be in a variety of patterns. These patterns are not, however, related to the CF of the fiber. The typical PST histogram of low-CF fibers to click stimulation recorded from the auditory nerve fibers is not as prevalent in the responses of higher-order fibers even as "low" in the system as the cochlear nucleus. The lack of the multiple peaks related to the CF of the unit (i.e., peaks at times equal to 1/CF) is interesting because this temporal information is evidently not usually transmitted beyond the cochlear nucleus in its original form. Figure 15.6 shows a comparison of PST histograms from low-CF fibers in both the auditory nerve and the cochlear nucleus in response to click stimulation. Such data imply that a simple relaying of information from one low-CF nerve to another does not occur in this part of the cochlear nucleus.

Neural Circuits

The study of neural circuits in the CANS is difficult. Almost all electrophysiological research involves measuring from one nerve fiber at a time. Studies of the activity of several neurons in the intact animal measured at the same time is very hard given the difficulties in

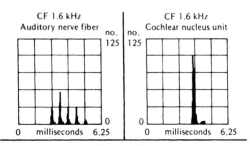

Click rate:10/sec

FIGURE 15.6 Comparison of PST histograms to click stimulation for two fibers of the same characteristic frequency (1.6 kHz): auditory nerve fiber (left) and cochlear nucleus fiber (right). Low-CF fibers in the cochlear nucleus typically do not show the modulated pattern related to 1/CF that is characteristic of auditory nerve fibers. The click is presented at time equal to 0. Adapted with permission from Kiang (1965).

developing multi-electrode recording techniques. One way to study neural circuits is to use what is called the *slice preparation*. This technique involves excising a segment of neural tissue from an animal (sometimes a very young animal) and placing it in a Petrie dish with the appropriate chemicals that allow the neural tissue to continue to function. While this is a difficult procedure, segments of different neural centers can be preserved and studied using the slice preparation technique. While "sound" cannot be delivered to such tissue slices, various neural fibers can be electrically stimulated and their function altered by chemical means. And, one can record neural activity from fibers in the slice preparation. Thus, one fiber can be stimulated and neural responses can be recorded from other fibers to help determine how the nerves in this section of tissue process information. While such *in vitro* (in glass) experiments have provided important information about the function of the CANS, caution must always be exercised in making generalization to the *in vivo* state (in the living organism).

There is one type of neural circuit that might play an important role in the central processing of sensory information. This circuit or network is called a *lateral inhibitory network*. A schematic diagram of a simple lateral inhibitory network is shown in Figure 15.7. Fibers that are adjacent or lateral to an excitatory fiber send inhibition across the network, as shown in Figure 15.7. One important conse-

quence of a lateral inhibitory network is that it "sharpens" the information presented at the input to the network. That is, contrasts in the neural excitation that occur in the input are amplified by the lateral inhibitory network, thus making the changes in contrast that occur across the network more noticeable. For instance, a change in the amplitude between one region of the spectrum and another region

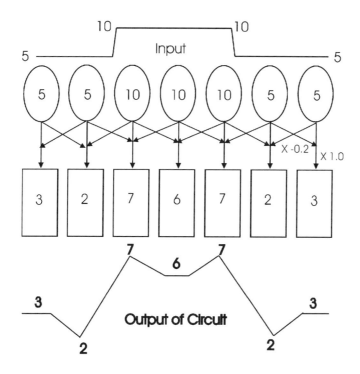

FIGURE 15.7 A schematic diagram of a lateral inhibitory network is shown. Input neurons shown as circles send information to output neurons shown as squares. The input neurons send an excitatory signal (X1) to the output neuron directly below it, and inhibitory signals to the output neurons to the left and right (X–0.2). The excitation is equal to the activity in the input neuron (i.e., multiply the input activity times 1, X1.0), while the inhibition is equal to negative 0.2 times the activity in the input neuron (i.e., multiply the input activity by –0.2, X–0.2). The amount of neural activity in an output neuron is the sum of the excitation and the two paths of inhibition. An input stimulus is shown along the top such that the stimulus increases in magnitude from 5 to 10 and back to 5 across the neuronal array. Consider the output neuron on the left; it receives 5 units of excitation from the input neuron, –1 units of inhibition (–0.2 × 5) from the right, and –1 units of inhibition from the left (assuming the network continues to the left). The result is 3 units of excitation for this output neuron (5 – 1 – 1 = 3). Similar mathematics will yield the units of output activity shown along the bottom. Note that the input contrast change at each edge of the input stimulation is made more obvious in the output due to this lateral inhibitory network.

(as occurs for the formants of speech, see Chapter 14), might be enhanced and made more obvious if the spectral information is processed by a neural network that had lateral inhibitory properties. That is, in Figure 15.7 the input stimulus might be the amplitude spectrum of a sound that changes in amplitude from 5 to 10 at one frequency region and back to 5 at another frequency region. If each input neuron was tuned to a particular frequency and each input neuron fed its outputs to this lateral inhibitory network, the spectral contrast might be enhanced, allowing for neural coding of subtle spectral changes.

Evoked Potentials

In our discussion of the CANS, we will stress that many CANS locations (e.g., the cortex) are relatively inaccessible and that discrepancies in the data from various investigations can often be attributed to differences in anesthesia. This being the case, responses from the auditory cortexes of awake-behaving animals (especially humans) become extremely significant to the contribution of knowledge about how we hear. A method for recording CANS activity that can be used with awake adult humans is to place electrodes on the scalp and record variations in electrical activity that occur in conjunction with the presentation of an auditory stimulus. When responses are recorded in this manner, the electrical activity of interest, called the *auditory-evoked response* (AER), is often quite small relative to other recorded activity. This activity reflects changes in the *electroencephalogram* (EEG), which is related to acoustic stimulation. By a method known as *signal averaging* (see Appendix E), the desired responses, which appear in conjunction with the presentation of an auditory stimulus, can be separated from the unwanted activity. These

potentials are usually recorded from the scalp at the center of the top of the head, known as the *vertex*. The AER, a complex response, occurs at the abrupt onset or termination of an acoustic signal. Its waveform has a negative peak about 100 msec after presentation of an appropriate stimulus and a later positive peak. The response is highly variable from person to person, increasing in amplitude, and decreasing in latency (time between stimulus onset and response onset) with increases in stimulus level. Its variability may even be great from trial to trial for the same subject. Different cortical sources have been postulated to account for the early (about 50 to 100 msec) and later (about 300 msec) components of the waveform. Correspondingly, it has also been suggested that the earlier components are more affected by variations in the acoustic stimulus, whereas the later components are more affected by nonacoustic variables such as attention, expectancy, significance, decision, and contingency.

The waveform in Figure 15.8 shows the auditory-evoked response to a 1000-Hz 300-msec tone presented at 20, 40, and 60 dB SL. The AER was obtained by presenting the tone 64 times. The EEG activity following each stimulus presentation was then summed. This method of signal averaging tends to cancel any parts of the EEG activity that are random (not correlated) with respect to the signal and to enhance any EEG activity correlated with the signal. The peaks and valleys are referred to as P_1 (positive peak), N_1 (negative peak), P_2 (positive peak), and N_2 (negative peak 2), etc. These peaks almost always occur in the AER, but their amplitude and time of occurrence (latency from stimulus offset) can vary as a function of stimulus variable (e.g., level) or variables dealing with attention of the subject.

In Chapter 8, we discussed the recording of the AP from humans. The AP has a latency from stimulus onset of about 1 msec, depend-

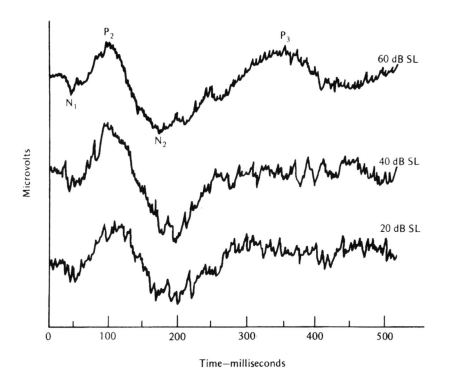

FIGURE 15.8 Auditory-evoked response (AER) to 64 repetitions of a 1000-Hz 30-ms tone burst presented at 20, 40, and 60 dB SL. Points N_1, P_2, N_2, and P_3 refer to particular points in the waveform (see text). Figure courtesy of Dr. Donald C. Teas, University of Florida, Gainesville.

ing on the level of the stimulus. The AERs just discussed have latencies of 50 msec or more, depending on which particular response is being measured. The AP is known to be generated by the synchronous firing of primarily high-frequency cochlear nerve fibers, and the AER is thought to be a cortical response. Are there electrophysiological responses with intermediate latencies that involve brainstem auditory pathways that can be recorded from humans? The answer appears to be yes. The recording of these responses is accomplished with a vertex electrode. These responses are called *brainstem-evoked responses* (BSERs) because they are thought to be generated pre-dominantly by nuclei of the brain stem. Figure 15.9 shows a brainstem-evoked potential in response to a click stimulus presented binaurally. The latencies of these responses are from 1 to 8 msec. The responses are low in voltage and require signal averaging (often over 1000 recordings are averaged) to make them "stand out" from the background electrical activity. The source of the first wave, wave I, is the auditory nerve. The places where the other potentials originate are not as well known, but it is thought they are generated at the different sites in the ascending auditory pathways.

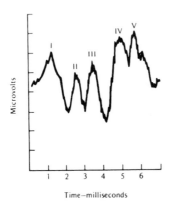

Time—milliseconds

FIGURE 15.9 Brainstem-evoked potential response (BSER) to a click presented 1000 times binaurally. The response was recorded from the vertex referenced to the right mastoid. Roman numerals represent the labeling of the peaks according to Jewett and Williston, 1971 (see text). Recording courtesy of Dr. Donald C. Teas, University of Florida, Gainesville.

There are also evoked potentials that can be measured with latencies between those of wave V (approximately 5–7 msec after stimulation) and the N_1 of the AER (approximately 60–90 msec after stimulation). These potentials are often referred to as *middle-latency* potentials. For the AER, BSER, and the middle-latency potentials, both the amplitude of the evoked response and its latency are used to relate the response to the stimulus or the state of the listener. For the BSER, latency is the measure most often used. Figure 15.10 shows the decrease in the latency of wave V that occurs with an increase in click stimulus level. Given the very short latencies of these responses, a click stimulus must be used to evoke the BSER. A change in the latency of wave V implies a change in the speed with which the neural information reaches the site of generation of wave V (perhaps at the level of the inferior colliculus). The latency of wave V may also vary as a function of changing the spectral content of the click.

As normative data are established for these responses and the source or sources for each wave identified, deviations from these norms can be used to detect possible deficiencies in the auditory pathways of the brainstem. Thus, the study of electrophysiology of the auditory pathways is producing data for the study and diagnosis of human hearing from the cochlea to the cortex. We will describe a few of the centers in the CANS that have been studied the most.

Brain Images

An x-ray provides an image of our internal organs. Similarly, various imaging techniques (see Appendix E) can be used to image the brain. Some image techniques provide an anatomical view of the brain. Other imaging techniques can provide a functional image, that is,

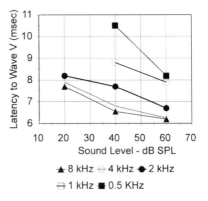

FIGURE 15.10 The latency from stimulus onset to the peak of wave V of the brainstem-evoked response is shown as a function of level and frequency content of the click stimulus. The latency decreases with increases in stimulus level and as the frequency content of the click is restricted to higher and higher frequencies. The frequency content of the click was controlled by filtering a brief acoustic transient with the center frequencies shown in the figure legend. Reprinted with permission from Klein and Teas (1978).

an image that represents those neural circuits that were active during a particular stimulus events. Figure 15.11 displays a PET (*positron emission tomography*) image of the cortex of a human when they are presented a stimulus like that shown in Figure 13.10 (the stimulus, iterated rippled noise, produces a complex pitch perception, whose pitch strength can be systematically varied). PET images result from the way in which cells of a neural circuit that are actively discharging while processing information take up a radioactive marker that has been injected into the subject. Only those cells that are highly active while the stimulus is present will have a high concentration of this radioactive marker. The PET scanner forms an image of the brain allowing one to see where this high concentration is found. Computer and statistical methods are used to make sure that the cortical area of high concentration results from the stimulus presenta-tion and is not just a result of random neural activity. The image in Figure 15.11 suggests that a particular cortical region (in this case, the region in the left and right *temporal planum* area of the auditory cortex) is active when this stimulus is present, and its activa-tion is enhanced when the strength of the pitch associated with the stimulus in-creases. This implies that this region of the cortex plays a role in processing pitch strength for these types of complex sounds. Functional imaging such PET and fMRI (see Appendix F) may provide valuable informa-tion about how neural circuits and sites in the human brain and brainstem process acoustic information.

We will now turn to a discussion of several neural centers in the ascending CANS. The various strategies discussed above are being used to better understand the role each center plays in auditory processing.

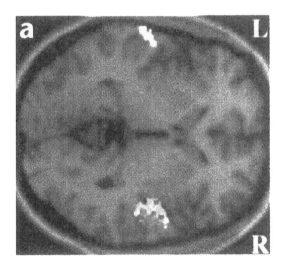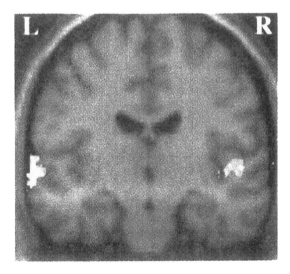

FIGURE 15.11 A positron emission tomography (PET) image showing increased brain activity (light areas) in the left and right superior planum areas at the level of the primary auditory cortex in response to iterated rippled noise that produces a complex pitch (see Chapter 13). Reprinted with permission from Griffiths *et al.* (1998).

COCHLEAR NUCLEUS

Figure 15.12 shows schematically the projection of cochlear location and cochlear nerve fibers on the cochlear nucleus. Upon entering the cochlear nucleus, each nerve fiber separates and goes to three separate regions within the cochlear nucleus. The three regions are the *anteroventral cochlear nucleus* (AVCN), the *posteroventral cochlear nucleus* (PVCN), and the *dorsal cochlear nucleus* (DCN). From this diagram we can see that in each division of the cochlear nucleus the cochlear partition is completely represented (tonotopically represented) from the base to apex. Electrophysiological investigations of the cochlear nucleus confirm this representation by demonstrating that the CFs of fibers within each region duplicate the CFs expected on the basis of the audi-

tory nerve innervation of that region. An example of an electrode penetration from one such investigation is given in Figure 15.13. This figure shows how the CFs of the various units are arranged in a tonotopic manner from low CF to high CF in both the AVCN and DCN.

Figure 15.14 summarizes many of the physiological findings from different neuronal types in the different regions of the cochlear nucleus. The center portion of Figure 15.14 shows the variety of cell types and schematic diagrams of the neural cell morphology: *spherical bushy* (SB) cells, *stellite* (St) cells, *globular bushy* (GB) cells, *multipolar* (M) cells, *octopus* (O) cells, *globular* (G) cells, and *fusiform* (F) cells, reading from left to right in the figure. These different neuronal types are distributed in the different subareas of the cochlear

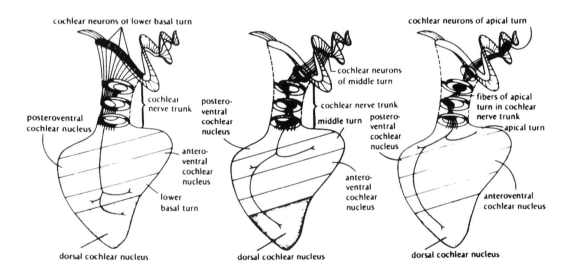

FIGURE 15.12 Schematic diagrams showing that basal turn (left), middle turn (center), and apical turn (right) auditory nerve fibers innervate different areas of the three main regions of the cochlear nucleus. The spatial separation of frequency from base to apex in the cochlea is reflected in the characteristic frequencies of the fibers that leave the cochlea and is maintained within each of the three main regions of the cochlear nucleus. Based on a similar diagram by Schuknecht (1974). Adapted with permission.

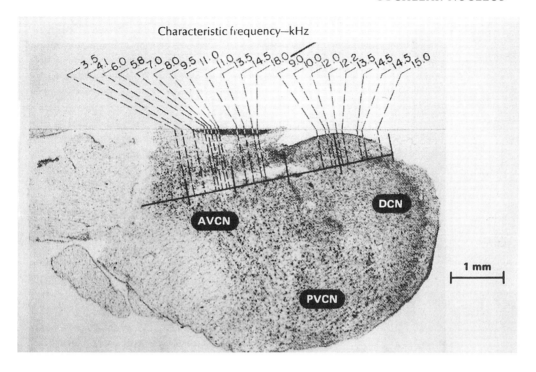

FIGURE 15.13 Cross-section of the cochlear nucleus showing the track made by an electrode penetration. The characteristic frequencies of neurons recorded from various points within the anteroventral cochlear nucleus (AVCN) and dorsal cochlear nucleus (DNC) show that the spatial separation of frequency is maintained within those two divisions of the cochlear nucleus. PVCN = posteroventral cochlear nucleus. This tonotopic organization is maintained throughout the central auditory system. Adapted with permission from Rose, Galambos, and Hughes (1959).

nucleus, and the arrows heading upward depict the fact that these neurons send information up the brainstem to other nuclei in the CANS. The nerve coming in from the bottom of the figure is an auditory nerve fiber (a.n.). The typical tuning curve, rate-level functions, period histogram, and PST histogram for auditory nerve fibers are shown at the bottom from left to right, respectively. Notice how these various physiological measures vary from neuronal type to neuronal type within the rest of the cochlear nucleus. Additional explanation is provided in the figure caption. This diagram suggests the enormous variation and complexity that exists in the anat-

omy and physiology of the cochlear nucleus. Similar complexity exists at each level of the CANS.

Although it isn't yet clear what function the cochlear nucleus plays in auditory processing, the large interconnections and the complex physiology suggests that the cochlear nucleus is refining the code for sound provided by the auditory periphery. The fibers coming from the cochlear nucleus appear to innervate most of the other nuclei in the auditory brainstem. The neural circuits of the cochlear nucleus appear capable of making neural comparisons between neurons that are tuned to approximately the same frequency as well as among

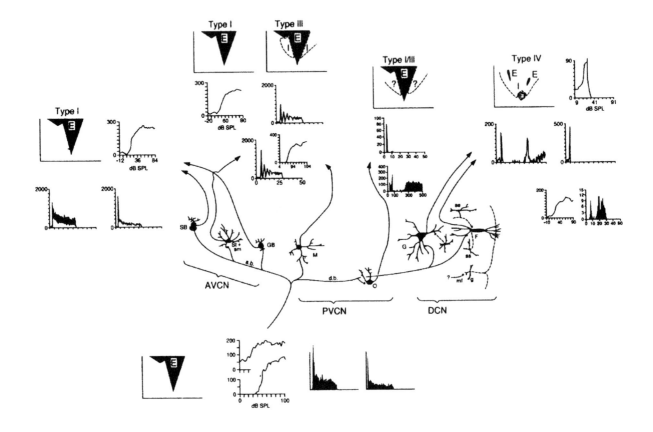

FIGURE 15.14 Cochlear nucleus cell types and the variety of physiological measures are shown. The cell types are the spherical bushy (SB) cell, globular bushy (GB) cell, stellite (St), mulitipolar (M) cell, octopus (O) cell, globular (G) cell, and fusiform (F) cell. The typical tuning curve, rate-level functions, period histogram, and PST histogram for auditory nerve fibers are shown at the bottom from left to right, respectively. In other areas of the cochlear nucleus, the tuning curves exhibit areas of excitation (E), Inhibition (I), and responses that may be a little of both (?). All of the histogram types shown in Figure 15.5 are found in the PST histograms of cochlear nucleus fibers. Sometimes more than one type of PST histogram is found in one region, such as is shown on the left for the SB cells of the AVCN and on the right for the DCN. Additional circuitry is shown as the dotted lines for the DCN. Figure courtesy of Bill Shofner, Parmly Hearing Institute, Loyola University of Chicago, adapted with permission from the work of Young (1984).

fibers that have different CFs. There is also evidence that many cells in the dorsal cochlear nucleus react in a manner that suggests a lateral inhibitory network. Comparisons among fibers of the same CFs might be used to help refine the code for stimulus level or in sup-

pressing the information from echoes. Comparisons among fibers with different CFs might aid in processing the spectral profile of complex stimuli. The presence of a lateral inhibitory network could help sharpen the neural representation of spectral information that is present in a complex sound field, as discussed above.

SUPERIOR OLIVARY COMPLEX

Central sites in the auditory system process information coming from the two ears. Two ears are crucial for our ability to localize acoustic events in space, as described in Chapter 12. Many neurons and nuclei in the ascending pathways have been studied as a function of presenting interaural level and temporal differences, the crucial stimulus values for azimuth localization. The neurons in the olivary complex are the first brainstem neurons to receive strong inputs from both cochlea.

The olivary complex is composed of at least four nuclei: the *lateral superior olive* (LSO), the *medial superior olive* (MSO), the *trapezoid body*, and the *preolivary nuclei*. The MSO and LSO are usually referred to as the olivary complex, and the trapezoid body is often shown as a separate nucleus (see Figure 15.3). Sometimes the preolivary nuclei are included with the MSO and LSO. The LSO and the MSO receive inputs from both cochlear nuclei.

Information from the ipsilateral inputs to the LSO are usually excitatory in increasing the discharge rate of the neuron. Contralateral stimulation of the LSO is usually inhibitory. Thus, stimulation from both ears may decrease the firing of the neuron relative to the firing that occurs when only the ipsilateral ear receives the sound. Most LSO units have high-frequency CFs. Differences in the level at the two ears (and occasionally, time of arrival of the stimulation at the two ears) can change the nature of the excitatory–inhibitory (E–I cells). This type of unit discharges with few spikes when there is approximately equal stimulation at both ears, and then the discharge rate increases as a function of changing the interaural level difference. These binaural units indicate the types of interactions that occur within the LSO; therefore, the LSO might form a network for processing interaural level differences used to determine the location of sound sources.

Units of the MSO tend to be excitatory–excitatory (E–E) types in that inputs from both sides cause an increase in the firing rate of the MSO unit. MSO units tend to have lower CFs than LSO units. As the stimulation at one ear comes on earlier in time than that at the other ear, the discharge rate of many MSO units increases, as shown in Figure 15.15. In many cases, the unit will discharge with the greatest rate when there is a particular interaural delay. In these cases, the unit is said to display a *characteristic delay*, in that this unit may indicate the existence of a particular interaural delay. This change in discharge rate with changes in interaural time by MSO units might aid the nervous system in determining the location of a sound in space (see Chapter 12).

INFERIOR COLLICULUS

The inferior colliculus receives inputs from both the olivary complex and the cochlear nucleus and has three major areas: the *central nucleus*, about which most is known; the *dorsal cortex of the inferior colliculus*; and the *paracentral nuclei*. Units in the inferior colliculus tend to be excitatory–inhibitory (E–I), although there are E–E cells as well, appear to be tonotopically organized in sheets of cells, and have a wide variety of neuronal types, as was seen in the cochlear nucleus. Cells in different parts of the inferior colliculus are "monaural," in that they respond to the input of only one ear, while others are "binaural," responding to bilateral stimulation.

Both the spectral processing that appears to take place within the cochlear nucleus and the

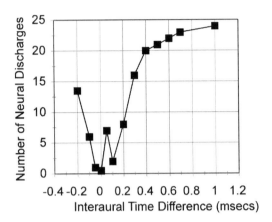

FIGURE 15.15 Number of neural discharges from a superior olive neuron as a function of the interaural time difference between clicks presented to both ears. Note that, as one ear receives the click slightly before the other ear, the neuron increases in number of discharges. Adapted with permission from Moushegian, Rupert, and Whitcomb (1972).

binaural processing that occurs in the olivary complex are seen in the inferior colliculus. Units in the central nucleus respond very accurately to differences in interaural time and level, perhaps refining the processing of sound source location that occurs in the MSO and LSO. In addition, there is evidence that units in the inferior colliculus respond differentially to changes in the vertical location of sound, which presumably is related to the spectral characteristics of the complex changes in sound produced by the head-related transfer function (HRTF, see Chapter 12). These data suggest that one possible role of the inferior colliculus is to combine information from different processes that occur lower in the brainstem, such as combining interaural information with spectral information to provide processing of sound in two or three dimensions (see Chapter 12).

Figure 15.16 displays the neural output of an inferior colliculus (IC) unit in response to a

broadband noise stimulus presented with interaural time delays ranging from –4000 µsec (4 msec) to 4000 µsec (4 msec). This IC cell with a low-frequency CF has a peak number of neural responses that occur as a periodic function of the interaural time delay. The peak closest to zero ITD is at approximately 400–500 µsec, which may be the interaural time difference that this IC cell is signaling. Such cells may be part of a neural circuit in the IC that helps process interaural time differences.

AUDITORY CORTEX

As shown schematically in Figure 15.1, the human *auditory cortex* is located in a deep groove or convolution in the brain called the *fissure of Sylvius*. The inaccessibility of the auditory cortex in the primate brain makes it difficult to study. In the cat's brain, the auditory cortex is more accessible, and therefore it is often used in research involving the cortex. The cat's cortex is often subdivided into vari-

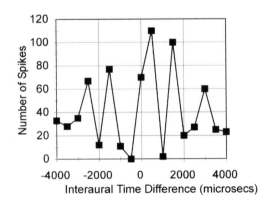

FIGURE 15.16 Response (number of spikes) of a neuron in the inferior colliculus with a low-frequency center frequency responding to a wideband noise presented with different interaural time differences (in µsec). Adapted with permission Yin and Chan (1988).

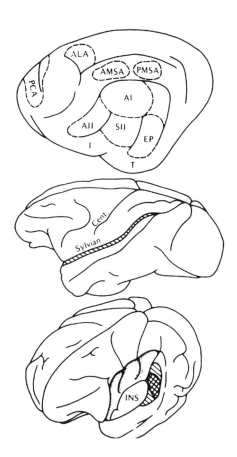

FIGURE 15.17 **Top**: Auditory cortex of the cat. The cat auditory cortex, which is easily accessible, is often subdivided into the various regions indicated. The primary auditory regions are AI, AII, SII, and EP. Other areas that receive auditory input are shown. **Center**: The auditory cortex of the monkey, which is like the human one, lies within the Sylvian fissure. **Bottom**: The monkey auditory cortex is shown by opening the Sylvian fissure. Adapted with permission from Elliott and Trahiotis (1972).

ous areas, as illustrated in Figure 15.17, which compares the cortexes of cat and monkey. The human auditory cortex is more like the monkey's. The bottom diagram is drawn with the Sylvian fissure opened to expose the auditory cortex. The projections to the cortex contain

bilateral information. Thus, each cochlea has an input to each hemisphere.

Early studies of the tonotopic organization of the auditory cortex revealed that frequency maps of the cortex could be made by stimulating small portions of the exposed cochlea and demonstrating a point-to-point projection of the cochlea onto the primary cortex (tonotopic organization). Frequency maps of the cortex could also be made by using tone bursts and recording the resulting activity at the cortex. The studies were made with the animal under deep anesthesia; in some cases, *strychnine* was applied to various areas of the cortex to raise the excitability of these cortical areas. Later studies showed that under reduced or no anesthesia less well-defined tonotopic organization could be obtained in the cortex. However, more recent studies using monkeys have produced consistent results from both anesthetized and awake animals. From studies of both anesthetized and awake animals, evidence now exists that the cortex is organized in columns of cells. Each column is perpendicular to the surface of the cortex and has cells with similar CF. The tuning curves for cortical neurons vary in shape, some being as narrow as those found in the auditory nerve and others being broader. This type of cortical organization might allow the auditory system to arrange information for complex auditory pattern recognition.

The time pattern of responses of cortical neurons seems to be especially sensitive to changes in stimulation. One-third of the cortical neurons can be stimulated only by tones of changing frequency or by more complex sounds. Very few units maintain a discharge rate above the spontaneous level for the duration of the stimulus, as was seen in auditory nerve discharge patterns. Figure 15.18 shows PST histograms illustrating various discharge patterns of cortical neurons to stimuli at their CFs and at various levels of intensity. Sinusoi-

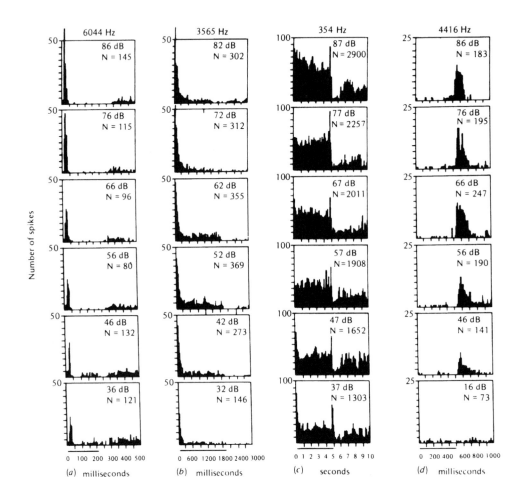

FIGURE 15.18 PST histograms to tone bursts recorded from cortical units. Bar at bottom of each column = duration of the tone burst. Columns = variations in the patterns caused by changes in frequency. Rows = changes in level. Note that most of the units respond to changes (at onset or offset) in stimulation, rather than to the total duration of the stimulus. Adapted with permission from Brugge and Merzenich (1973).

dal tone bursts are *not* often the appropriate stimuli to investigate the response patterns of many cortical neurons.

By using stimuli that vary in their frequency content as a function of time (frequency modulation, see Chapter 4), interesting response patterns have been detected. Some neurons will respond only to an in-

crease in frequency, others only to a decrease in frequency, and still others only to an increase in frequency of low-frequency stimuli and to a decrease in frequency of high-frequency stimuli. Units in the auditory cortex and other neural centers can also be investigated using amplitude-modulated stimuli. *Neural temporal modulation transfer functions,* like those obtained psychophysically (see Chapter 10), can be obtained in which neural discharge rate is determined as a function of stimulus modulation rate. Although some units respond in a manner similar to the psychophysical results, others appear to respond best to certain modulation rates as if these units were tuned to a particular amplitude modulation rate. Figure 15.19 displays the neural response of an auditory cortical unit tuned to about 10 kHz when a 10-kHz tone was sinusoidally amplitude modulated (100% depth of modulation) at different rates (from 1 to 30 Hz). The vector strength of this cell was greatest for modulation rates of approximately 5 Hz, as if this unit might be tuned to low modulation rates of approximately 5 Hz. Vector strength is a measure of how well the unit's discharge rate is synchronized to the period of the modulation envelope (see Chap-

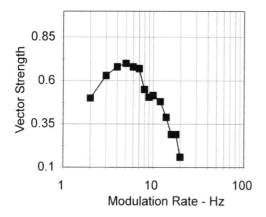

FIGURE 15.19 Vector strength of a neuron in the auditory cortex to a sinusoidally amplitude modulated 10-kHz carrier tone. The CF of the neuron was approximately 10 kHz. Vector strength is a measure of the ability of the neuron to phase lock to the modulated envelope of the amplitude-modulated sound (see Chapter 9). A vector strength of 1.0 indicates maximal phase locking. This neuron appears to phase lock best to a modulation rate of approximately 5 Hz. Adapted from with permission Schreiner and Langner (1988).

ter 9). A vector strength of 1.0 means that the neural unit only produces spikes at one phase (e.g., at the peak) of the modulation envelope. Low modulation rates are those that produce the largest amount of MDI and CMR (see Chapter 14). Thus, cortical amplitude and frequency modulation–sensitive neurons might underlie the perception of some of the more complex stimuli to which we are sensitive (see Chapter 14).

Another method of studying the function of the auditory system is to cut, remove, or render unresponsive part of the system and observe differences in an animal's ability to analyze acoustic stimulation. Typically, animals are trained to make a response that indicates they have detected or recognized some auditory stimulus; a part of the cortex is then removed or altered, and the animal is retested to see whether there is a change in its ability to perform the original task. If so, the missing part of the cortex is assumed to be responsible for the animal's ability to analyze the stimulus for this auditory task. Experiments utilizing such *lesions* and *behavioral techniques* are difficult to summarize because of differences in the site and size of the lesions, the testing techniques, and the nature of the stimulus. Generally, simple tasks are less affected by cortical lesions than are more complex tasks. Pure tone intensity and frequency discrimination tasks (see Chapter 10) may be disrupted after the lesion, but they can be relearned. More difficult tasks based on patterns (such as order of presentation or localization of a sound in space) suffer more after a lesion of the auditory cortex than do simpler tasks. Thus, it appears probable that failure on different types of tasks reflects different deficits in central processing of the stimulus. On the other hand, failure to find a deficit, temporary or permanent, does not necessarily indicate that the lesioned structure did not take part in analysis of the stimulus tested in the normal animal.

The animal behavioral studies, in agreement with the electrophysiology, appear to indicate that the cortical centers are organized more to process complex stimulus situations than to encode the basic parameters of the stimulus.

SUMMARY

In this chapter, we have attempted to describe briefly the anatomy and physiology of the central auditory system. The anatomical description was limited to the main nuclei of the ascending and descending pathways and to the auditory cortex in general. Both pathways were shown to have complicated innervation and bilateral representation at almost every nucleus. The interconnections within each nucleus were not discussed in detail, but their existence is evident. There is a wide variety in the neural morphology of CANS neurons that most likely leads to a variety of physiological responses. Neural circuits within each nucleus appear to operate on excitatory and inhibitory interactions. Neural circuits may be studied using the slice preparation technique, and such circuits may exhibit properties such as lateral inhibition. The major nuclei of the central auditory nervous system are the cochlear nucleus (subdivided into the DCN, PVCN, and AVCN), superior olive (subdivided into the MSO and LSO), inferior colliculus (subdivided into the central nucleus, dorsal cortex, and paracentral nuclei), medial geniculate, and auditory cortex (subdivided into AI, AII, SII, and EP). Tonotopic organization was said to be maintained at every level of the system, and tuning curves retained their sharpness for most neurons at all levels. The descending pathways perform a control function of both an excitatory and inhibitory nature on the incoming information. Time patterns of neurons in the central auditory system, both in the nuclei and the cortex, reflect the anatomical complexities of the nuclei, the pathways, and the cortex. Cells in the olivary complex and higher (e.g., the inferior colliculus) in the system discharge differentially to changes in interaural differences. Auditory-evoked responses are a means of obtaining some information concerning central auditory processing.

SUPPLEMENT

A great deal of the material in this chapter is covered in Chapters 6, 7, and 8 of the book by Pickles (1988), in books by Webster *et al.* (1992) and Fay and Popper (1992), and in a volume edited by Altschuler *et al.* (1989). The functional study of the CANS is in its infancy, with the major effort being devoted to describing the basic anatomical structures and physiological properties of these structures. As has been pointed out many times, both the anatomy and physiology are complex, making it difficult to determine the functional role of any particular neural site. Most of what we know about what CANS function comes from studies of animals such as the bat (Suga, 1988) and the barn owl (Konishi *et al.*, 1988). Articles by Wickesberg and Ortel (1990) and Neti, Young, and Schneider (1992) provide insights into some possible functional properties of the cochlear nucleus. Langner (1992) provides a similar review of modulation processing in the CANS. Shamma (1985) should be consulted for a review of one role that lateral inhibition might play in auditory processing.

Determining the various anatomical subareas of a nucleus often changes over time as more is learned about the structure. This applies especially to the inferior colliculus, where the subareas defined in this chapter are based on the work of Morest and Oliver (1984).

Auditory brainstem electric responses have proven to be popular and useful tools in clinical assessment of the auditory system. Interested students might read textbooks by Moore (1983) and Glattke (1983).

A potentially important aspect of the study of the central nervous system involves how the nervous system develops. The neural connections must develop in early life, and the way in which they develop may play a very important role in how well the auditory system functions. For instance, deprivation of certain kinds of acoustic information at a critical stage in neural development may alter the type of structure or function occurring within the central nervous system. These and other topics can be found in articles by Brugge *et al.* (1981) and Lippe and Rubel (1983), and in edited books by Ruben *et al.* (1986), Rubel *et al.* (1997), and Werner and Rubel (1992).

One tool and set of data we have not covered in this book is *genetics* (see Chapter 16 and Appendix F). The genetic code plays a crucial role in determining how the auditory system functions. By probing for aspects of the genetic makeup of auditory structures, hearing scientists are learning a great deal about the structure and function of the auditory system, as well as understanding about auditory deficits that are largely determined by heredity. Understanding modern biological genetic tools is important, but covering this topic would require providing background information beyond the intended scope of this book. The interested reader should first study any basic genetics textbook.

CHAPTER 16

The Abnormal Auditory System

The previous chapters have dealt with the structure and function of the normal auditory system. Over one's lifetime these normal systems can become abnormal for many reasons. In these cases, one may experience a hearing loss or in extreme cases deafness. While knowing about the normal auditory system is crucial for understanding the causes of hearing loss, it is also important to understand the abnormal auditory system. In addition, knowledge about the abnormal system can provide valuable information about the normal auditory system. Many things can be done to reduce the likelihood that the auditory system will be damaged, and a variety of sensory aids and rehabilitation strategies are available to help people who have a hearing loss. Medical interventions exist that can eliminate or reduce the severity of a hearing loss. Chapter 1 provided a general description of the clinical fields of *otology* and *audiology* and the practitioners in these fields. This chapter will not cover the treatment of hearing loss, but it will provide an overview of some of the things that can happen to the structure and function of the auditory system when it is damaged.

DAMAGE TO THE AUDITORY SYSTEM

In general, hearing loss can be caused by *sound exposure*, *ototoxic drugs* (drugs and chemicals that are poisonous to auditory

structures), *aging, diseases and infections, accidents*, and *heredity*. Most is known about how the inner ear can be damaged, and a great deal of this knowledge is based on studies of *noise-induced hearing loss* (NIHL). NIHL refers to changes in normal auditory function that occur as a consequence of exposure to loud levels of sound. In this context, sound at these high levels is considered unwanted because it causes hearing loss. Noise is used as a general term (as opposed to the more specific definitions provided in Chapters 4 and 11) to refer to any sound that is unwanted. In addition to causing a hearing loss, sound can become unwanted (a noise) because it interferes with work or play, interferes with communication, interferes with sleep, or is annoying.

EFFECTS OF NOISE ON THE INNER EAR

In Chapters 7 and 8, we discussed the important role of the haircells and their stereocilia in converting or transducing the mechanical and hydrodynamic forces within the cochlea into neural impulses in the auditory nerve. Obviously, damage to the haircells would drastically affect the transduction processes and change the ability to hear. Unfortunately, haircells appear to be most vulnerable to overstimulation that we call noise.

The amount of inner ear damage is related to the exposure level of the noise, among other factors. Exposure to a very high level of noise can cause the type of inner ear destruction illustrated in Figure 16.1. In this example, the noise destroyed the entire organ of Corti in one section of the cochlea. It appears that the primary cause of inner ear damage is that the elastic limits of the organ of Corti were exceeded, and it was virtually torn apart. Noises with very high levels can also cause middle ear damage, such as rupture of the tympanic membrane. For lower levels of noise exposure, the outer haircells may remain in place, but their stereocilia may swell, fuse, or otherwise be distorted, as shown in Figure 16.2. The stereocilia, tectorial membrane, and basilar membrane may all be structurally changed if the noise exposure is sufficiently intense.

Since each of these structures plays an important role in the normal transduction of sound to neural impulses, their damage from noise exposure can produce significant hearing loss.

The effects of noise on hearing may be temporary or permanent. In mammals, if the haircells are severely damaged, they will not recover or be replaced by new haircells (see Haircell Regeneration). If haircell damage is slight, haircells can recover, and hearing will return to normal. Mechanical destruction of the organ of Corti caused by exceeding its elastic limits will result in permanent loss of the haircells. However, less severe haircell damage is probably caused by various physicochemical processes such as those associated

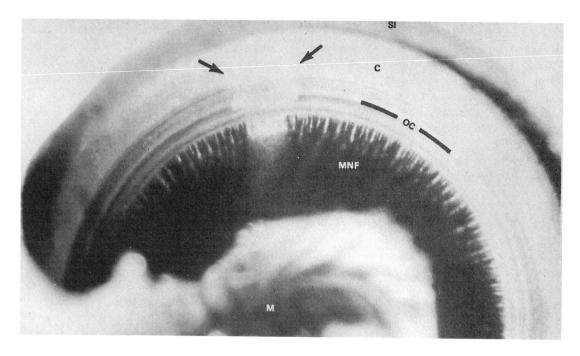

FIGURE 16.1 (a) Low-power light micrograph of a large portion of the cochlea. The curvature of the cochlea is obvious. The innervation of the organ of Corti (OC) by the myelinated nerve fiber (MNF) leaving the modiolus (M) is clearly seen. The arrows indicate an area in which the organ of Corti is missing due to a lesion or disruption caused by acoustic overstimulation, that is, a loud sound. Chinchilla photographs courtesy of Dr. Ivan Hunter-Duvar, Hospital for Sick Children, Toronto.

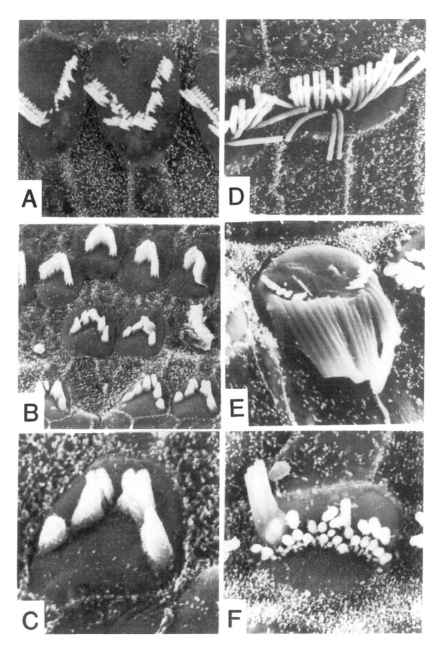

FIGURE 16.2 Various types of stereocilia distortion due to noise exposure. Micrographs A–C are OHCs and D–F are IHCs. (**A**) Missing OHC stereocilia a few days after a pure-tone exposure. The remaining stereocilia are erect. (**B**) Three rows of OHCs showing various degrees of fusing of stereocilia due exposure. (**C**) Higher magnification of fused OHC stereocilia from B. (**D**) Floppy IHC stereocilia due to pure-tone exposure. This situation is probably not reversible. (**F**) IHC with many stereocilia missing and some remaining stereocilia fused due to noise exposure. This situation is not reversible. Chinchilla photomicrographs courtesy of Dr. Ivan Hunter-Duvar, Hospital for Sick Children, Toronto.

with the metabolic activity (which takes place in the stria vascularis, see Chapter 8) of the overexerted haircell. It is also possible that moderate noise exposure alters the sensitive biomechanical connections of the inner ear, especially those that might be maintained by the motile responses of the outer haircells (see Chapter 8).

Hearing Loss Due to Noise Exposure

Hearing may be affected by exposure to noise in several ways. One that has received the most attention is a change in hearing sensitivity or threshold. An increase in auditory thresholds because of exposure to noise is called a *noise-induced threshold shift* (NITS). If over time—be it minutes, hours, or days—the threshold returns to its preexposure level (i.e., there is no NITS), it is called a *noise-induced temporary threshold shift* (NITTS), or simply a *temporary threshold shift* (TTS). If, however, the threshold does not return to its preexposure value, it is called a *noise-induced permanent threshold shift* (NIPTS), or simply *PTS* (*permanent threshold shift*). In many experiments, the amount of TTS reaches a limit at an asymptotic value; in these cases, the term *asymptotic threshold shift* (ATS) is used to describe the upper limit of TTS. In order to measure TTS or PTS, the threshold (see Appendix D and Chapter 10) is measured twice, once before the exposure and then again after the exposure.

After exposure to noise, the amount of hearing loss depends greatly on when the postexposure measurement is made. For most situations in which the noise exposure is moderately high, some hearing will be restored as time passes after the exposure is terminated. For many laboratory TTS studies, threshold measurements are made about 2 or 4 minutes after the listener is removed from the noise

(labeled TTS_2 or TTS_4; the subscript refers to the time in minutes after noise offset). This is done because it takes some finite amount of time to measure a threshold by most psychophysical techniques, especially when animals are the test listeners. Figure 16.3 shows how TTS_2 increases as the time of exposure and level of the exposing noise increases.

Even in PTS studies, after the exposure has ceased, some hearing is restored. Thus, measures of hearing loss made immediately at the end of an exposure may consist of two components: a temporary component that will recover to some degree (i.e., TTS) and a permanent component that remains for a lifetime (i.e., PTS). Hearing losses that combine TTS and PTS are called *compound threshold shifts*, or *CTS*.

TTS as a Function of Level, Duration, Spectral Content, and Temporal Pattern

Several exposure stimulus factors affect the amount of TTS or PTS: *level, duration, spectral*

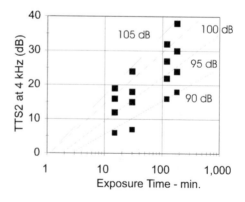

FIGURE 16.3 Growth of TTS_2 at 4 kHz after exposure to a bandpass noise (1200–1400 Hz) as a function of exposure time. Growth functions are shown for four different noise-exposure levels. Data of Ward *et al.* (1959), with permission.

content, and *temporal pattern*. Also, individual differences among people and species differences greatly affect the size of the threshold shift. The relationship between hearing loss and each of these factors is complex because no one factor can be considered apart from the others.

For continuous exposures of moderate intensities (approximately 75 to 105 dB SPL) lasting less than 8 hours, hearing loss (in decibels of threshold shift) as reflected in TTS measurements grows approximately linearly with increases in sound pressure level of the exposure stimulus. For low intensities below about 70 or 75 dB SPL, there is no measurable TTS. At very high intensities, above about 130 dB SPL, the effects on hearing loss are more erratic.

For moderate noise levels (80 to 105 dB), TTS is approximately proportional to the logarithm of the exposure time up to 8 or 12 hours, after which the amount of TTS appears to remain relatively constant. For instance, if a 15-minute exposure produces a TTS_2 of 10 dB, and a 30-minute exposure produces a TTS_2 of 15 dB, then a 1-hour exposure will result in a TTS_2 of 20 dB (i.e., for this example each doubling of the exposure time results in 5 dB of TTS_2). Figure 16.3 illustrates the growth of TTS_2, measured at 4000 Hz, as a function of the duration of the exposure to a 1200–2400 Hz bandpass-filtered noise of various intensities.

A number of "rules" have been suggested to establish the amount of threshold shift that a noise might generate as a function of exposure duration. One rule is called the *equal energy rule of noise exposure*, and it states that sounds of the same energy produce the same threshold shift. Thus, for each doubling of a sound's duration, the sound power can be reduced by 3 dB (see Chapters 3 and 10) to maintain the same energy and thus the same amount of TTS (thus, the equal energy rule is sometimes called the *3-dB rule*). Some data suggest that the sound power needs to be decreased by 5 or 4 dB for each doubling (the *5-dB* or *4-dB rules*) of duration to produce a constant TTS.

The amount of measured TTS also depends on the frequency spectrum of the fatiguing sound and on the frequency at which threshold is being measured. For exposure levels of less than 80 dBA or for short exposure durations that produce thresholds shifts lasting less than 2 minutes, the maximum TTS occurs at the frequency of the exposure stimulus. TTS decreases equally along the spectrum on both sides away from the exposure frequency that produces the maximum TTS. For higher-level exposures that produce longer-lasting TTS, the maximum effect does *not* occur at the frequency of the exposure sound. Also, TTS is not produced equally for exposure stimuli of different frequencies even though the level of the exposure maybe the same at all frequencies. In general, more TTS is produced at higher frequencies, at least up to approximately 4 to 6 kHz. Thus, when the exposure stimulus is a broadband noise, maximum TTS is found at frequencies between 3000 and 6000 Hz (Figure 16.4). When the exposure is a pure tone, TTS is found to increase as the frequency of the exposure tone increases. Furthermore, maximum TTS occurs at frequencies above the exposure frequency, with progressively higher frequencies showing the maximum TTS as the level of the exposure is increased. For high levels of exposure to a pure tone, the maximum TTS is produced one-half to one octave above the exposure frequency (sometimes called the "*half-octave shift*"). The spread of TTS to higher frequencies is also seen for bands of noise where the maximum TTS is produced one-half to one octave above the upper cutoff frequency of the exposing noise band. The exact relationship between the frequency region of maximal TTS (or PTS) and

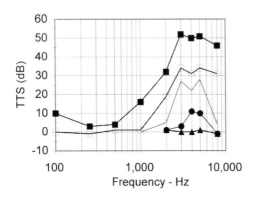

FIGURE 16.4 Amount of hearing loss in dB as a function of signal frequency with a 20-min 115-dB SPL white, broadband Gaussian noise as the exposure stimulus. The curves represent temporary threshold shifts (TTSs) measured at various times after the noise was turned off. Some hearing loss is still evident 24 hours after exposure to the noise. Adapted with permission from Postman and Egan (1949).

the parameters of the exposing stimulus is complex and depends on the details of the situation in which the exposure takes place.

The relation between hearing loss and the temporal pattern of the exposure stimulus is extremely complex, and a detailed explanation is beyond the scope of this text. For instance, when an exposure is intermittent, hearing may recover somewhat during the time the noise is turned off, increase when the noise comes on again, or stimulate the middle ear reflex (helping to reduce the effect of the noise on the inner ear, see Chapter 6), etc. The situation is further complicated if the exposure stimulus has high peak levels of short duration (e.g., the sound of a gun shot). Such stimuli are called *impulses*, and their effects on hearing are difficult to study because they can vary in many ways, such as peak pressure, pulse duration, rise and decay time, direction of pressure change (rarefaction or condensa-

tion), repetition rate, and the number of impulses in a given exposure period.

Recovery of Hearing after Noise Exposure

In many cases, hearing loss will disappear after the exposing noise is turned off. Figure 16.5 shows how the hearing loss builds up ("Growth") as the duration of the noise exposure increases and then decreases ("Decay") as the time after the cessation of the noise increases. The functions shown in the figure caption represent curves fit to data from several studies. As can be seen, the growth of threshold shifts occurs over a 5- to 10-hour period and then reaches its maximum; the decay of threshold shifts is exponential with

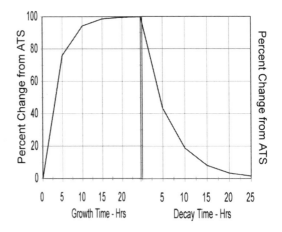

FIGURE 16.5 Growth and decay of TTS as a function of exposure time and time after exposure. The amount of TTS is plotted in terms of percentage of ATS; that is, a percentage of the maximum amount of TTS. The curves are a fit to a wide variety of data. The Growth curve is fit by the equation $y = D(1 - \exp(-t/T))$, and the Decay curve by, $y = C(\exp(-t/T))$, where D and C are constants, T is the growth or decay constant, t is the exposure time or time after exposure, and "exp" is the exponential argument. Adapted with permission from Melnick (1991).

time, which means that, if threshold shift in decibels is plotted on a logarithmic axis as a function of time, the data would be fit by a straight line. Notice that by approximately 24 hours hearing is back to normal (i.e., threshold shift is 0%).

OTOTOXIC DRUGS

Many drugs and chemicals are poisonous to the tissues of the auditory system, and as such they are *ototoxic*. Many antibiotics, especially several *aminoglycosides*, such as *kanamycin*, *neomycin*, and *streptomycin*, can be toxic to the stria vascularis and to haircells. These antibodies are often used in animal research to selectively destroy haircells (aminoglycosides destroy primarily outer haircells), allowing the researcher to better understand the role of haircells in both normal and abnormal auditory processing. Other drugs such as *diuretics* (*ethacrynic acid* and *furosemide*) affect the stria vascularis (see Chapter 7) as do the aminoglycosides.

Several drugs can lead to the sensation of *tinnitus* (see Supplement to Chapter 8), or "ringing" in the ears. Tinnitus is the subjective report of sounds, usually high-frequency sounds, that occur inside the head and cannot be measured by any objective external method. In many cases, the "loudness" of tinnitus can completely debilitate an individual. Tinnitus often, but not always, is accompanied by a hearing loss. *Salicylate*, an active ingredient in aspirin, and *quinine* both taken in large doses can lead to tinnitus and sometimes hearing loss. Thus, people with arthritis who take large doses of aspirin often experience noticeable tinnitus and can develop a hearing loss. Tinnitus can occur for many different reasons, including aspirin use, and there is currently no "cure" for tinnitus.

AGING

We all lose our hearing abilities as we grow older. Hearing loss that is due to aging is called *presbycusis*. Hearing loss probably starts when a person is in their twenties, but does not become noticeable (or even measurable) until much later in life (e.g., after 50), although there is a great deal of variability in the age at which measurable hearing loss occurs, and men tend to have more old-age hearing loss than women. The amount of hearing loss also depends on the history of noise exposure people experience over their lifetime. The more noise a person has been exposed to, the greater the hearing loss, so that people exposed to noise will lose their hearing at an earlier age than those who have not been exposed.

Presbycusis starts with hearing loss at the extremely high frequencies and progresses to lower frequencies as a person ages. Sounds of all frequencies cause the stereocilia of the haircells at the base of the cochlea to bend (shear), while only low frequencies cause such stereocilia bending for haircells at the apex (see Chapter 8). Thus, the haircells at the base of the cochlea "wear out" before those at the apex. Since haircells at the base carry information about high-frequency sounds and those at the apex carry information about low frequencies (see Chapters 8 and 9), hearing loss progresses from high to low frequencies as a function of aging. The progressive loss of hearing from high to low frequencies as a function of aging is one more example of how the place theory of hearing helps explain both normal and abnormal hearing.

DISEASES AND INFECTIONS

One of the most common childhood diseases is "middle-ear infection" or *otitis media*. Otitis media is an infection of the middle ear that can cause pain and hearing loss. Often the middle ear contains fluid (*otitis media with effusion*), and this fluid also causes hearing loss. The eustachian tube can become blocked, and this too can lead to hearing loss, since a blocked eustachian tube does not allow for the proper pressure balance across the tympanic membrane (see Chapter 6). If the infection is not stopped, there is a chance that it can spread to the inner ear and thereby lead to the possibility of *meningitis*, which is a very serious condition. Young children are more prone than adults to otitis media due to, among other things, the position of the eustachian tube relative to the middle ear, allowing infections from the nasal passages to more easily enter the middle ear of young children. Children who suffer frequent or long bouts of otitis media, and therefore have long periods of hearing loss, may be delayed in developing speech skills. While there is no "cure" for otitis media, antibodies can help deal with the infection (but the antibodies may not actually reduce the conditions that *cause* otitis media).

Several diseases, especially *otosclerosis*, cause the ossicular chain to calcify or become impaired in other ways. Without an operating ossicular chain, the sound level that gets to the inner ear is significantly reduced (between 30 and 60 dB), since the amplification provided by the ossicular chain is reduced or eliminated (see Chapter 6). It is also possible that, when the ossicular chain is inoperable, sound reaches both the round and oval windows (instead of just the oval window via the stapes), making it difficult for vibrations to be transmitted to the inner ear. The ossicular chain or part of the chain can be surgically replaced with an "artificial" chain in an operation called a *stapedectomy*. While the new stapes can restore much of the conductive hearing loss, a mild hearing loss may still exist.

A condition called *Ménière's disease* refers to a class of symptoms involving the inner ear and/or the vestibular system. Recall from Chapter 7 that the inner ear is closely connected to the vestibular system and that the auditory and vestibular nerves are close together in the VIIIth nerve bundle as it ascends toward the brainstem. In many cases of Ménière's disease, the organ of Corti appears to swell (a condition referred to as *hydrops*). This can result in hearing loss (sometimes the amount of the hearing loss fluctuates over time), a loss of balance, and a loss of the sense of equilibrium. There is not a good cure for Ménière's disease.

The most common condition that affects neural function is the formation of tumors. Tumors can cause all kinds of hearing abnormalities as a consequence of cutting off the transmission of neural information. In removing tumors, it is often necessary to remove part of the auditory nerve, brainstem, or cortex, and significant hearing and other neural difficulties often result. The use of *magnetic resonance imaging* (MRI; see Appendix F) has made it much easier for otologists and neurosurgeons to accurately locate tumors. The use of electrophysiological techniques (borrowed from those used in research) to monitor the status of the nerves that are being affected by the tumor also aids the surgical team in successfully removing tumors without destroying the nerves.

HEREDITY

Hearing loss can be hereditary. The revolution in cell and molecular biology is making it possible to understand the genetic basis of hearing (see Appendix F). As such, a better understanding of many forms of inheritable causes of hearing loss is currently underway. The gene(s) that are most likely responsible for conditions like *Usher's syndrome* and *Waardenburg syndrome*, which are two different inherited conditions that each include hearing loss and possible deafness, are now being identified. Other forms of deafness, including sensory–neural hearing loss, are currently being studied for their genetic causes. While knowing what genes trigger these disorders is an important discovery, knowing the genetic cause of a disorder does not yet mean that there is a cure. Some have argued that *gene therapy* will be a major way to "cure" inheritable disorders. Gene therapy involves inserting or injecting into the body new genes or proteins that have the ability to repair defective genes that one has inherited. However, gene therapy is in its infancy and is still a very controversial method, with limited demonstrated successes. Genes dealing with cochlear function have been implanted into animal models, with limited success in having the genes express themselves in a functional manner in the implanted animal. A major way to study the influence of genes on the development of the auditory system and hearing disorders is the use of transgenetic mice (see Appendix F).

RELATION BETWEEN INNER EAR DAMAGE AND HEARING LOSS

We have seen that noise or drug exposure or other variables can lead to inner ear damage and resultant hearing loss. We might therefore expect that measuring threshold shift would indicate the amount of inner ear damage. Several studies have investigated this proposition using animals that can be trained to respond to the presence or absence of a sound. Threshold sensitivity curves can be measured with these animals by various psychophysical techniques (see Appendix D). After their normal hearing has been assessed, the animals are exposed to the noise or some other agent that can cause hearing loss; then their hearing recovery curves are calculated, and the amount of PTS is determined. Once PTS is established across the frequency range, the inner ears are prepared for histological observation by various microscopic techniques (see Appendix F). The histological examination assesses the condition of the organ of Corti along the whole length of the basilar membrane, with particular attention to the condition of each haircell. The orderly arrangement of haircells allows the investigator to count the existing haircells and also those that are missing. Such plots are called *cytocochleograms* or simply *cochleograms*; one is shown in the bottom of Figure 16.6. The top of Figure 16.6 shows PTS for this animal as a function of the frequency tested. Note the correspondence between the way in which the hearing loss is distributed from low to high frequencies and the way the outer haircell loss is distributed from apex to base along the cochlea. Such a correspondence supports the place theory of frequency encoding (Chapter 8), which states that each place along the cochlear partition codes for a particular frequency.

The correspondence between haircell loss and threshold shift is not always as straightforward as Figure 16.6 suggests. The amount and pattern of haircell damage caused by the same noise, drug exposure, or other insult can vary greatly between individuals. Also, the

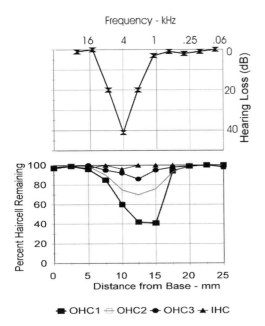

FIGURE 16.6 A cochleogram is shown on the bottom figure for an animal exposed to a signal with frequencies in the 4000-Hz region of the spectrum. As can be seen, between 10 and 15 mm from the base there are missing outer haircells (OHC1, 2, and 3 for the three rows), but few missing inner haircells. Permanent threshold shifts (top figure) were measured in a spectral region from 1 to 16 kHz. The figure shows the typical correspondence between the frequency region of a hearing loss and a localized region of outer haircell loss. Adapted with permission from Moody, Stebbins, and Johnsson (1976).

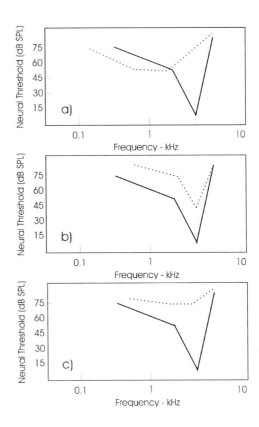

FIGURE 16.7 A comparison of a normal auditory nerve tuning curve (solid dark line) with its tip at CF and low-frequency tail with tuning curves from animals with severely damaged or missing haircells (dotted curve). (**a**) Comparison when outer haircells are severely damaged showing the loss of the highly tuned tip of the tuning curve, loss of sensitivity, and movement of the tip toward high frequencies. (**b**) Comparison when inner haircells are severely damaged showing that the tuning curve maintains its overall shape, but there is a loss of sensitivity. (**c**) Comparison when both inner and outer haircells are severely damaged showing a major loss in sensitivity and a significant change in the shape of the tuning curve.

hearing loss and haircell damage do not always correspond with each other. A further complication is that different animal species differ in susceptibility to acoustic overstimulation, drug dose, etc. Thus, it is difficult to generalize from research on animals to the effects on humans.

As reviewed in Chapters 8 and 9, the inner and outer haircells have different functions. Thus, it is reasonable to assume that the consequences of damage to each type of haircell will differ. Figure 16.7a shows a normal tun-

ing curve (solid line) and a neural tuning curve after the outer haircells were destroyed but the inner haircells remained relatively intact (dotted line). Figure 16.7b shows a similar comparison, but when the inner haircells have received more damage than the other hair-cells. Figure 16.7c shows a comparison of tuning curves when both the outer and inner haircells are damaged.

As can be seen, the loss of outer haircells (Figure 16.7a) results in a significant loss in neural sensitivity in the region of the "tip" of the tuning curve, that is, in the frequency region to which the nerve fiber is tuned. There appears to be a shift toward higher frequencies for the lowest sensitivity (CF) of the tuning curve after outer-haircell damage. There is a small change in the low-frequency "tail" of the tuning curve after outer-haircell damage, and the difference in the "tip-to-tail" sensitivity changes by about 40 dB after outer-haircell damage. Inner-haircell loss (Figure 16.7b) tends to preserve the general shape of the tuning curve, but results in a 40- to 50-dB loss in the sensitivity of the nerve. When both types of haircells are damaged (Figure 16.7c), the shape of the tuning curve is altered as it was for the condition when just the outer haircells are damaged and there is a very large decrease (75–90 dB) in neural sensitivity. Changes in the sensitivity and shape of the tuning curves suggest that both hearing sensitivity and frequency resolution would be severely altered when outer haircells are destroyed. Damage to just the inner haircells would probably result in hearing loss as measured by threshold shift, but since the shape of the tuning curve is not significantly altered, frequency resolution may not be affected as much.

Other changes occur in the function of the auditory nerve following haircell loss. For instance, when outer haircells are damaged, the nonlinear (compressive) input–output relationship between sound level input and either basilar membrane displacement (see Chapter 7, Figure. 7.16) or auditory nerve discharge rate (see Chapter 9, Figure 9.2) becomes much more linear, especially for signals whose frequencies are near the CF of the fiber. The loss of the compressive nature of coding sound level is probably responsible for several manifestations of hearing impairment. For instance, loudness recruitment (see Chapter 13, Figure 13.2) can be partially explained by the loss of nonlinear compression due to outer haircell loss.

HAIRCELL REGENERATION

As mentioned previously, mammalian haircells do not regrow (*regenerate*) after they have been damaged. As a result, damaged haircells lead to a permanent loss of hearing. However, in some species (birds and fish) haircells appear to regenerate after being destroyed from exposure to intense noise or ototoxic drugs. In birds it has been shown that, not only do haircells regenerate after damage, but they appear to regain their physiological function of generating nerve impulses. In some bird species, auditory thresholds recover to near normal levels after intense noise or ototoxic drug exposure. Thus, for birds (and perhaps fish) haircells can regenerate, and as a result these species may not experience permanent hearing loss due to noise or ototoxic drug exposure. Figure 16.8 shows a *basilar papilla* from a bird (zebra finch) immediately after a major noise exposure and the basilar papilla 90 days following the exposure. The basilar papilla is the sensory receptor surface in birds that contains auditory haircells. The basilar papilla 90 days later is near normal, indicating that the haircells have regenerated.

The reason why the haircells of fish and birds regenerate and those of mammals do not is incompletely understood. The inner ear of birds and fish are different from that of mammals. For instance, neither species has the coiled cochlea that characterizes mammalian inner ears. It is also the case that the haircells

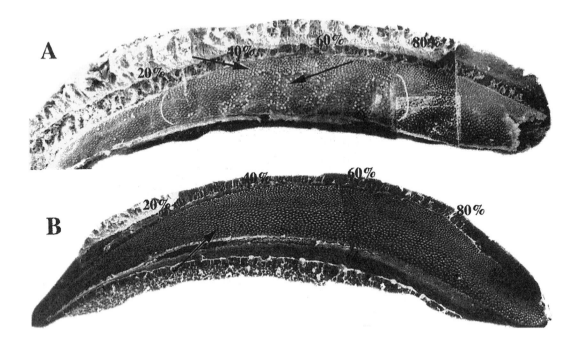

FIGURE 16.8 (a) The basilar papilla from a bird (zebra finch) obtained immediately after 24 hours of exposure to an intense sound. A large area of abnormal or missing haircells are noted in the region marked by the white parenthesis and highlighted by the arrows. (b) The basilar papilla from a bird exposed to the same sound, but allowed to recover 90 days before the basilar papilla was studied. The haircells appear normal or near normal throughout the entire basilar papilla, suggesting that the haircells have regenerated after damage. Reprinted with permission from Ryals *et al.* (1999).

that exist in fish, and in most birds, are not excited by a membrane that vibrates with traveling wave properties. The arrangement of the haircells also differs in birds and fish from that found in most mammals. Several other differences in the basic anatomy and genetic makeup of the haircells of birds and fish offer suggestions as to why haircells may regenerate in birds and fish but not in mammals.

AUDITORY CENTRAL NERVOUS SYSTEM CHANGES

While most is known about the effects of noise and other agents on the inner ear, sev-

eral important changes have also been studied in the brainstem and cortical regions responsible for auditory processing. An important concept in investigating central nervous system changes that result as a function of exposure to environmental conditions is *plasticity*. Plasticity refers to changes in the anatomy and physiology of neural structures that occur after the structures receive neural inputs that differ from those that they previously received, especially because of environmental influences (e.g., exposure to sounds). For instance, in vision, if one eye is blinded, then the anatomical structure and physiological response of neurons in cortical regions that would normally receive inputs from that eye (retina) or from both retinae will change so that they will be more responsive to the "seeing" eye. Thus, plasticity is often a way in which the central nervous system "compensates" for a loss in one form of processing. Plasticity is most readily seen in developing animals, but there is ample evidence that the adult central nervous system is also capable of such reorganization. Learning may be a result of plasticity.

While there is evidence for plasticity in the auditory system, such evidence is not as abundant as it is for visual and *somatosensory* (sense of touch) systems. Haircell regeneration, discussed above, is one form of neural plasticity. After prolonged exposure to moderate levels of noise that would *not* cause haircell damage, some cells in the cochlear nucleus (and at the periphery) appear to become more resistant to damage (these cells have become "toughened") when a very loud noise is presented. It is as if the moderate level of previous noise exposure is protecting against additional detrimental consequences of the loss of neural function. This form of toughening is not well understood and is not always found, even in the same experiment.

Cells in kittens' *auditory cortex* that appear to be involved with sound localization are different if the kittens grow up with only one ear able to send neural signals to the auditory cortex. That is, parts of the auditory cortex responsible for sound localization appear to be organized differently for kittens who are deprived of binaural inputs as opposed to those that received normal inputs from both ears during early life. This reorganization might aid cats in sound localization after they have suffered monaural deafness. If a normal-hearing person goes through a period of time when they are deprived of normal binaural information (e.g., they wear an earplug in one ear for a long period of time), they will initially lose some of their abilities to localize sounds (especially in the vertical direction), but over time they can regain some of this loss. That is, it is as if the nervous system has adapted to the new form of binaural information allowing for near normal sound localization.

The cochlear implant offers another example of the possible plasticity of auditory function. The cochlear implant or prosthesis is a wire containing up to 22 electrodes that is inserted into the cochlea (following surgery) of people with significant hearing loss that is of a cochlear origin. A sound transducer translates sound into electrical current that stimulates the different electrodes, which in turn can stimulate surviving auditory nerve fibers. Each electrode carries information about a specific frequency region of the sound according to the way in which the transducer operates. Since each electrode is at a different place along the cochlea, presumably the frequency information is distributed in the cochlea in a manner similar (but by no means identical) to the way a normal cochlea operates. The current follows the approximate amplitude envelope of the sound. Many patients fit with a cochlear implant can learn to function very well with sound, often gaining significant ability to understand speech and even music. In many cases, people who had no or little ability to hear throughout most of their lives learn to use sound when fitted with a cochlear implant. This suggests that the auditory system can "learn" to process this new form of information. If so, then the auditory system is exhibiting strong evidence for neural plasticity.

SUMMARY

Intense sounds, aging, ototoxic drugs, disease and infections, accidents, and heredity

may lead to an abnormal auditory system. Hearing loss caused by high sound levels can be permanent (causing PTS) or temporary (causing TTS). The level, duration, spectral content, and temporal pattern of the exposing sound all influence the amount of PTS and TTS and the way in which TTS decreases after cessation of the exposing sound. Ototoxic drugs (such as aminoglycosides and diuretics) damage different tissues of the auditory system, often in a selective manner. Other drugs (such salicylates and quinine) can cause tinnitus or hearing loss. Hearing loss due to aging (presbycusis) progresses from high to low frequencies as we age. Many diseases or conditions can lead to hearing loss (e.g., otitis media, otosclerosis, Ménière's disease, tumors). Many forms of hearing loss are also inheritable. Cochleograms reveal the relationship between haircell loss and the frequency region of PTS. Auditory nerve tuning curves change in different ways depending on whether inner or outer haircells are damaged. While the haircells of mammals do not regenerate, those of many birds and fish appear to regrow, and perhaps regain their function after being damaged or destroyed. The anatomy and physiology of the central auditory nervous system often change (plasticity is demonstrated) as a result of insult, injury, or other forms of alterations in how acoustic input reaches the central nervous system.

SUPPLEMENT

The book by Moore (1995), *Perceptual Consequences of Cochlear Damage*, provides a through review of the existing data on inner-ear damage and its effects on hearing. The book by Rubel, Popper, and Fay (1999) covers many aspects of neural plasticity including haircell regeneration. Van de Water, Popper,

and Fay's (1996) book covers material related to clinical issues, including aspects of the genetic causes of hearing loss. Pickles (1988) provides a good overview of many ototoxic drugs, especially how they are used to better understand auditory structure and function. The website (www.nih.gov/nidcd) of the National Institutes on Deafness and Other Communication Disorders (NIDCD) of the National Institutes of Health (NIH) often provides up-to-date material on clinical issues in hearing and how they relate to the basic hearing sciences.

Interest in the effects of noise on hearing had a resurgence during World War II, and research on the subject has continued since then. As a result, there is a large body of literature and several published summaries. Two texts, one edited by Henderson *et al.* (1976), and the other edited by Hamernik *et al.* (1982), and a series of articles in volume 90(1) of the *Journal of the Acoustical Society of America* (Clark, 1991a,b; Melnick, 1991; Patterson, 1991; Rosowski, 1991) represent an outstanding collection of current research relevant to the effects of noise on hearing.

As a result of noise exposure studies on humans and animals, criteria have been established for the levels and durations of exposures likely to cause PTS. Called *damage risk criteria* (DRC), these data show the duration and level combinations that can be expected to cause permanent hearing loss if we are exposed to them on a regular basis such as at work. The damage risk criteria were first developed by the Committee on Hearing, Bioacoustics, and Biomechanics (CHABA) of the National Research Council, the action organization of the National Academy of Sciences. Recent standards developed by the Environmental Protection Agency (EPA) and the Occupational Safety and Health Agency (OSHA) recommend that people not be exposed in the workplace to sound levels that exceed 85 dBA in any 8-hour day. Other standards for permissible noise level exist for community noise, airport noise exposure, and noise in the workplace (see Harris, 1991).

A major goal of the work on TTS was to use TTS to predict what might lead to PTS. For obvious reasons, studies of PTS are difficult to do in humans, yet it is crucial to understand what noise exposure conditions might lead to PTS. By studying TTS and ATS, it was hoped that reliable predictions for PTS could be obtained. The CHABA DRC was largely based on TTS data and assumptions concerning the ability of TTS to predict PTS. A great deal of data suggest that many of these assumptions are not valid, indicating that caution is required in using TTS and ATS measures as predictors of PTS (see Melnick, 1991). Clark (1991a) provides a review of the damaging effects of everyday sounds. Impulse noise (brief but intense noise) presents several special problems for the accurate assessment of noise and its effects on hearing. These short-duration sounds have a broad spectrum (see Chapter 4), and both the level of the sound and its spectral content determine the amount of threshold shift (see Patterson, 1991).

Understanding why fish and bird haircells regenerate and mammals do not is an important area of study. If a way can be found to regenerate haircells, then many forms of hair loss may be overcome by whatever intervention causes haircell regeneration (see, for instance, *Science*, Vol. 1992). Canlon *et al.* (1988) have studied the neural "toughening" that is thought to occur following moderated exposure to sound. Brugge (1988) in cats and Hofman *et al.* (1998) in humans have studied localization or the physiological basis for localization following sound localization deprivation or alteration of the cues used for localization. Miller and Spelman (1990) should be consulted to learn more about the cochlear prosthesis.

Animals have been used in noise research since Pavlov's work in the 1920s. Most systematic current studies using animal models can trace their origins to the study of Miller *et al.* in 1963, which used the cat as the animal model. In 1963, Miller introduced the chinchilla as an animal for use in hearing research. Later studies by Miller (1970), Carder and Miller (1972), and Clark (1991b) indicate that the chinchilla is a widely used animal model for hearing.

Sinusoids and Trigonometry

*I*n addition to describing harmonic motion or vibration, sinusoids describe the relationship between sides and angles of triangles. The sinusoidal function is one of many *trigonometric functions*. Figure A.1 is a triangle with sides *A*, *B*, and *C* and angles *a* (opposite side *A*), *b* (opposite side *B*), and *c* (opposite side *C*). The following equations describe some of the trigonometric relationships among sides and angles:

$$\sin a = A/B, \cos a = C/B, \tan a = A/C, \quad \text{(A.1)}$$

where sin is the sine function, cos the cosine function, and tan the tangent function.

Another way to view these three trigonometric relationships is seen in Figure A.2. The circle is a *unit circle* (radius of 1); "sin *a*" describes how far the point *P* has moved along the unit circle from the zero (or starting) position in the vertical direction, and "cos *a*" describes how far point *P* has moved along the unit circle from the zero (or starting) point in the horizontal direction. Note that the sine function starts at 0, goes to 1, and back through 0 to –1, then to 0 again as angle *a* increases from 0 to 360°. The cosine function, however, starts at 1, goes to 0, then to –1, back to 0, and then to 1 as angle *a* increases from 0 to 360°.

The tangent function is formed from the sin and cos functions:

$$\tan a = \sin a / \cos a. \quad \text{(A.2)}$$

This can easily be seen from equation (A.1):

$$A = B \sin a \text{ and } C = B \cos a,$$
$$\tan a = A/C = (B \sin a) / (B \cos a)$$
$$= \sin a / \cos a. \quad \text{(A.3)}$$

Given equation (A.2), we see that the tangent takes on values from minus infinity to positive infinity as the angle *a* is varied.

Some of the relationships among the trigonometric functions that provide powerful tools in analyzing acoustical signals are as follows:

$$\sin a = \cos (90° - a)$$

or

$$\cos a = \sin (90° - a). \quad \text{(A.4)}$$

$$\sin^2 a + \cos^2 a = 1. \quad \text{(A.5)}$$

Equations (A.4) and (A.5) can be derived directly from the unit circle shown in Figure A.2. Other relationships are

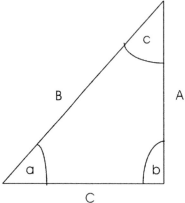

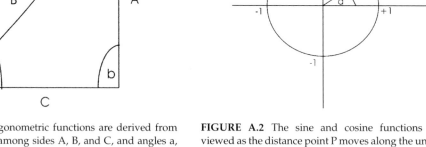

FIGURE A.1 Trigonometric functions are derived from the relationships among sides A, B, and C, and angles a, b, and c of a right triangle.

FIGURE A.2 The sine and cosine functions can be viewed as the distance point P moves along the unit circle.

$$\sin (a \pm b) = \sin a \cos b \pm \cos a \sin b. \quad (A.6)$$

$$\cos (a \pm b) = \cos a \cos b \mp \sin a \sin b. \quad (A.7)$$

$$\sin a + \sin b = 2\{\sin [(\tfrac{1}{2})(a + b)] \\ \times \cos [(\tfrac{1}{2})(a - b)]\}. \quad (A.8)$$

$$\sin 2a = 2 \sin a \cos a. \quad (A.9)$$

$$\sin a \sin b = (\tfrac{1}{2})[\cos (a - b) \\ - \cos (a + b)]. \quad (A.10)$$

Many more relationships can be derived from those listed above. For instance, we may derive $\cos 2a$ from equation (A.7):

$$\cos 2a = \cos (a + a) \\ = \cos a \cos a - \sin a \sin a \\ = \cos^2 a - \sin^2 a.$$

Rearranging equation (A.5):

$$\sin^2 a = 1 - \cos^2 a,$$

and substituting for $\sin^2 a$ in the equation above for $\cos 2a$, we then obtain

$$\cos 2a = \cos^2 a - (1 - \cos^2 a) \\ = 2 \cos^2 a - 1. \quad (A.11)$$

USING SINUSOIDS TO DESCRIBE ACOUSTIC EVENTS

These few ideas concerning sinusoids and trigonometry can help solve and describe a great variety of acoustic situations, for example: (1) to determine the instantaneous amplitude of a sinusoid at different times and for different frequencies and starting phases, and (2) to investigate nonlinearities.

Suppose that there is a 100-Hz sinusoid with a peak amplitude A and a starting phase of 0°, and we would like to know the instantaneous amplitude one-thousandth of a second (1 msec) after the sinusoidal vibration begins. We can use the equation for a sinusoid (A.12) to obtain this result. The equation $y = A \sin (2\pi f t + \theta)$ is the definition of a sinusoid when the sinusoidal terms are expressed in radians.

We can change the equation for terms expressed in degrees by using the following:

$$y = A \sin (360° \; 1/T \; t + \theta), \qquad \text{(A.12)}$$

where T is the period of the sinusoid in seconds ($1/T$, of course, equals f), θ is the starting phase, A is the peak amplitude, and y is the instantaneous amplitude. In our example, $T = 1/100$ Hz (10 msec), $t = 1/1000$ second (1 msec), and $\theta = 0°$. Thus,

$$
\begin{aligned}
y &= A \sin (\; 360° \times 1/(1/100) \times 1/1000 + 0) \\
&= A \sin (360° \times 100 \; / \; 1000) = A \sin 36°.
\end{aligned}
$$

The sin of $36°$ is 0.59, so 1 msec after a 100-Hz vibration begins, the instantaneous amplitude is $0.59A$ in magnitude. Thus, if the peak amplitude was 100 dyne/cm^2 of pressure, the instantaneous amplitude at 1 msec would be 59 dyne/cm^2.

If θ were $44°$ instead of $0°$, then the equation would be written as

$$
\begin{aligned}
A \sin (36° + \theta) &= A \sin (36° + 44°) \\
&= A \sin (80°) = 0.98A,
\end{aligned}
$$

or, in our example, where A is 100 dyne/cm^2, the instantaneous amplitude of a 100-Hz vibration with starting phase of $44°$ would be 98 dyne/cm^2 at 1 msec after it started.

NONLINEARITY

A linear system, as we stated in Chapter 4, is one described by a straight-line relationship between the input to the system and its output. The general equation for a straight line is

$$y = mx + b, \qquad \text{(A.13)}$$

where m is a slope constant, b an intercept constant, x the input, and y the output.

A nonlinear system is one described by a relationship between an input and output that is not represented by a straight line. One such simple nonlinear equation is

$$y = x + x^2. \qquad \text{(A.14)}$$

If we assume that the input to the system is the sum of two sinusoids, $A \sin (2\pi f_1 t) + A \sin (2\pi f_2 t)$, then

$$x = A \sin (2\pi f1 t) + A \sin (2\pi f2 t),$$

according to equation (A.14). (Throughout this discussion we shall assume that all phase angles, θ, are zero.)

Since $y = x + x^2$, the value of y is as follows:

$$
\begin{aligned}
y &= A \sin (2\pi f_1 t) + A \sin (2\pi f_2 t) \\
&\quad +[A \sin (2\pi f_1 t) + A \sin (2\pi f_2 t)]^2 \\
&= A \sin (2\pi f_1 t) + A \sin (2\pi f_2 t) \\
&\quad + [A^2 \sin^2 (2\pi f_1 t) \\
&\quad + 2A^2 \sin (2\pi f_1 t) \sin (2\pi f_{2} t) \\
&\quad + A^2 \sin^2 (2\pi f_2 t)]. \qquad \text{(A.15)}
\end{aligned}
$$

Let us look at the last three terms of equation (A.15) and use trigonometric identities.

First, using equation (A.10),

$$
\begin{aligned}
A^2 \sin^2 (2\pi f_1 t) &= A^2[\sin (2\pi f_1 t) \sin (2\pi f_1 t)] \\
&= (A^2/2)[-\cos (2\pi(2 f_1) t) + \cos (2\pi(0) t)] \\
&= (A^2/2)[-\cos (2\pi(2 f_1) t) + 1] \\
&= -(A^2/2)\cos (2\pi(2 f_1) t) + A^2/2 \\
&= A^2/2 - (A^2/2)[\cos (2\pi(2 f_1) t)]. \qquad \text{(A.16)}
\end{aligned}
$$

Second, using equation (A.10),

$$
\begin{aligned}
2A^2 \sin &(2\pi f_1 t) \sin (2\pi f_2 t) \\
&= A^2 \cos (2\pi(f_1 - f_2) t) \\
&\quad - A^2 \cos (2\pi(f_1 + f_2) t). \qquad \text{(A.17)}
\end{aligned}
$$

Third, using equation (A.16) but substituting f_2 for f_1,

$$A^2 \sin^2 (2\pi f_2 t) = -A^2/2 \cos (2\pi(2f_2)t) + A^2/2$$
$$= A^2/2 - A^2/2 \cos (2\pi(2f_2)t). \quad (A.18)$$

By combining these three parts with the results of equation (A.15), we have

$$y = A \sin (2\pi f_1 t) + A \sin (2\pi f_2 t)$$
$$- A^2/2 \cos (2\pi(2f_1)t)$$
$$- A^2/2 \cos (2\pi(2f_2)t)$$
$$+ A^2 \cos (2\pi(f_1 - f_2)t)$$
$$- A^2 \cos (2\pi(f_1 + f_2)t) + A^2. \quad (A.19)$$

This in turn means that when f_1 and f_2 are the input frequencies of x, the following frequencies described in equation (A.19) constitute the output y if the output y is from the nonlinear system described in equation (A.14). These frequencies are $f_1, f_2, 2f_1, 2f_2, f_1 - f_2$, and $f_1 + f_2$. They may be the aural harmonics ($2f_1, 2f_2$) and combination tones ($f_1 - f_2, f_1 + f_2$) described in Chapters 6, 12, and 13.

An easy method of calculating the resultant amplitude and starting phase uses *vector addition*. In vector addition, each unit to be added has two values: a magnitude and a direction, or phase value. For the addition of sinusoids the peak amplitude is the magnitude of the vector, and the starting phase is the direction or phase of the vector. Thus, for the example shown in Figure A.3 the vector diagram in Figure A.4 may be used to determine the sum or resultant peak amplitude and starting phase. The vector diagram of Figure A.4 is constructed by first plotting the amplitude of the sinusoid with the smallest starting phase angle as a horizontal line with a length proportional to its peak amplitude (A_1). A line representing the other sinusoid is drawn at angle α to the line A_1 and with a length proportional to its peak amplitude (A_2). The angle α is the difference in the starting phases between the two sinusoids. The resultant vector is determined by completing the triangle. Thus, the resultant sinusoid has a peak amplitude R and a starting phase of θ, as shown in Figure A.4. Rather than having to construct vector diagrams, some trigonometric identities may be used to solve for the resultant

VECTORS

In the previous examples, sinusoids of different frequencies were added together. In adding sinusoids of the same frequency, both the amplitudes and starting phases of the sinusoids must be considered. Notice the examples shown in Figure A.3. The two original sinusoids (A1 and A2) have the same frequency and amplitudes, but the starting phases are different. The resultant sinusoid has a different phase from either of the original starting phases, and the resultant amplitude is not the simple algebraic sum of the original two amplitudes.

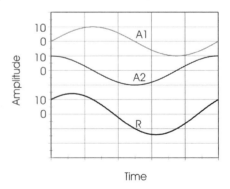

FIGURE A.3 The sum of two sinusoids (A1 and A2) (see text).

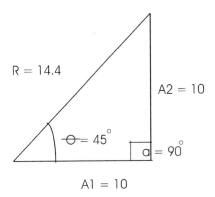

R = 14.4

A2 = 10

$\theta = 45°$

$\alpha = 90°$

A1 = 10

FIGURE A.4 A vector diagram for the condition shown in Figure A.3. The lengths of the vectors are the amplitudes of the two sinusoids, and the phase α is the starting phase difference between the two sinusoids. The resultant amplitude and phase can be determined from this right triangle or from equations (A.21) and (A.22).

vector's magnitude (peak amplitude) and phase (starting phase):

$$R^2 = A_1^2 + A_2^2 + 2A_1A_2 \cos(\alpha), \quad (A.20)$$

and

$$\theta = \cos^{-1}[(R^2 + A_1^2 - A_2^2)/(2RA_1), \quad (A.21)$$

where A_1 and A_2 are the peak amplitudes of the two sinusoids being added together, such that A_1 represents the sinusoid with a small starting phase angle, α is the difference in the starting phases of these two sinusoids, R is the resultant peak amplitude, θ is the resultant starting phase, and $\cos^{-1}$ means the angle whose cosine equals the ratio $(R^2 + A_1^2 - A_2^2)/(2RA_1)$. Equations (A.21) and (A.22) are called the *law of cosines*.

For example, in Figure A.3 and shown in the vector diagram of Figure A.4, the resultant peak amplitude and starting phase using the law of cosines are

$$R^2 = 10^2 + 10^2 + 2 \times 10 \times 10 \cos(90°)$$
$$= 100 + 100 + 200 \times (0) = 200 = 14.14,$$

and

$$\theta = \cos^{-1} + 10^2 - 10^2)/(2 \times 14.14 \times 10)]$$
$$\theta = \cos^{-1}(200/282.80)$$
$$\theta = \cos^{-1}(0.707)$$
$$\theta = 45°.$$

To find the inverse cos or $\cos^{-1}$ or arccos using a calculator or a computer program, look for keys or values that say arc, or $\cos^{-1}$, or arccos; enter the number (e.g., 0.707) and then the appropriate key or key sequence. On some calculators and programs, you may need to make sure that the calculator is in the "degree" rather than in the "radian" mode.

Vector diagrams and the law of cosines are useful tools for obtaining the magnitude (amplitude) and phase (starting phase) in situations involving values of objects having both a magnitude (amplitude) and a direction component (starting phase).

APPENDIX B

Logarithms

he logarithm is the solution to the following equation for x, that is, if

$$y = 10^x, \qquad (B.1)$$

then

$$x = \log_{10} y. \qquad (B.2)$$

In other words, $\log_{10} y$ or, for simplicity, $\log y$, is the logarithm to the base 10 of the number y. Thus, x is the number to which 10 must be raised in order to obtain the answer y. A set of simple relationships between x and $\log y$ can be demonstrated as follows:

$$0 = \log 1, \text{ since } 1 = 10^0,$$
$$1 = \log 10, \text{ since } 10 = 10^1,$$
$$2 = \log 100, \text{ since } 100 = 10^2,$$
$$3 = \log 1000, \text{ since } 1000 = 10^3. \qquad (B.3)$$

So, for instance, $\log 1{,}000{,}000$ is $\log 10^6 = 6$.

Two simple rules concerning logarithms are used to solve most problems.

Rule 1:
$$\log (x\, y) = \log x + \log y. \qquad (B.4)$$

Rule 2:
$$\log (x/y) = \log x - \log y. \qquad (B.5)$$

A special case of rule 1 is rule 1a:

$$\log x^n = n \log x,$$

since,

$$\log x^n = \log (x\, x\, x, \ldots\ldots , n \text{ times})$$
$$= \log x + \log x +, \ldots\ldots, n \text{ times}$$
$$= n \log x.$$

These two rules are useful in dealing with problems in which x is not a whole number or y is not some integer multiple of 10. For instance, what is the log of 412 or what is $10^{0.63}$?

Tables of logarithms usually have the logarithms of numbers between 1 and 10; so if you use these tables to find log 412, you are really looking up log 4.12, which is 0.615. Any answer such as 0.615, when found in the logarithm tables, is called a *mantissa*. Rules 1 and 2 can help determine logarithms of numbers besides those from 1 to 10, such as 412: $\log 412 = \log (4.12 \times 100) = \log 4.12 + \log 100 = 0.615 + 2 = 2.615$ (remember from equation (B.3), log $100 = 2$). The number 2 is called the *charac-*

teristic of the logarithm. Additional examples are as follows:

$$\log 2.6 = 0.335$$

[obtained directly from a logarithm table],

$$\log 2165 = \log (2.165 \times 1000)$$
$$= \log 2.165 + \log 1000$$
$$= 0.335 + 3 = 3.335,$$
$$\log 0.31 = \log 3.1/10 \text{ (using rule 2)}$$
$$= \log 3.1 - \log 10$$
$$= 0.491 - 1 = -0.509,$$
$$\log 0.0031 = \log 3.1/1000$$
$$= \log 3.1 - \log 1000$$
$$= 0.491 - 3 = -2.509.$$

To solve the equation $y = 10^{0.63}$, we simply go backward through the table, that is, log 4.26 is 0.63, so $10^{0.63}$ is equal to 4.26; thus, $y = 4.264$.

DECIBELS

The primary use for logarithms in audition is in working with the decibel.

To find decibels from a ratio (r).

1. If the ratio is one of pressure, then dB = 20 log r.

2. If the ratio is one of intensity, power, or energy, then dB = 10 log r.

To find the ratio from a decibel number:

1. Determine if the ratio is pressure or intensity/power/energy.

2. If the measurement is pressure, divide the decibel number by 20; if it is intensity, power, or energy, divide the decibel number by 10 (let's call this number z, i.e., $z = $ dB/20 or dB/10).

3. Then the ratio is equal to 10^z.

Recall that for ratios between 0 and 1 (measured value less than the reference value) the decibel result will be a negative number. And, for negative decibels, the ratios will be between 0 and 1.

Fourier Analysis

A complete description of an acoustic stimulus requires it to be defined in both the time and frequency domains. The auditory system is also capable of processing sound in both domains. We have introduced the concept of Fourier analysis of time and frequency domains in Chapters 2 and 4. In these discussions, the reconstruction of the time-domain waveform was described from a complete description of the phase and amplitude spectra (see Figure 4.2 for this description). Appendix A also shows how, with the use of trigonometric identities, the equations for summing sinusoidal functions could be written and solutions obtained. We have not been able to describe how the frequency domain description can be derived from knowledge of the time-domain waveform. We did point out in Chapter 5 that filters can be used in practical situations to estimate the amplitude spectrum of a sound. However, an analytic definition for equating the frequency domain to the time domain has not been presented. This appendix is intended to help the student, in a general way, understand Fourier analysis. From this discussion the student should learn something about the processes by which the spectrum of a waveform can be derived from a definition of the stimulus in the time domain.

In general, $f(t)$ will represent the time-domain waveform. For instance, for a pure tone,

$$f(t) = A \sin (2\pi f\, t).$$

We will use $F(w)$ as our representation of the frequency domain description, where $w_0 = 2\pi f$, f is the frequency in Hertz, and t is the time in seconds.

The equation for *Fourier series* (C.1) is

$$f(t) = \tfrac{1}{2}a_0 + \sum_{n-1}^{\infty} [a_n \cos (nw_0t) + b_n \sin (nw_0t)], \quad (C.1)$$

where n are integers from 1 to ∞ that represent the different number of frequencies in the frequency domain. Notice that the equation states that $f(t)$, the time-domain description, is equal to a sum of sinusoidal (and cosinusoidal) terms. This is the essence of *Fourier's theorem*, that the time-domain waveform can be constructed from a sum of sinusoidal components.

To use equation (C.1), we need to know the values of a_0, a_n, b_n, and w_0. They are defined as follows:

$$a_o = 2/T \int_{-T/2}^{2/T} f(t)\, dt = 0, \qquad \text{(C.2)}$$

$$a_n = 2/T \int_{-T/2}^{2/T} f(t) \cos(nw_o t)\, dt, \qquad \text{(C.3)}$$

$$b_n = 2T \int_{-T/2}^{2/T} f(t) \sin(nw_o t)\, dt, \qquad \text{(C.4)}$$

$$w_o = 2\pi T.$$

T is the period of the waveform in the time domain. This version of Fourier analysis is therefore limited to periodic time-domain waveforms, and, as can be seen from equation (C.1), for situations in which the spectra are discrete (only an integer number of frequencies exist in the spectra). The sinusoids in the frequency domain of the waveform consist of the *fundamental frequency* (w_o or $2\pi/T$) and any of the higher harmonics (integer multiples of the fundamental frequency) that are present after equations (C.2)–(C.4) are solved. The fundamental frequency is the frequency at which the periodic time-domain waveform repeats.

The $\int$ is the symbol for integration. In general, it means to find the total area under the function over the limits shown on the integration sign. In other words, it is the continuous sum of the magnitude of the value of the function over the limits of integration. Thus, in equations (C.2)–(C.4) we are to find the area under the functions from minus half a period ($-T/2$) to plus a half a period ($T/2$) of the periodic time-domain function.

By using trigonometric identities (see Appendix A) equation (C.1) can be rewritten as

$$f(t) = C_o + \sum_{n=1}^{\infty} C_n \cos(nw_o t - {}_n), \qquad \text{(C.5)}$$

where $C_o = \tfrac{1}{2}a_o$, and

$$C_n = \sqrt{a_n^2 + b_n^2}, \qquad \text{(C.6)}$$

$$\theta_n = \tan^{-1}(a_n/b_n). \qquad \text{(C.7)}$$

The term C_n is used to define the amplitude spectrum of the waveform, and θ_n is used to define the phase spectrum of the waveform. The term C_o is often referred to as the dc (for direct current) term of the Fourier series. It indicates the amount of asymmetry the time-domain waveform has about the zero amplitude value. It is also sometimes called the amplitude of the zero Hertz component in the amplitude spectrum.

To solve for these spectra, perform the following steps. The time-domain waveform would be defined. Then equations (C.2)–(C.4) would be solved. The resulting vales of a_o, a_n, and b_n would be used in equations (C.3)–(C.7). The relationship between C_n and nw_o could be plotted as the amplitude spectrum, and the relationship between θ_n and nw_o could be plotted as the phase spectrum, such as in Figure 4.8b,c.

Let us consider the waveform (a square wave) shown in Figure 4.8. This is a complex waveform, because it is not a sinusoid and it is periodic with a period of T msec. Thus, we can obtain its spectra following the steps outlined above.

The time-domain description is

$$f(t) = \begin{cases} 0, & \text{for } -T/2 \le t \le 0, \\ 10, & \text{for } 0 < t \le T/2. \end{cases} \qquad \text{(C.8)}$$

We now solve for a_o:

$$a_o = 2/T \int_{-T/2}^{T/2} f(t)\, dt = 5. \qquad \text{(C.9)}$$

The solution for a_n is

$$a_n = 2/E \int_{-T/2}^{T/2} f(t) \cos(nw_o t)\, dt$$

$$= 2/T \int_{-T/2}^{0} -\cos(nw_o t)\, dt$$

$$+ \int_{0}^{T/2} \cos(nw_o t)\, dt$$

$$= 0 \qquad \text{(C.10–C.11)}$$

as long as n is not equal to zero. Thus, *all the a_n's are zero*, since all the integrals (areas) are zero.

The solution for b_n is

$$b_n = 2/T \int_{-T/2}^{T/2} f(t) \sin(nw_o t)\, dt \qquad \text{(C.12)}$$

$$= 2/T \left[\int_{-T/2}^{0} -\sin(nw_o t)\, dt \right.$$

$$+ \int_{0}^{T/2} \sin(nw_o t)\, dt \qquad \text{(C.13)}$$

$$= 2/n\pi(1 - \cos(n\pi)]. \qquad \text{(C.14)}$$

Thus, $b_n = 0$ when n is an even number and $b_n = 4/n\pi$ when n is an odd number. Thus, only the odd harmonics of the fundamental exist in the spectrum of the square wave.

For equation (C.1) we have the following Fourier series for the square wave:

$$f(t) = \frac{4}{\pi}(\sin(w_o t) + \frac{1}{3}\sin(3w_o t)$$

$$+ \frac{1}{5}\sin(5w_o t) + \ldots). \qquad \text{(C.15)}$$

Now we combine these to obtain C_n and θ_n:

$$C_n = \sqrt{a_n^2 + b_n^2} = \sqrt{b_n^2} = b_n = \frac{4}{n}\pi \qquad \text{(C.16)}$$

for all $n = 1, 3, 5, 7, \ldots$, and $a_n = 0$ for all n, and

$$\theta_n = \tan^{-1}(a_n/b_n) = \tan^{-1}(0) = 90°, \qquad \text{(C.17)}$$

since $a_n = 0$, and the tangent (tan) of 90° is zero.

The amplitudes (C_n) of the odd harmonics ($n = 1, 3, 5, 7, \ldots$) of w_o are: $4/\pi$, $4/3\pi$, $4/5\pi$, $4/7\pi$, etc. All phases are 90°. Only odd harmonics of the fundamental frequency, $w_o = 2\pi/T$, exist. Figure 4.8b,c shows the amplitude and phase spectra of this waveform.

A similar set of equations is used to solve for the spectra of time-domain waveforms that are not periodic. The Fourier series is replaced by the *Fourier integral*, which involves the use of *complex numbers*. A complex number is defined as:

$$a + bi,$$

where $i = \sqrt{-1}$.

Because the square root of –1 does not have a solution, a set of mathematical rules (called *complex math*) exists to help solve equations in which this term (square root of –1) (*i*) must appear. The term *a* is called the *real value* of the complex number, and *bi* is the *imaginary value* of the complex number.

The Fourier integral for time-domain waveforms that are not periodic is defined as follows:

$$f(w) = \int_{-\infty}^{\infty} f(t)\, e^{-wt}\, dt, \qquad \text{(C.18)}$$

and, conversely, the time-domain waveform can be defined as

$$f(t) = \tfrac{1}{2}\pi \int_{-\infty}^{\infty} f(w)\, e^{iwt}\, dw, \qquad \text{(C.19)}$$

where e is the exponential base with a value of 2.718..., and i is $\sqrt{-1}$. Equations involving complex math enable one to compute the amplitude and phase spectra for nonperiodic time-domain waveforms.

The use of computers to solve problems of Fourier analysis often involves a computer algorithm called the *Fast Fourier Transform* (FFT). In general, the FFT involves dividing the domain waveform into n equal intervals. The number of intervals n must be greater than $2WT$, where W is the highest frequency component that is represented and T is the duration of the stimulus. The n instantaneous amplitudes of the time-domain waveform form the inputs to the FFT program. The FFT program provides, in essence, the values of a_o, a_n, and b_n. These values can then be used with equations (C.3) and (C.4) to compute the amplitude and phase spectra. The FFT algorithm provides a fast and efficient method for solving the Fourier series on a computer, but its limitations must be fully understood before it is used.

This appendix is not intended to make you an expert in Fourier analysis. Rather, we hope to have provided an indication of how the spectra are obtained from the time-domain description of a waveform. We also hope that the terms, definitions, and concepts will make it easier for you to understand other descriptions of Fourier analysis.

Psychophysics

*I*n its broadest sense, *psychophysics* is the study of the relationship between the psychological and physical aspects of a stimulus. Historically, the methods of psychophysics have centered around two general approaches. One approach focuses on *discrimination*. The listener is presented with two or more stimuli—for instance, sinusoids of two different intensities—and then is asked whether the stimuli are different. The discrimination procedures are viewed as an "indirect" means of obtaining an answer to the following question: What *can* the listener respond to? The late S. S. Stevens believed that an experimenter could obtain a direct estimate of what the listener *does* respond to and, often, what the listener *can* respond to. Thus, the second general class of psychophysical procedures involves directly asking the listener about the stimulus. These are usually called *scaling procedures*.

In discrimination tasks, the experimenter is interested in obtaining an estimate of the smallest differences in a stimulus parameter (e.g., sound level) to which the auditory system is sensitive. That is, to what difference in sound level *can* the listener respond? However, an experimenter who is not careful will find out *only* to what difference the listener *does* respond when asked to make the discrimination. Another way of stating the problem is that the experimenter is interested in the listener's *sensitivity* to the stimulus change and not in the ability to respond in the experimental situation (*response proclivity or response bias*). The listener might have a particular *bias* toward responding one way, and the experimenter does not want this bias to influence the measure of the auditory system's sensitivity. Discrimination tasks are designed to estimate sensitivity.

The *scaling* procedures are generally designed to obtain information about various subjective or psychological aspects or dimensions. For instance, as the level of a sinusoidal sound is changed, what does a listener report? If one responds that its loudness is changing, then scaling procedures attempt to measure how loudness changes.

CLASSICAL PSYCHOPHYSICS

Many psychophysicists are interested in measuring the smallest value of some stimulus that a listener can detect (*absolute limen* or *absolute threshold*). In addition to investigating the absolute threshold, investigators study the *difference limen* or *difference threshold*, that is, the smallest difference between two values of a stimulus dimension. Two of the *basic* classical methods were the *method of limits* and *method of constant stimuli*, used to estimate absolute and difference thresholds. A variation of the method of limits, the *method of adjustment*, was also used.

Many of the classic methods, like the method of constant stimuli, were designed to obtain a *psychometric function* directly from a listener's responses. The experimenter chooses 5 to 10 values of a stimulus parameter, for instance, seven values of level between 12 and 18 dB SPL, of a 1000-Hz tone, and then presents each value approximately 100 times (in the example, this would mean 700 total trials). The listener is asked to indicate on each trial "Yes, I detected the tone" (Y) or "No, I did not detect the tone" (N). The experimenter then tabulates the proportion of times the listener said "yes" for each stimulus value. These data are then used to plot a psychometric function as in Figure D.1. In general, a psychometric function relates a measure of a listener's performance (such as proportion of "yes" responses) to a value of the stimulus (such as sound level).

The estimate of either the absolute or difference threshold is obtained from a point on a psychometric function, as shown in Figure D.1. Some arbitrary value of the proportion of "yes" responses is chosen. Usually this value is 50% of the total range possible for the proportion of "yes" responses. The psychometric function is then used to obtain that value of the stimulus parameter that would yield the 50% value for the proportion of "yes" responses. In Figure D.1 that value is 14.5 dB SPL, and thus 14.5 dB SPL is labeled the *absolute threshold* or *absolute limen*.

The method of constant stimuli has an advantage over the method of limits in that the experimenter can get an estimate of the listener's bias in the experiment by including "catch" or "blank" trials in the sequence. That is, occasionally the experimenter presents no stimulus to the listener. The listener does not know this has happened, and so the response is still either "yes" or "no." If there is a bias toward either response, then the proportion of "yes" responses on the blank trials will reflect this bias.

All of the classical procedures, however, are still affected by response bias. In most experimental contexts, instruction, experience, and feedback can be used to control the effect of bias. We will now study a psychophysical procedure in which a measure of sensitivity is obtained that does not change as a function of the listener's response criterion.

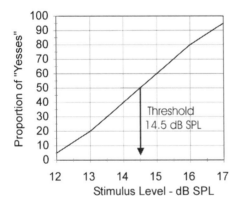

FIGURE D.1 Psychometric function relating proportion of "yes" responses to a stimulus value (intensity in dB SPL in this example). A threshold is obtained by noting the intensity (14.5 dB SPL) that yields 50% "yes" responses.

THEORY OF SIGNAL DETECTION (TSD)

The TSD procedure is similar to the method of constant stimuli. The experimenter might present 100 trials to the listener; 50 trials contain a tone and 50 trials contain nothing (i.e., "catch" trials). On each trial, the listener responds either "Yes, a tone was presented," or "No, a tone was not presented." Of course, only half the trials contained a tone.

There are four possible combinations of stimulus presentations and listener responses: (1) the stimulus is presented and the listener can say "yes" (a *hit*); (2) the stimulus is presented and the listener can say "no" (a *miss*); (3) the stimulus is not presented and the listener can say "yes" (a *false alarm*); and (4) the stimulus is not presented and the listener can say "no" (a *correct rejection*). The proportion of times the listener had a hit, false alarm, miss, and correct rejection is entered in a *response table*, as shown in Table D.1. In the theory of signal detection, the values of the stimulus parameters remain constant for all trials; then another block of trials is presented in which another value of a stimulus parameter (e.g., sound level) is used. In one experiment there might be four or five response tables, each one representing the results from presenting a different level.

The more intense the stimulus, the easier it is for the listener to detect the tone, and so the hit proportion will increase and the false alarm proportion will decrease (correct rejections must increase). That is, the listener will detect the tone more often when it is pre-

TABLE D.1 Four Stimulus Response Tables (Parts A and B show data for a case when a tone was high in intensity and therefore easy to detect. The two tables differ in the response bias of the listener. Tables C and D show data for a case when a tone was low in intensity and therefore hard to detect. The two tables again differ in the response bias of the listener. The ROC curves for these data are shown in Figure D.2. The entries are the number of responses; since based on 100 trials, also percentages.)

A — High intensity "Yes" bias	"Yes"	"No"
Signal	Hit 90	Miss 10
No-signal	False alarm 30	Correct rejection 70
$P(C) = 80\%$		

B — High intensity "No" bias	"Yes"	"No"
Signal	Hit 70	Miss 30
No-signal	False alarm 10	Correct rejection 90
$P(C) = 80\%$		

C — Low intensity "Yes" bias	"Yes"	"No"
Signal	Hit 75	Miss 25
No-signal	False alarm 45	Correct rejection 55
$P(C) = 65\%$		

D — Low intensity "No" Bias	"Yes"	"No"
Signal	Hit 55	Miss 45
No-signal	False alarm 25	Correct rejection 75
$P(C) = 65\%$		

sented and correctly state "no" when the tone is not presented. The listener might be told to change their response bias. That is, in case 1, the listener could be told to say "yes" only when absolutely sure that the tone was heard; in case 2, the listener could be told to say "yes" even if the tone is not clearly heard. In this situation, the tone will not become "easier to hear," but the listener will say "yes" more in case 2 than in case 1. The hit proportion will be larger in case 2 than in case 1 because the listener is simply saying "yes" more often in case 2. The false alarms must also be larger in case 2 than in case 1 because the false alarms also represent the "yes" responses. Therefore, changes in sensitivity, *when the response criterion of the subject does not change*, increase (or decrease) hits and decrease (or increase) false alarms; whereas changes in bias, *when sensitivity does not change*, increase (or decrease) both hits and false alarms. Four examples of these types of stimulus–response tables are shown in Table D.1.

We can illustrate these effects in what is called a *receiver operator characteristic* (ROC) curve, as shown in Figure D.2. Each datapoint in the ROC curve is taken from the corresponding hit and false alarm proportions of Table D.1. If we vary (or a listener varies) the bias of a listener, then the datapoints will fall along curves like the two shown in Figure D.2. Notice that, as you follow one curve from the lower left to the upper right, the false alarms and hits increase. This is what should happen if bias is being altered and sensitivity has not changed. The difference between the two curves represents the listener's sensitivity to the difference in the level of the tone. For any two similar points on the two curves, the hit rate for the upper curve is greater than that for the lower curve. In general, changes in response criterion will move the hits and false alarms along one curve, whereas changes in

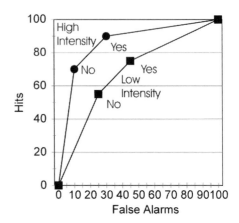

FIGURE D.2 Receiver Operator Characteristic (ROC) curves for the data in Table D.1. The upper curve shows the two bias conditions (A and B) when the tone's intensity was high (table, parts A and B), while the lower curve shows the two bias conditions (C and D) when the tone's intensity was low (table, parts C and D). The area under the upper curve is greater than that under the lower curve.

sensitivity will yield a point on another curve in the ROC space. Therefore, the *ROC curve enables us to separate bias effects from sensitivity effects.* Notice that the area under the upper curve is larger than the area under the lower curve, and this area cannot change as a function of bias. Thus, *the area under any ROC curve is a measure of sensitivity, which is unaffected by response bias.*

Another measure of sensitivity is computed by using this equation:

$P(C)$ = (hit proportion)
   × (proportion of times signal presented)
   + (1 – false alarm proportion)
   × (proportion of times signal not presented).

If the signal is present half the time and no signal the other half, then

$$P(C) = \frac{p(\text{hit}) + (1 - p\ (\text{false alarm})}{2,}$$

that is, the percentage correct ($P(C)$) is similar to the area under the ROC curve, although $P(C)$ is slightly affected by response bias. Other measures are derived from the TSD procedure that depend only on sensitivity and not on response bias.

A psychometric function can then be derived using the area under the ROC curve or $P(C)$. For each value of a stimulus parameter (such as sound level), we obtain a response table; the response table is used to compute an ROC area or $P(C)$. As the stimulus level changes, the area under the ROC curve and $P(C)$ will vary between 50 and 100%, and the psychometric function will be similar to that obtained with the other psychophysical methods. We can then use the midpoint of this function (in this case, 75%) to estimate a threshold in the same way as in the method of constant stimuli (see Figure D.1).

In all three psychophysical procedures, the first and most important function that is obtained is the psychometric function. After the psychometric function is derived, an estimate of the threshold is obtained. The value of the threshold is arbitrary because the choice of a value of the listener's performance, such as $P(C)$, was arbitrary. The thresholds, therefore, are a function of both the stimulus and the listener's performance.

DIRECT SCALING

Magnitude estimation, ratio comparison, and *cross-modality matching* are three scaling techniques in which the listener is asked to judge the magnitude of a subjective attribute such as loudness, timbre, space, volume, and the like. From the techniques, a *scale* relating the perceived magnitude (e.g., loudness) to the physical stimulus value (e.g., level in decibels) is obtained to describe a scale for that subjective attribute.

These procedures are usually reliable and valid. When the data are plotted as the logarithm of the magnitude of the subjective measure versus the logarithm of the stimulus value, the data are almost always fit by a straight line. The fact that a straight line fits data that are plotted on log–log coordinates means that the magnitude estimate is related to the stimulus dimension by a *power function*.

The power function is given by

$$P = kS^n, \tag{D.1}$$

where P is the subjective unit, S the physical unit, k the proportionality constant, and n the power variable. Knowing the power function, we can compute such information as the number of decibels required to double the loudness of a tone (see Chapter 13).

Notice that, if in Equation (D.1) we take the logarithm of both sides of the equation (see Appendix B for the use of logarithms), we have

$$\log P = n \log S \pm \log k. \tag{D.2}$$

In Equation (D.2), $\log k$ is still a constant and $\log P$ is now linearly related to $\log S$. That is, if we plot $\log P$ versus $\log S$ we generate a straight line, as exists over a large portion of that shown in Figure 13.2. The slope of this line is the number n, or the *power value*. The power value is often used in relating the psychological aspects to the physical aspects of any stimulus.

MATCHING PROCEDURES

In the matching paradigm, two stimuli are presented to the listener, who is asked to match their equality along some stimulus attribute, such as loudness. With the matching procedures the two waveforms to be matched usually differ in many physical dimensions, and the subject is asked to adjust one of these physical dimensions to make a subjective match.

The matching procedures allow the experimenter to decide which aspect of a stimulus makes it appear equal, according to some subjective attribute, to another stimulus. Thus, although tones of different frequencies might have equal physical intensities, they may not be judged equal in loudness. In Chapter 13, we show how matching procedures allow us to determine the phon scale and equal loudness contours.

Neural Anatomy and Physiology

ANATOMY

Anatomy is the science that studies the structure of the animal body and the relationship of its parts. In order to discuss the relationship among various parts of the body, it is necessary to develop an appropriate vocabulary. For instance, terms like *upper* and *lower* are relative and, hence, meaningless unless you know in what position the animal is placed. Figure E.1 illustrates how confusing this situation can be when attempts are made to compare the anatomy of an upright two-legged animal with that of a horizontally oriented four-legged animal. The anatomist has developed a vocabulary that differs for animals carried in the horizontal versus the vertical position. For illustrative purposes, consider the human and the dog shown in Figure E.l. The head of the human is the *superior* end, whereas that of the dog is termed the *anterior* end or *rostral* (which literally means "having a beak"). The terms *cephalic* and *cranial* may be applied to the head end of an animal carried in either position. The terms *superior* and *anterior* have additional meanings: *superior* is often used as a relative term meaning a structure occupying a higher position than another structure. For example, if you are standing, your nose is superior to your mouth; but if

you are lying on your back, your nose is superior to your ear. *Inferior* is the opposite of *superior* and, thus, is used to describe a structure occupying a lower position than another structure. *Anterior* refers to the forward part of an organ or body; thus, for horizontally oriented animals it is the head end, but for vertically oriented animals it refers to the belly surface of the body. The more general use of the term *anterior* is to describe a structure that is situated in front of another structure. *Ventral* means "toward the belly" and is, therefore, synonymous with *anterior* for humans. *Posterior* can also refer to the tail end of a horizontally oriented animal. *Dorsal* refers to the back side and is, therefore, the upper side of horizontal animals and the same as the posterior of humans.

Other anatomical terms do not have these dual meanings. In the text, we have made an attempt to use these less ambiguous terms and to define them when they are used for the first time. For quick reference the following definitions should be helpful:

Medial means the middle, or toward a line drawn lengthwise through the middle of the body (midline).

Lateral means the sides of the body (right and left) or away from the midline.

Peripheral means at or near the surface.

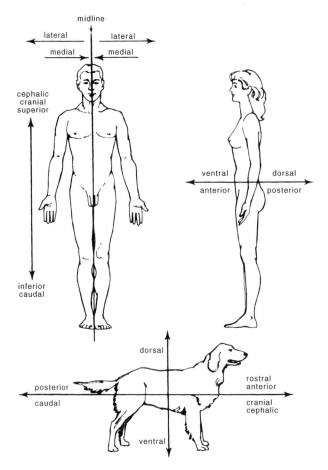

FIGURE E.1 Basic anatomical terminology. Terms as used for vertically and horizontally oriented animals.

Central means situated near the middle or center of the body. (In this text it is often used to mean the part situated toward the center of the head or cortex.)

Proximal means near and is therefore used with a particular structure as a reference.

Distal means away from some reference structure.

External means toward the outer surface.

Internal means toward the inner surface.

The auditory anatomist is mainly concerned with *microanatomy*—that is, the details of the structures as seen through the microscope—rather than the relationship of the structures to the body as a whole. When studying the structures of the ear with a microscope, one often concentrates on the minute structure of the tissue. This is called *histology*. The histologist may study normal or abnormal tissues with any of a number of microscopic techniques.

NEURAL ANATOMY

The anatomy of the auditory system central to the cochlea consists of neurons. The branch of anatomy that investigates the nervous system and nerve tissue is *neuroanatomy*. There are many kinds of neurons: *receptor neurons*, which receive stimulation from sensory receptor cells (such as haircells) and in turn excite other neurons; *interneurons*, which receive stimulation from one neuron and pass it to the next; *motorneurons*, which receive stimulation from a neuron and in turn stimulate a muscle; and various other neurons depending on the method of classification. For our present purposes, it will suffice to consider the receptor neuron in Figure E.2 to illustrate the general anatomy of a neuron. The neuron consists of three main parts: the *cell body* or *soma*, the *dendrites*, and the *axon*. The wall of the cell body is called the *cell membrane*. Inside the cell body is the *nucleus*. The nucleus is surrounded by intracellular fluids called *cytoplasm* and is responsible for the metabolic activities of the cell. The dendrites are specialized for receiving excitation from another cell, receptor cell, or nerve. The axon usually is elongated and is specialized for transmitting the excitation out of the cell. Most neurons have several dendrites to receive stimulation and only one axon to deliver neural impulses. Axons may, however, have several branches called *collaterals*, and they may also have many endings called *telodendria*. Transmission of stimulation from one neuron to another takes place across an interspace between neurons called a *synapse*. When observed by direct electron microscopy, the end of the axon can be shown to contain *vesicles* at a synapse and a thickened cell membrane. These synaptic vesicles, which appear as little round sacs, contain chemicals, *neurotransmitters*, used in transmitting stimulation across the synaptic gap or junction to the next neuron. Axons may be covered with

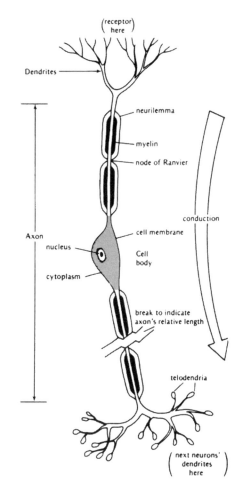

FIGURE E.2 Basic structure of a receptor neuron such as those found in the auditory system.

two coverings or sheaths: the *neurilemma* and the *myelin sheath*. The neurilemma is a very thin sheath found on the outside of the peripheral nerve fibers that aids the axon in regeneration when it is injured. Neurons in the central nervous system generally do not have a neurilemma. The myelin sheath is a relatively thick, fatty substance that surrounds the axon and serves as sort of an insulator. This myelin sheath is interrupted at regular intervals called the *nodes of Ranvier*. These nodes assist the neuron in the transmission of neural impulses called *action potentials*. The myelin sheath does not cover the cell body. Groups of neural cell bodies in the central nervous system are called *ganglia*. A similar grouping of cell bodies in the central nervous system is referred to as a *nucleus*. A bundle of nerve fibers or axons in the periphery is usually referred to as a *nerve*. In the central system such a bundle may be called a *tract*.

PHYSIOLOGY

The study of the *function* of body structure is called *physiology*. This appendix deals primarily with *neurophysiology* or *neurology*, the physiology of nerves and nuclei, and will explain how a nerve discharge begins in a nerve cell and how it is propagated down the nerve's axon to the next nerve.

NEURAL PHYSIOLOGY

The nerve's cell membrane separates the *intracellular* and *extracellular fluids*. The intracellular and extracellular fluids contain water and chemicals whose mixture causes *ions* to be formed. Ions take on positive or negative electric charges; the charges of the ions differ between the intracellular and extracellular fluids of the neuron. An electrical difference in charge between two ions produces an *electric potential*. That is, the potential exists for the flow of electric current. For instance, the two poles of a battery are of opposite charge because each pole consists of different chemicals with opposite ionic charges. Thus, a potential

difference exists between the poles of the battery. Electric current will flow between the poles if a conductor, such as a metal wire, is connected between the poles of the battery.

Because a difference in the charge of the ions exists between the intracellular and extracellular fluids, a potential difference exists across the cell membrane. This potential difference always exists whether or not the nerve is excited, and so this permanent potential difference is called a *resting potential*.

The intracellular fluid consists primarily of *potassium* and has a *negatively charged ionic field*; the extracellular fluid is primarily *sodium* and has a *positively charged ionic field*. The resting potential difference between the two fluids (across the cell membrane) is between 50 and 90 millivolts (mV). See Figure E.3.

The nervous system, thus, possesses the potential for the flow of electric current. The most widely accepted theory of how this flow is initiated concerns a change in the permeability of the cell membrane to potassium and sodium as a function of excitation of the nerve. The chain reaction is outlined as follows.

First, some *adequate stimulus* (such as the shearing of the haircell cilia) changes the permeability of the cell membrane that allows for sodium to flow into the cell. This exchange of chemicals, and hence of ions, changes the potential difference from a large negative value to a less negative value, that is, to a positive value (this process is called *depolarization*). When the potential difference reaches a *critical potential difference*, then the cell membrane immediately in front of the initial point of change also changes in permeability. An ion exchange occurs at this point along the nerve. Meanwhile, the ions move toward the resting state back at the initial region of excitation (the term *hyperpolarization* is used to describe the process whereby the cell returns to a more negative charge). At this initial place of excita-

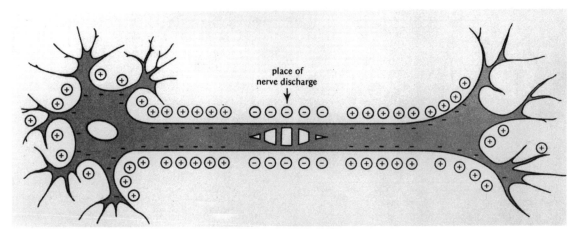

FIGURE E.3 Positive (outside of cell) and negative (inside of cell) ions and their charges as a function nerve stimulation. The ions are transported through the cell wall at the point of the nerve discharge.

tion, the sodium begins to flow (sometimes referred to as a sodium pump) from the cell until the resting potential is reached at this point along the cell. An ion exchange is propagated down the nerve. Meanwhile, the membrane behind the propagated chemical exchange is being restored to its initial state.

This change in the transfer of ions can be measured by noting the electric potential change at *one point* along the nerve (see Figure E.4). The resting potential is approximately −45 mV. Next, a positive increase (*A*) in the potential occurs when the critical voltage is reached. The increase to +40 mV at time *B* is a result of the flow of sodium into the cell. Then as the cell membrane goes back to its resting state and the ions flow the other way, there is a decrease in the potential with some undershoot to −70 mV at time *C*. Finally, at time *D* the potential difference is back at its resting state. The time between *B* and *C* is called the *absolute refractory period* of the nerve impulse. During this time, no new discharge can occur because the flow of ions must reach a resting state before they can move again. The time

between *C* and *D* is the *relative refractory period*, because it is difficult (but possible) for a new discharge to occur during this period. The refractory periods indicate the limit as to how often a nerve impulse can be propagated down the nerve (the limit is about 1000 impulses per second, although this rate is rarely achieved). One important point concerning the nerve discharge should be emphasized. Once a discharge begins, it retains the same amplitude and has the same waveform shape (Figure E.4) independent of the parameters of the stimulus that initially excited the nerve. This is referred to as the *all-or-none law of nerve conduction*. The all-or-none law poses a real challenge to understanding how the nervous system encodes the multitude of events occurring in the complex world.

Once the propagated discharge has reached the end foot of the axon, the discharge must cross the synapse before it can begin the chain of propagation down the next nerve fiber. Small chemical packets called *vesicles* (they rest in the end foot of the axon) aid in transferring the chemical action across the synapse.

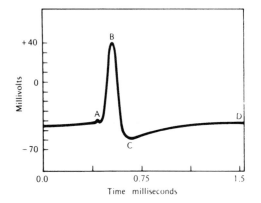

FIGURE E.4 Voltage changes during stimulation of nerve. From the resting potential there is an increase (*depolarization*), A, to a positive voltage, B, then a return to negative voltage (*hyperpolarization*) with a negative undershoot, C, and back to the resting potential, D. This is the voltage change that exists at the point of nerve discharge shown in Figure E.3. The time between A and D is the *total refractory period* of the neural discharge.

The vesicles help release *neurotransmitters*, chemicals that are transmitted across the synapse to the receiving dendrites, into the synapse. When the neurotransmitter reaches the receiving dendrite, it either excites (depolarizes) or inhibits (hyperpolarizes) the cell membrane of the receiving dendrite, depending on whether or not the neurotransmitter has a chemical make-up that causes polarization (excitation) or hyperpolarization (inhibition).

ELECTROPHYSIOLOGY

Electrophysiology is the study of the electrochemical changes in the nervous system. Its main technique involves implanting an electrode into neural tissue and measuring the potential difference between the change in the tissue and some other point. The other point (reference electrode) might be another part of the nervous system or some neutral location. Each time a spike discharge is propagated down the nerve, the electrode will respond to the potential difference. This potential difference is then amplified and fed to a recording instrument.

The auditory electrophysiologist is concerned with the potential differences existing not only in nerves but also in the structures of the inner ear. No matter where these difference exist, the electrode always measures the differences in electric charge between two points. As described earlier, this difference has the potential for electric current flow.

The physiological measures most often used are electrical. Sometimes electrical signals such as the cochlear microphonic (CM) can be recorded at a distance from the ear. However, because of the distance, the CM amplitude is very small relative to the other background electrical activity associated with other bodily functions. Presumably, most of this background activity is uncorrelated with the presentation of a signal to the ear and is thus considered noise. The CM, of course, is perfectly correlated with the occurrence of the acoustic signal because the CM always occurs when the signal occurs. Thus, if the signal is presented many times and the electrical responses of the distant electrode following each stimulus presentation are summed, the uncorrelated background noise should tend to cancel and the CM associated with the signal presentation is enhanced. This technique is called *signal averaging*. A computer is required for sampling the CM waveforms and performing the addition. Many different types of weak electric potentials are recorded in this manner, especially those associated with auditory-evoked changes in the EEG (electroencephalogram) or brain waves recorded from the top of the head. The recording of the brainstem-evoked potential (BSER, see Chapter 15) usually requires thousands of EEG waveforms to be averaged for the investigator to discern a definite pattern in the averaged response.

Techniques and Tools Used to Study Hearing

A wide variety of techniques and tools have been used to acquire the data described in this book. We have already mentioned a few of them: psychophysical techniques in Appendix D; mathematical techniques in Appendices A, B, and C; and signal processing tools in Appendix C. This appendix describes some of these in more detail and briefly introduces some additional techniques and tools.

SIGNAL PROCESSING

Computers are usually used today to generate, measure, and control the sounds that are presented to experimental subjects. The sounds are described in equation form in a computer program, and the resulting numbers are converted to analog voltages by a *digital-to-analog converter* (*D-to-A converter* or DAC). These voltages are lowpass filtered, amplified, and sent to loudspeakers or headphones. *Digital signal processing* software (e.g., FFT computer routines, as explained in Appendix C) is used to generate the signals so that the signal arriving at the loudspeaker or

headphones is the sound the experimenter intended to present. An *analog-to-digital converter* (ADC) is a piece of computer hardware used to convert analog sound signals into digital numbers that can be processed by a computer.

Computers are useful in modeling auditory processes. Models of auditory processing can be actual physical models such as those used by von Bekesy, mathematical formulations of systems such as filters, or simulations of many operations that might occur in an auditory process. Figures 9.15 and 14.1 are examples of the output of a *computational auditory model* based on computer simulations of several stages of auditory processing.

MICROSCOPY

Appendix E describes the terminology of anatomy. The main anatomical tool is the *microscope*, which provides *photomicrographs* (photographs taken via the microscope). We will briefly explain a variety of microscopic techniques. There are two main types of microscopy: light and electron. The most widely used tools of light microscopy in the study of the anatomy of the auditory system are the *standard light transmission* microscope and the *phase contrast light* microscope. With the standard light microscope, a thin slice or section of the specimen of interest is placed in the scope. A strong light is passed through it from one side, and on the opposite side the investigator observes the light that passes through the specimen with a system of magnification lenses. Figure F.1 shows a block diagram of the standard transmission light microscope and two of the other microscopes to be discussed. The plane in which a specimen is sectioned may be referred to by various terms such as *cross-sectional* or *longitudinal*. The exact meaning of such terms is relative to whatever the investigator considers the "normal" orientation of the tissue of interest. The important point is that an investigator has an endless possible number of planes in which to section the specimen.

Another method of light microscopy is phase-contrast microscopy. The phase-contrast microscope takes advantage of the fact that light passes through various tissues with different speeds, thus differentially affecting its phase. The phase-contrast microscope changes these phase differences (which are not visible to the eye) into amplitude differences, which appear as differences in brightness (i.e., light-dark contrast). The phase-contrast microscope is often used in conjunction with the *whole mount* or *surface preparation technique*. With this technique, whole portions of the organ of Corti and basilar membrane of 2 to 5 mm in length can be placed in the microscope for observation, and a sizable number of adjacent haircells can be viewed. Another advantage of this technique is that we can focus at different levels within the specimen. Thus, by adjusting the microscope to be in focus at various levels within the specimen we can obtain a visual "slice" of the specimen without

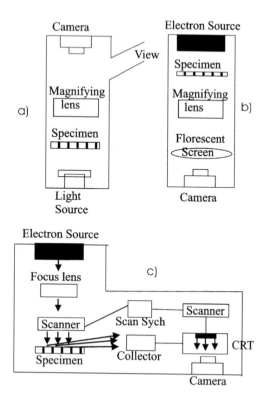

FIGURE F.1 Simple block diagrams of three types of microscopes: (**a**) standard light transmission microscope, (**b**) direction electron microscope, (**c**) scanning electron microscope. The first two types are similar in that they both transmit the signal (either light or electrons) through the specimen. The scanning electron microscope reflects electrons off the surface of the specimen.

the inherent difficulties normally encountered in attempting mechanical slicing of the very delicate organ of Corti.

Since the 1950s, the *transmission* or *direct electron microscope* has gained in popularity because of its remarkable resolving power. The specimen is sliced in sections and placed in the scope. Instead of passing light through the specimen, electrons are transmitted in a beam through the specimen in a manner similar to the way the light beam is diffracted in the light microscope. The electronic beam is magnified by electromagnetic lenses and then directed toward a fluorescent screen, where the image is produced and photographed. By comparing the transmission electron micrographs in Figure 7.7 with the light micrograph of Figure 7.5a, we can obtain an idea of the resolving capabilities of this instrument. Figure 7.7 shows a great deal more detail than Figure 7.5a.

The *scanning electron microscope* (SEM) has a better resolving power than the light microscope, but it does not have the same level of capability as the direct electron microscope. The SEM has the advantage of having a greater depth of field. This means that a larger portion of the viewed specimen will be in focus. The SEM does not use a beam of electrons transmitted through the specimen. Instead, it uses an electron beam, focused on the specimen, which loosens electrons from the surface of the specimen. These loosened electrons and the reflected electron beam are collected and used to form an image on a screen that resembles a television set. This process allows viewing of only the surface of the specimen. The SEM derives its name from the fact that the electron beam scans over the specimen in a systematic fashion. The specimen is usually prepared to emit electrons by prior coating with a very thin layer of gold or other soft metal. The scan of the microscope's electron beam is synchronized with the scan of the electron beam in the cathode-ray tube (similar to a TV picture tube). The collected electrons that are loosened from the specimen, or reflected by it, modulate the electron beam of the picture tube to create the image. Figures such as Figure 7.2 and 7.6 are SEM photomicrographs.

NEURAL STAINS AND MARKERS

The anatomist, along with the physiologist and biochemist, has a large arsenal to probe the structure and function of neural tissue. In general terms, the tissue is exposed to some chemical, pharmacological, or radioactive agent. These agents may stain, mark, or label certain parts of the neural tissue (that is, cell bodies as opposed to axons). The stained parts can be differentiated from other tissue (e.g., the axon from the cell body) under the microscope. An older technique involved destroying some part of a neural location (*lesions*). The nerves then die, and, under the microscope, the dead nerve fibers appear different from live nerves. Both the new chemical and the older lesion techniques allow the anatomist to trace the pathway of some particular part of the nervous system. In other preparations, the chemical agent may label the tissue (or part of the tissue) only when the neuron discharges. That is, the agent interacts with the chemistry of the neuron because of the chemical or metabolic changes that occur when a nerve spike is initiated. This interaction then causes the tissue to be labeled, and the labeled parts can be studied with a microscope or with some other instrument that is sensitive to the chemical or radioactive agent.

One example of such chemical agents is *horseradish peroxidase* (HRP). HRP is a protein that is taken up by a neural cell and goes to the cell body. After HRP is injected into some neural tissue in a live animal, the HRP is taken up by neural cells in the region of the injection. When this neural region is bathed in the proper chemicals and examined under a microscope, the fibers that have taken up the HRP can be seen by the dark stain the process gives to these fibers. Figure F.2 is an example of a fiber that has been marked using the HRP technique. Using HRP enables the neuroscien-

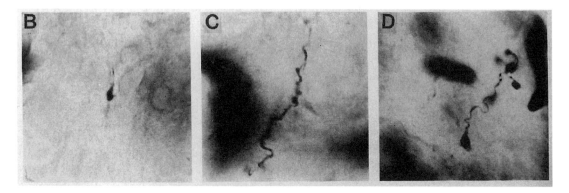

20 μm

FIGURE F.2 Photomicrographs of HRP-injected brain slices showing individual nerve fibers in the cochlear nucleus of the hamster. Reprinted with permission from Schweitzer and Cecil (1992).

tist to trace the path of fibers within a local region of the nervous system.

Immunohistochemical techniques enable one to study how individual molecules are distributed in a section of neural tissue. For instance, an investigator may wish to determine which parts of some neural tissue have a certain type of neurotransmitter. In some cases, the investigator might want to know whether the neurotransmitter molecules in the tissue are ones that excite neurons to produce neural discharges or ones that inhibit neuronal discharging. The *immune system* of the body produces *antibodies* in response to *antigens*, which are *peptides* such as those that reside on the surface of viruses. In normal function, when antibodies, produced by white blood cells, come into contact with the appropriate antigen, the antibody attaches to the antigen and seeks to destroy the virus. However, antibodies can be developed for a wide variety of antigen molecules. When an antibody is made for a particular molecule (antigen), it will attach itself to that molecule in neural tissue. By attaching a marker chemical to the antibody

and bathing neural tissue in the substance, the antibody will attach itself to the antigen molecule (e.g., an excitatory neurotransmitter molecules), and the marker or dye that was mixed with the antibody will then show up on the microscope under ultraviolet light, revealing those neural structures that contained the antigen molecule (e.g., an excitatory neurotransmitter). Figure F.3 shows a photomicrograph from an immunohistochemical procedure.

Some techniques allow the scientist to determine those neural sites that were active for processing some particular stimulus condition. One such technique uses radioactive *2-deoxyglucose* (2-DG). 2-DG is like glucose (a sugar), which cells need to maintain their metabolism. If 2-DG is present in some neural tissue, those neurons that are active (are discharging) will take up the 2-DG to supply the energy necessary to maintain the firing rate. By marking the 2-DG with a radioactive substance, one can see those areas of the tissue that have taken up the 2-DG (i.e., have been neurally active) on an *autoradiograph* of the tissue. In an experiment, radioactive 2-DG is injected into the neural tissue of interest and the animal is presented with the test sound for a fairly long period of time. The animal is killed and the tissue fixed and prepared for investigation. The autoradiograph will reveal which neural structures were involved with processing the information associated with the sound. Figure F.4 shows an autoradiograph from a 2-DG study.

ELECTROPHYSIOLOGY

Appendix E briefly outlines the way in which electrophysiological measurements are made. Most of the data shown in this textbook have been obtained with *extracellular recordings* made with single electrodes. That is, the electrode was near the neuron, but not actually inside it. *Intercellular* recordings, which involve implanting an extremely small elec-

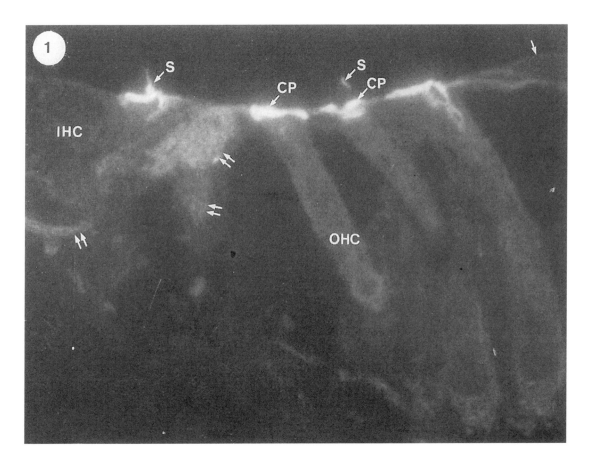

FIGURE F.3 Autoradiograph of the organ of Corti of a Guinea pig injected with an antibody against actin (actin is a protein that is involved with haircell motility, see Chapter 8). The light areas on the top of the haircells (OHC and IHC) show the areas where actin is found near the stereocilia (S) and cuticular plate (CP). The double arrows indicate regions where actin is present near the supporting cells. Reprinted with permission from Slepecky and Ulfendahl (1992).

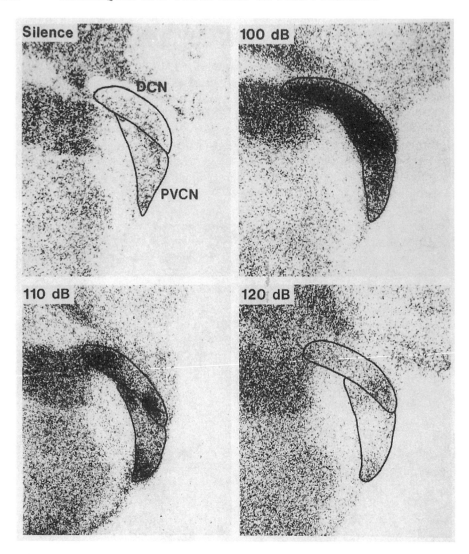

FIGURE F.4 Patterns of 2-DG uptake in the dorsal (DCN) and posterior ventral (PVCN) cochlear nucleus of a gerbil. The four panels indicate the level of uptake (the darker the area, the more 2-DG uptake, and thus the more neurally active the area) to four intensity levels of 2-octave (1414–5356 Hz) noise. The amount of activity is greatest at 100 dB, decreases in some regions of the nucleus at 110 dB, and is much less at 120 dB. Some form of suppression might be occurring that results in lowering of activity at high noise levels. Reprinted with permission from Ryan, Axelsson, and Wolf (1992).

trode inside of a neuron, can be made (see Chapter 8), but they are much more difficult to perform.

Multi-electrode techniques are used for at least two reasons. In some experiments, a number of electrodes are implanted, with some electrodes used for recording neural activity and others used to inject dye or chemicals (e.g., HRP) into the tissue. This makes it possible for the investigator to anatomically investigate the neurons that were recorded from, since the dyes would mark these neurons. In some cases, a multi-electrode can be used to record from and/or stimulate a number of different neurons.

IMAGING

The evoked-potential technique was briefly described in Appendix E as one means of providing an "image" of the neural activity of the brain and/or brainstem. Other imaging techniques—such as *x-ray*, *magnetic resonance imaging* (MRI), *functional magnetic resonance imaging* (fMRI), *positron emission tomography* (PET), or *computerized tomography* (CT)—have also been used to map the neural activity of the brain. X-ray, CT, fMRI, and MRI use either x-rays or magnetic fields to map the brain, with computer reconstruction of the image being an important part of the CT, MRI, and fMRI techniques. In PET, the patient receives an injection of radioactive 2-DG and, as explained above, the radiograph will reveal those areas of the brain that were active in processing information during the time of the test. These imaging techniques can measure large sections of the brain, but the resolution is not as great as that obtained with microscopic of electrophysiological techniques. Since brain imaging requires averaging neural activity over time, these techniques are also not very sensitive to events that will change over a short period of time. Figure 15.11 is a PET image.

BIOPHYSICAL TECHNIQUES

Since such membranes as the tympanic membrane, oval and round window membranes, and basilar membrane all vibrate, techniques have been developed to study their vibrations. However, because the amplitude of vibration is so incredibly small, only recently has modern technology provided the means for direct measurement. Four of these tools are: *real-time holography*, the *Mössbauer technique*, *optical heterodyne spectroscopy*, and *laser interferometry*. Holography, laser interferometry, and spectroscopy take advantage of the laser beam to produce a light source that reflects off the membranes or off microscopic reflecting surfaces placed on the membranes so that the vibrations can be measured. The Mössbauer technique involves placing a radioactive source on the membrane and measuring the change in emitted radioactivity that occurs as a function of the membrane's movement.

Since the late 1970s, new biophysical techniques allow tissue to be excised from an organism and placed in an artificial environment for study. This *in vitro* experiment (as opposed to *in vivo*, or in the organism) allows the scientist to examine the anatomy, electrophysiology, or biochemistry of the tissue under a variety of controlled situations. The chemical and metabolic environments of the tissue can be varied to determine the interaction of these variables with neural action.

The study of isolated haircells (Chapter 8) is done with *in vitro* procedures in which the organ of Corti is removed from an animal, while the haircells are chemically excised from this tissue and placed in a bath that allows the haircells to remain biologically active. One powerful technique (*patch clamping*) involves attaching an electrode directly to the neural cell and controlling the electrical state of the neuron to allow one to carefully measure either the voltage or current flow within this isolated cell. Other manipulations of the haircell are also possible, such as physiologically recording from a haircell while individual stereocilia are moved. Such experiments are highly technical and require great experimenter technique and skill. When the tissue is taken from a developing organism, the *in vitro* study enables the experimenter to investigate the growth of neural tissue as well as aspects of its genetic code.

GENETICS

Every cell in the body contains *genes* that specify traits of the cell and how the cell functions. Thus, all of the parts of the ear are controlled by genes. Not all genes are normal, as some become mutant; these *mutant genes* may cause a cell to function abnormally. Thus, in a condition like Usher's syndrome (see Chapter 16), different genes (probably as many as eight) need to occur in certain combinations involving some mutant genes in order for the syndrome and its accompanying deafness to occur. These combinations of genes occur based on inheriting genes, since each parent controls half of the genes for their child. If the parents carry the particular genes including mutant genes, and if they are combined in the child in the proper manner, the child will have Usher's syndrome, and he or she will carry some of the genes that make it likely (but not certain) that they could have a child with the syndrome.

Since genes control the way in which all cells work, knowing the genetic makeup of a cell can be very useful in understanding its biological function. It is possible that introducing "new" genes into a cell may allow that cell to operate in a different manner. For instance, perhaps mammalian haircells would regenerate if the proper genes could be introduced into the cells. Thus, genetics assists in understanding the basis of many inheritable conditions, genetics can provide basic information about how biological structures function, and altering the genetic makeup of cells may allow them to operate in different, but important, ways.

List of Abbreviations and Symbols

CHAPTER 1

Au.D.	Doctor of Audiology
CV	confounding variable
DV	dependent variable
ENT	ear, nose, and throat
IV	independent variable

CHAPTER 2

a,x	acceleration
A	peak amplitude
A_{rms}	root-mean-square (rms) amplitude
d	distance
$D(t)$	instantaneous amplitude (sometimes ai or $a(t)$)
F	force
f	frequency in Hertz
Hz	Hertz
m	mass
Pr	period (sometimes T)
rms	root mean square
s	stiffness
t	time
v	velocity
w	$\theta/t = 2\pi f$; angular velocity
θ	starting phase angle

CHAPTER 3

Ar	area
Ab	absorption
a	acceleration
c	speed of sound
dB SL	decibels in sensation level
dB SPL	decibels in sound pressure level
E	energy
F	force
f	frequency
fo	fundamental frequency
I	sound intensity
l, L	length
m	mass
P	power
p	pressure
$p(t)$	instantaneous pressure
po	density of a medium
R	resistance
RT	reverberation time
r	radius
s	stiffness
T	duration
t	time
V	velocity
Vol	volume
v	voltage
x	distance

X	reactance
Xm	mass reactance
Xs	stiffness or spring reactance
Z	impedance
Zc	*characteristic impedance*
λ	wavelength
λ_o	wavelength of f_o

CHAPTER 4

$A, A(t)$	Amplitude
AM	amplitude modulation
BW	bandwidth
D	duration
ERB	equivalent rectangular bandwidth
$e(t)$	envelope function
f_x	frequency number x
$f(t)$	fine-structure function
F_c	carrier frequency
$F_c \pm F_m$	side bands
F_m	modulation frequency
FM	frequency modulation
m	modulation depth
N_o	spectrum level (noise power per unit bandwidth)
Pr	Period
SAM	Sinusoidal Amplitude Modulation
TP	total power
$x(t)$	complex sound function
β	modulation depth for frequency modulation, m/F_m

CHAPTER 5

$Atten_{db}$	attenuation in decibels
BW	bandwidth
CF	center frequency
cut	cutoff frequency
ERB	equivalent rectangular bandwidth
f	frequency

$f_i,$	where i = 1, 2, 3, 4 … different frequencies
f_r	resonant frequency
in	input
m	mass
out	output
Q	indicator of relative bandwidth, CF/BW
Q_{10}	Q at 10-dB downpoints
roll	roll off in db/octave
s	stiffness

CHAPTERS 6–9 (ANATOMY)

an	auditory nerve
A	afferent nerve
BC	border cell
BM	basilar membrane
C	Claudius's cell
COCB	crossed olivocochlear bundle
Cn	cochlear nerve
CP	cuticular nerve
D	Deiters cell
E	efferent nerve fibers
Fn	facial nerve
Gc	ganglion cell
H	Hensen's cell
HC	haircell
HP	habenula perforata
HS	Hensen's stripe
IAM	internal auditory meatus
IHC	inner haircell
IP	inner pillar
IS	inner spiral bundle
ISC	inner sulcus cell
m	microvilla
M	modiulus
Mc	mitochondria
MNF	myelinated nerve fibers
n	nerve
N	nucleus
OC	organ of Corti

OCB	olivocochlear bundle
OHC1	outer haircell, row 1
OHC2	outer haircell, row 2
OHC3	outer haircell, row 3
OP	outer pillar
OS	outer spiral fibers
OW	oval windows
PC	pillar cell
PP	phalangeal process
PST	poststimulus time histogram
R	radial fibers
Rm	Reissner's membrane
RW	round window
S	stapes
Sc	sterocilia
SI	spiral ligament
SL	spiral lamina
Sm	scala media
SM	stapedial muscle
SN	space of Nuell
St	scala tympani
Sv	stria vascularis
TB	temporal bone
TC	tunnel of Corti
TL	tympanic layer
Tm	tectorial membrane
TR	tunnel radial fibers
TSB	tunnel spiral bundle
tt	tensor tympani
UCO-CB	uncrossed olivocochlear bundle
VIIIth nerve	eighth cranial nerve

CHAPTER 6–9 (PHYSIOLOGY)

ac	aleternating current
ALSR	average localized synchronized rate
AP	action potential
BF	best frequency
CF	characteristic frequency
CM	cochlear microphonic

dc	direct current
DPOAE	distortion product otoacoustic emission
EP	endolymphatic potential
EOAE	evoked otoacoustic emission
$f1, f2$	frequencies
HRTF	head-related transfer function
kHz	kilohertz
OAE	otoacoustic emission
PST	poststimulus time histogram
SOAE	spontaneous otoacoustic emission
SP	summating potential
TEOAE	transient evoked otoacoustic emission

CHAPTER 10

AM	Amplitude modulation
ANSI	American National Standards Institute
dB HL	decibels in hearing level
E	Energy
f	frequency
F_m	modulation frequency
I	intensity
jnd	just-noticeable difference
m	depth of amplitude modulation, and masker [subscript of P (power) or V (voltage)]
M	masker
MAF	minimal audible field
MAP	minimal audible pressure
$n(t)$	noise waveform
$P(C)$	percent correct
s	signal [subscript of P (power) or V (voltage)]
Δf	frequency difference (usually a jnd)
DI	temporal difference (usually a jnd)
DT	temporal difference (usually a jnd)
P	power
RETSPL	reference equivalent threshold sound pressure level
SAM	Sinusoidal Amplitude Modulation
S	signal

SPL	sound pressure level
T	time (duration)
TMTF	temporal modulation transfer function
DS/S	Weber fraction
α	phase angle of addition

CHAPTER 11

CBW	critical bandwidth
CR	critical ratio
E	energy
ERB	equivalent rectangular bandwidth
E/N_o	signal-to-noise ratio
$f, f1, f2$	frequencies
FM	forward masking
I	intensity
DI	temporal difference (usually a jnd)
M	masking tone
No	noise power per unit bandwidth (spectrum level)
p, g, f, f_o	parameters of the ROEX filter
P	power
Ps	masked signal power
Pncb	power of noise in critical band
r, b, f, f_o	parameters of the gammatone filter
r, f, f_o, b, n	parameters of the gammatone filter
SM	simultaneous masking
SU	suppressor tone
$W(g)$	filter weighting function

CHAPTER 12

BMLD	binaural masking level difference
HRTF	head-related transfer function
MAA	minimal audible angle
MLD	masking level difference
DILD	interaural level difference
DIPD	interaural phase difference

DITD	interaural time difference
S_x, M_x	MLD conditions, $x = 0$, no difference; m = one ear; π:interaural difference

CHAPTER 13

ABLB	alternating binaural loudness balance
dBA	decibels on the A scale
$f1 - f2$	difference tone
$2f1 - f2$	cubic difference tone
IRN	iterated ripple noise
RIS	regular interval stimuli

CHAPTER 14

CMR	comodulation masking release
$F1, F2, F3$	speech formants
MDI	modulation detection interference
TMTF	temporal modulation transfer function

CHAPTER 15

an	auditory nerve
AI, AII, SII	auditory cortex
ABLB	alternating binaural loudness balance
Ach	acetylcholine
AER	auditory evoked response
AVCN	anteroventral cochlear nucleus
BSER	brain stem electric (or evoked) response
CANS	central auditory nervous system
CF	center frequency
DCN	dorsal cochlear nucleus
E	excitation, as in $E–I$ and $E–E$ cells
EEG	Electroencephalogram

F	Fusiform cells
fMRI	functions magnetic resonance imaging
G	Globular cells
GB	Globular bushy cells
GABA	gamma-aminobutyric acid
HRTF	head-related transfer function
I	inhibition, as in *E–I* cells
LSO	lateral superior olive
M	Multipolar cells
MSO	medial superior olive
O	Octopus cells
PET	positron emission tomography
PST	poststimulus histogram
PVCN	posteroventral cochlear nucleus
SB	spherical bushy cells
St	stellite cells

CF	center or characteristic frequency
CTS	compound threshold shift
DRC	damage risk criteria
EPA	Environmental Protection Agency
NIDCD	National Institute on Deafness and Other Communication Disorders
NIH	National Institutes of Health
NIHL	noise-induced hearing loss
NIPTS	noise-induced permanent threshold shift
NITTS	noise-induced temporary threshold shift
OSHA	Occupational Safety and Health Agency
PTS	permanent threshold
SEM	scanning electron micrograph
TTS	temporary threshold shift
TTS_n	temporary threshold shift *n* minutes after cessation of exposure stimulus

CHAPTER 16

ATS	aysmptotic threshold shift

Any symbols used uniquely in the Appendices are defined there.

References and Author Index

Secondary Sources

As mentioned in the Preface, this book provides a restrictive references list. The following secondary sources should be consulted for a more complete list of references, especially the textbooks by Moller (2000), Moore (1989), Pickles (1988), and Rosen and Howell (1991):

Altschuler, R., Hoffman, D., Bobbin, R., and Clopton, B., eds. (1986). *Neurobiology of Hearing: The Cochlea*. Raven, New York.

Altschuler, R., Hoffman, D., Bobbin, R., and Clopton, B., eds. (1989). *Neurobiology of Hearing: The Central Nervous System*. Raven, New York.

Blauert, J. (1997). *Spatial Hearing*. MIT Press, Cambridge.

Dallos, P., Popper, A. N., and Fay, R. R., eds. (1996). *The Cochlea*. Springer-Verlag, New York.

Edelman, G. W., Gall, W. E., and Cowan, M., eds. (1988). *Auditory Function: Neurobiological Bases of Hearing*. Wiley, New York.

Fay, R. R., and Popper, A. N., eds. (1992). *The Auditory Pathway: Neurophysiology*. Springer-Verlag: New York.

Fay, R. R., and Popper, A. N., eds. (1994). *Comparative Mammalian Hearing*. Springer-Verlag, New York.

Gilkey, R. H., and Anderson, T. A. (1997). *Binaural and Spatial Hearing in Real and Virtual Environments*. Lawrence Erlbaum, Hillsdale, NJ.

Green, D. M. (1976). *An Introduction to Hearing*. Erlbaum, Hillsdale, NJ.

Gulick, W. L., Gescheider, G. A., and Frisina, R. D. (1989). *Hearing: Physiological Acoustics, Neural Coding, and Psychoacoustics*. Oxford University Press, New York.

Hartmann, W. M. (1998). *Signal, Sounds and Sensation*. Springer-Verlag, New York

Moller, A. R. (2000). *Hearing: Its Physiology and Pathophysiology*, Academic Press, New York.

Moore, B. C. J. (1986). *Frequency Selectivity in Hearing*. Academic Press, London.

Moore, B. C. J. (1995). *Perceptual Consequences of Cochlear Damage*. Oxford University Press, Oxford.

Moore, B. C. J. (1997). *An Introduction to the Psychology of Hearing*, 3rd ed. Academic Press, London.

Pickles, J. O. (1988). *An Introduction to the Physiology of Hearing*, 2nd ed. Academic Press, London.

Rosen, S., and Howell, P. (1991). *Signals and Systems for Speech and Hearing*. Academic Press, London.

Rossing, T. (1990). *The Science of Sound*, 2nd edition. Addison-Wesley, Reading, MA.

Van de Water, T., Popper, A. N., and Fay, R. R., eds. (1996). *Clinical Aspects of Hearing*. Springer-Verlag, New York.

Webster, D., Fay, R. R., and Popper, A. N., eds. (1992). *The Auditory Pathway: Neuroanatomy*. Springer-Verlag, New York.

Yost, W. A., and Gourevitch, G., eds. (1987). *Directional Hearing*. Springer-Verlag, New York.

Yost, W. A., and Watson, C. S., eds. (1987). *Auditory Processing of Complex Sounds*. Erlbaum, Hillsdale, NJ.

Yost, W. A., Popper, A. N., and Fay, R. R., eds. (1993). *Human Psychoacoustics*. Springer-Verlag: New York.

Zwicker, E., and Fastl, H. (1991). *Psychoacoustics: Facts and Models*. Springer-Verlag, Berlin.

References

The figure numbers is in italics at the end of a reference for the source of a figure.

Abel, S. M., 158

Abel, S. M. (1971). Duration discrimination of noise and tone bursts. *J. Acoust. Soc. Amer.* **51**, 1219–1224. (*Fig. 10.8*)

Ades, H. W., 66, 228

Ades, H. W. (1959). Central auditory mechanisms. In *Handbook of Physiology*, Vol. 1: *Neurophysiology* (J. Field, H. W. Magoun, and V. E. Hall, eds). American Physiological Society, Washington, DC. (*Fig. 15.1*)

Ades, H. W., and Engstrom, H. (1974). Anatomy of the inner ear. In *Handbook of Sensory Physiology*, Vol. 5 (W. D. Keidel and W. D. Neff, eds.). Springer-Verlag, New York. (*Fig. 6.1*)

Adrian, E. D., 124

Adrian, E. D. (1931). The microphonic action of the cochlea in relation to theories of hearing. Report of a discussion on audition. Physical Society, London, 5–9.

Altmann, D. W.; *see* Mulroy; *see* Weiss

Altschuler, R. D., 102, 250

Altschuler, R., Hoffman, D.,. Bobbin, R., and Clopton, B., eds. (1986). *Neurobiology of Hearing: The Peripheral Nervous System*. Raven, New York.

Altschuler, R., Hoffman, D., Bobbin, R., and Clopton, B., eds. (1989). *Neurobiology of Hearing: The Central Nervous System*. Raven, New York.

Anderson, D. J.; *see* Hind; *see* Rose

Anderson, T. A.; *see* Gilkey

Angelborg, C., 103

Angelborg, C., and Engstrom, H. (1973). The normal organ of Corti. In *Basic Mechanisms in Hearing* (A. Moller, A., ed.). Academic Press, London.

Asanuma, A.; *see* Tanaka

Assmann, P. F.; *see* Darwin

Au, W. L., 192

Au, W. L. (1993). *The Sonar of Dolphins*. Springer-Verlag, New York.

Axelsson G. A.; *see* Ryan

Bacon, S. P., 159

Bacon, S. P., and Viemeister, N. F. (1985). Simulataneous masking by gated and continuous sinusoidal maskers. *J. Acous. Soc. Amer.* **78**, 1220–1230. (*Fig. 10.10*)

Bader, C. R.; *see* Brownell

Beranek, L. L., 36

Beranek, L. L. (1988). *Acoustics*. Acoustical Society of America, Melville, New York. (Originally published 1954.)

Bertrand, D.; *see* Brownell

Biddulph, R.; *see* Shower

Bilger, R. C., 125; *see* Jesteadt; *see* Reed

Bilger, R. C., Matthies, M. L., Hammel, D. R., and Demorest, M. E. (1990). Genetic implications of gender differences in the prevelance of spontaneous otoacoustic emissions. *J. Speech & Hear. Res.* **33**, 418–433.

Blauert, J., 191

Blauert, J. (1997). *Spatial Hearing*. MIT Press, Cambridge.

Bobbin, R.; *see* Altschuler

Borden, G. J., 225

Borden, G. J., and Harris, K. S. (1980). *Speech Science Primer*. Williams & Wilkins, Baltimore.

Borg, E.; *see* Canlon

Boring, E. G., 7

Boring, E. G. (1942). *Sensation and Perception in the History of Experimental Psychology*. Appleton-Century-Croft, New York.

Bray, C.; *see* Wever

Bregman, A. S., 7, 224

Bregman, A. S. (1990). *Auditory Scene Analysis: The Perceptual Organization of Sound*. MIT Press, Cambridge.

Brown, M. C.; *see* Liberman

Brownell, W. E., 125

Brownell, W. E., Bader, C. R., Bertrand, D., and de Ribaupierre, Y. (1985). Evoked mechanical responses of isolated cochlear outer hair cells. *Science* **227**, 194–196.

Brugge, J. F., 248, 251, 267; *see* D. J. Anderson; *see* Hind; *see* Rose

Brugge, J. F. (1988). Stimulus coding in the developing auditory system. In *Auditory Function: Neurobiological Bases of Hearing* (G. W. Edelman, W. E. Gall, and M. Cowan eds.). Wiley, New York.

Brugge, J. F., and Merzenich, M. M. (1973). Patterns of activity of single neurons of the auditory cortex in monkey. In *Basic Mechanisms in Hearing* (A. G. Moller, ed.). Academic Press, New York. (*Fig. 15.18*)

Brugge, J. F., Kitzes, L. M., and Javel, E. (1981). Postnatal development of frequency and intensity sensitivity of neurons in the anteroventral cochlear nucleus of kittens. *Hear. Res.* **5**, 217–229.

Buchel, C.; *see* Griffiths

Burns, W.; *see* Taylor

Cajal, 4

Canlon, B., 267

Canlon, B., Borg, E., and Flock. A. (1988). Protection against noise trauma by pre-exposure to a low level acoustic stimulus, *Hear. Res.* **34**, 197–200.

Cant, N. B., 232

Cant, N. B. (1992). The cochlear nucleus: Neuronal types and their synaptic organization. In *The Auditory Pathway: Neuroanatomy* (D. Webster, R. R. Fay, and A. N. Popper, eds.). Springer-Verlag, New York. (*Fig. 15.4*)

Carder, H. M., 267

Carder, H. M., and Miller, J. D. (1972). Temporary threshold shifts from prolonged exposure to noise, *J. Speech Hear. Res.* **15**, 603–623.

Carr, C. E.; *see* Konishi

Cecil, T.; *see* Schweitzer

Chandler, D. W., 191

Chandler, D. W., and Grantham, D. W. (1992). Minimum audible movement angle in the horizontal plane as a function of stimulus frequency and bandwidth, source azimuth, and velocity. *J. Acoust. Soc. Amer.* **91**, 1624–1636.

Cherry, C., 192

Cherry, C. (1953). Some experiments on the recognition of speech with one and with two ears. *J. Acoust. Soc. Amer.* **25**, 975–981.

Clark, L. F.; *see* Kiang

Clark, W. W., 266, 267

Clark, W. W. (1991a). Noise exposure from leisure activities: A review. *J. Acoust. Soc. Amer.* **90**, 175–182.

Clark, W. W. (1991b). Recent studies of temporary threshold shift (TTS) and permanent threshold shift (PTS) in animals. *J. Acoust. Soc. Amer.* **90**, 155–164.

Clopton, B.; *see* Altschuler

Colburn, S.; *see* Litovsky

Comis S. D.; *see* Pickles; *see* Osborne

Corti, 4

Carlyon, R. P., 224

Carlyon, R. P. (1991). Discriminating between coherent and incoherent frequency modulation of complex tones. *Hear. Res.* **41**, 223–236.

Cosell, W.; *see* Miller

Crawford, A. C.; *see* Fettiplace

Dallos, P., 66, 102, 124, 144

Dallos, P. (1973). *The Auditory Periphery: Biophysics and Physiology.* Academic Press, New York. (*Fig. 6.1*)

Dallos, P., Santos-Sacchi, J., and Flock, A. (1982). Intercellular recordings from cochlear outer hair cells. *Science* **218**, 582–584.

Dallos, P., Popper, A. N., and Fay, R. R., eds. (1996). *The Cochlea.* Springer-Verlag, New York.

Darwin C. J., 225

Darwin, C. J. (1981). Perceptual grouping of speech components differing in fundamental frequency and onset time. *Quart. J. Exp. Psychol.* **33A**, 185–207.

Darwin, C. J., and Ciocca V. (1992). Grouping in pitch perception: Effects of onset asynchrony and ear of presentation of a mistuned component. *J. Acoust. Soc. Amer.* **91**, 3381–3391.

Dau, T., 163

Dau, T. Kollmeier, B., and Kohlraush, A. (1997). Modeling auditory processing of amplitude modulation, II: Spectral and temporal integration in modulation detection. *J. Acoust. Soc. Amer.* **102**, 2906–2919.

Davis, H., 106, 109; *see* Pestalozza; *see* Saul; *see* Stevens; *see* Tasaki

Davis, H. (1956). Initiation of nerve impulses in the cochlea and other mechanoreceptors. Physiological triggers and discontinuous rate proceses., Bullock, T. D. (ed.). *Am. Physiol. Soc.* **11**, 60–71. (*Fig. 8.1*)

Davis, H. (1960). Mechanism of excitation of auditory nerve impulses. In *Neural Mechanisms of the Auditory and Vestibular Systems* (T. Rasmussen and W. F. Windle, eds.). Thomas, Springfield, IL. (*Fig. 8.5*)

Deiter, 4

Demorest, M. E.; *see* Bilger

Dent, M. L.; *see* Ryals

de Ribaupeirre, Y. ; *see* Brownell

Diamond, I. T., 228

Diamond, I. T. (1973). Neuronatamony of the auditory system: Report on a workshop. *Arch. Otolaryng.* **98**, 397–413. (*Fig. 15.1*)

Dolan, D. F.; *see* Nuttal

Dooling, R. J.; *see* Ryals

Dorland, A., 82

Dorland, A. (1965). *Dorland's Illustrated Medical Dictionary.* Saunders, London. (*Fig. 7.1*)

Dosanjh, D. S.; *see* Henderson

Durlach, N. I., 192

Durlach, N. I. (1972). Binaural signal detection: Equalization and cancellation theory. *Foundations of Modern Auditory Theory*, Vol. 2. (J. V. Tobias, ed.). Academic Press, New York.

Dye, R.; *see* Yost

Egan, J.; *see* Postman

Eldridge D.; *see* Tasaki

Elliott, D. N., 247

Elliott, D. N., and Trahiotis, C. (1972). Cortical lesions and auditory discrimination. *Psychol. Bull.* **77**, 198–222. (*Fig. 15.17*)

Engstrom, H.; *see* Ades; *see* Angelborg

Fastl, H.; *see* Zwicker

Fay, R. R., 102, 103, 124, 144, 192, 250; *see* Popper; *see* Webster; *see* Yost; *see* Rubel

Fay, R. R., and Popper, A. N., eds. (1992). *The Auditory Pathway: Neurophysiology.* Springer-Verlag, New York.

Fay, R. R., and Popper, A. N., eds. (1998). *Hearing in Fishes and Amphibians.* Springer-Verlag, New York.

Fechner, 4, 155

Fernandes, M. A.; *see* Hall

Fernandez, C.; *see* Davis; *see* Naughton; *see* Tasaki

Fettiplace, R., 145

Fettiplace, R., and Crawford, A. C. (1980). The origin of tuning in the Turtle cochlear hair cells. *Hear. Res.* **2**, 447–454.

Fletcher, H., 85, 170, 178, 202

Fletcher, H. (1924). The physical criterion for determining the pitch of a musical tone. *Phys. Rev.* **23**, 427–437.

Fletcher, H. (1953). *Speech and Hearing in Communication.* Van Nostrans, New York. (*Fig. 7.4*)

Fletcher, H., and Munson, W. A. (1933). Loudness: Its definition, measurement, and calculation. *J. Acoust. Soc. Amer.* **5**, 82–108.

Flock, A.; *see* Dallos, *see* Canlon

Fourier, J., 11

Frachowiak, R. S. J.; *see* Griffiths

Franks, J. R.; *see* Watson

Frisina, R. D.; *see* Gulick

Galambos, R.; *see* Rose

Gardner, M. B.; *see* Steinberg

Gengel, R. W.; *see* Watson

Geisler, C. D., 78, 124, 126, 144

Geisler, C. D. (1998). *From Sound to Synapse: Physiology of the Mammalian Ear.* Oxford Press, New York.

Gescheider G.; *see* Gulick

Giguere, C.; *see* Patterson

Gilkey, R. H., 191

Gilkey, R. H., and Anderson, T. A. (1997). *Binaural and Spatial Hearing in Real and Virtual Environments.* Erlbaum, Hillsdale, NJ.

Glasberg, B. R., 177; *see* Moore

Glasberg, B. R., and Moore, B. C. J. (1990). Deviation of auditory filter shapes from notched-noise data. *Hear. Res.* **47**, 133–138.

Glattke, T. H., 251

Glattke, T. H. (1983). *Short-Latency Auditory Evoked Potentials.* University Park Press, Baltimore.

Glorig, A.; *see* Ward

Gourevitch, G.; *see* Yost

Grantham, D. W., 162; *see* Chandler

Grantham, D. W., and Yost, W. A. (1982). Measures of intensity discrimination. *J. Acoust. Soc. Amer.* **72**, 406–411.

Green, D. M., 36, 160, 162, 164, 192, 211, 224; *see* Jesteadt; *see* Neff; *see* Weir

Green, D. M. (1971). Temporal auditory acuity. *Psychol. Rev.* **78**, 542–551.

Green, D. M. (1976). *An Introduction to Hearing.* Erlbaum Associates, New York.

Green, D. M. (1989). *Profile Analysis.* Oxford University Press, New York. (*Fig. 14.4*)

Green, D. M. (1993). Intensity processing. In *Human Psychoacoustics* (W. A. Yost, A. N. Popper, and R. R. Fay, eds.). Springer-Verlag, New York.

Green, D. M., and Yost, W. A. (1975). Binaural Analysis. In *Handbook of Sensory Physiology: Hearing* (W. D. Keidel and W. Neff, eds.). Springer-Verlag, New York.

Greenwood, D., 146

Greenwood, D. (1990). A cochlear frequency-position function for several species—29 years later. *J. Acoust. Soc. Amer.* **87**, 2592–2605.

Griffiths, T. D., 241

Griffiths, T. D., Buchel, C., Frachowiak, R. S. J., and Patterson, R. P. (1998). Analysis of temporal structure in sound by the human brain. *Nat. Neurosci.* **1**, 422–427. (*Fig. 15.11*)

Guinan, J. J., 78; *see* Warr

Guinan, J., and Peake, W. T. (1967). Middle ear characteristics of anesthetized cats. *J. Acoust. Soc. Amer.* **41**, 1237–1261.

Gulick, W. L., 7

Gulick, W. L., Gescheider, G. A., and Frisina, R. D. (1989). *Hearing: Physiological Acoustics, Neural Coding, and Psychoacoustics*. Oxford University Press, New York.

Guzman, S.; *see* Litovsky

Haggard, M.; *see* Hall

Hall, J. L., 201

Hall, J. L. (1975). Nonmonotic behavior of distortion product $2f_1$–f_2: Psychophysical observations. *J. Acoust. Soc. Amer.* **58**, 1046–1050. (*Fig. 13.6*)

Hall III, J. W., 218

Hall III, J. W., Haggard, M., and Fernandes, M. A. (1984). Detection in noise by spectro-temporal pattern analysis. *J. Acoust. Soc. Amer.* **76**, 50. (*Fig. 14.7*)

Hamernik, R., 266; *see* Henderson

Hamernik, R., Henderson, D., and Salvi, R., eds. (1982). *New Perspectives on Noise-Induced Hearing Loss*. Raven, New York.

Hammel, D. R.; *see* Bilger

Handel, S. (1989). *Listening: An Introduction to Perception of Auditory Events*. MIT Press, Cambridge.

Harris, K. S., 267; *see* Borden

Harrison, J. M., 228, 231

Harrison, J. M., and Howe, M. E. (1974a). Anatomy of the afferent auditory nervous system of mammals. In *Handbook of Sensory Physiology*, Vol. 5 (W. D. Keidel and W. D. Neff, eds.). Springer-Verlag, New York. (*Fig. 15.1*)

Harrison, J. M., and Howe, M. E. (1974b). Anatomy of the descending auditory system (mammalian). In *Handbook of Sensory Physiology*, Vol. 5 (W. D. Keidel and W. D. Neff, eds.). Springer-Verlag New York. (*Figs. 15.1 and 15.3*)

Hartmann, W. M., 19, 51, 61, 202, 225

Hartmann, W. M. (1988). Temporal fluctuations and the discrimination of spectrally dense signals by human listeners. In *Auditory Processing of Complex Sounds* (W. A. Yost and C. S. Watson, eds.). Erlbaum, Hillsdale, NJ.

Hartmann, W. M. (1998). *Signal, Sounds and Sensation*. Springer-Verlag, New York.

Hartmann, W. M., McAdams, S., and Smith B. K. (1990). Hearing a mistuned harmonic in an otherwise periodic complex tone. *J. Acoust. Soc. Amer.* **88**, 1712–124.

Hawkins, J. E.; *see* Moody

Hays, D.; *see* Jerger

Helmholtz, H., 4, 78; *see* Warren and Warren

Henderson, D. 266; *see* Hamernik

Henderson, D., Hamernik, R., Dosanjh, D. S., and Mills, J. H., eds. (1976). *Effects of Noise on Hearing*. Raven, New York.

Henning, G. B., 192

Henning, G. B. (1977). Detectability of interaural delay in high-frequency complex waveforms. *J. Acoust. Soc. Amer.* **55**, 84–90.

Hind, J. E., 136; *see* Anderson; *see* Rose

Hind, J., Anderson, D., Brugge, J., and Rose, J. (1967). Coding of information of pertaining to paired low-frequency tones in single auditory nerve fibers of the squirrel monkey. *J. Neurophysiol.* **30**, 794–816. (*Fig. 9.9*)

Hoffman, R. D.; *see* Altschuler

Hofman, P. M., 267

Hofman, P. M., van Riswick, J. G. A., and Van Opstal, A. J. (1998). Relearning sound localization with new ears. *Nat. Neurosci.* **1**, 417–422.

Hood, D. S.; *see* Watson

Howe, M. E.; *see* Harrison

Howell, P.; *see* Rosen

Hudspeth, A. J., 125

Hudspeth, A. J. (1983). The hair cells of the inner ear. *Sci. Amer.* **183**, 54–65.

Hughes, J. R.; *see* Rose

Javel, E.; *see* Brugge

Jerger, J., 78

Jerger, J., and Hays, D. (1980). Diagnostic applications of impedance audiometry: middle ear disorders; sensorineural disorders. In *Clinical Impedance Audiometry*, 2nd ed. (J. Jerger and J. Northern, eds.). American Electromedics, Acton, Massachusetts.

Jerger, J., and Northern, J., eds. (1980). *Clinical Impedance Audiometry*, 2nd ed. American Electromedics Corporation. Acton, MA.

Jesteadt, W., 157, 162, 203; *see* Weir

Jesteadt, W. (1980). An adaptive procedure for subjective judgments. *Percep. Psychophysiol.* **28**, 85–88.

Jesteadt, W., and Bilger, R. C. (1974). Intensity and frequency discrimination: One-tone and two-tone interval paradigms. *J. Acoust. Soc. Amer.* **55**, 1266–1279.

Jesteadt, W., Weir, C. C., and Green, D. M. (1977). Intensity discrimination as a function of frequency and sensation level. *J. Acoust. Soc. Amer.* **61**, 169–177. (*Fig. 10.7*)

Jewett, D. L., 240

Jewett, D. L., and Williston, J. S. (1971). Auditory evoke potentials for fields averaged from the scalp of humans. *Brain* **94**, 681–696. (*Fig. 15.9*)

Johnsson, L. G.; *see* Hawkins

Keith, A.; *see* Wrightson

Kemp, D. T., 125

Kemp, D. T. (1978). Stimulated acoustic emissions from within the human auditory system. *J. Acoust. Soc. Amer.* **64**, 1386–1391.

Khanna, S. M.; *see* Tonndorf

Kiang, N. Y.-S., 133, 137, 236; *see* Peake, *see* Liberman

Kiang, N. Y.-S. (1965). Stimulus coding in the auditory nerve and cochlear nucleus. *Acta Otolaryngol.* (Stockholm) **59**, 186–200. (*Fig. 15.6*)

Kiang, N. Y.-S., Watanabe, T., Thomas, E. C., and Clark, L. F. (1965). *Discharge Patterns of Single Fibers in the Cat's Auditory Nerve*. MIT Press, Cambridge. (*Fig. 9.7 and Fig. 9.10*)

Killion, M. C., 161; *see* Yost

Killion, M. C. (1978). Revised estimate of minimum audible pressure: Where is the "Missing 6 dB." *J. Acoust. Soc. Amer.* **63**, 1501–1505.

Kirikae, 78

Kistler, D. J.; *see* Wightman

Kitzes, L. M.; *see* Brugge

Klatt, D., 225

Klatt, D. (1982). *Review of the Science and Technology of Speech Synthesis*. Committee on Hearing and Bioacoustics Report. National Research Council, Washington, DC (*Fig. 14.10*)

Klein, A. J., 240

Klein, A. J., and Teas, D. C. (1978). Acoustically dependent latency shifts of BSER (wave V) in man. *J. Acoust. Soc. Amer.* **63**, 1887–1892. (*Fig. 15.10*)

Kohlrausch, A.; *see* Dau

Kollmeier, B.; *see* Dau

Konishi, M., 250

Konishi, M., Takahashi, T. T., Wagner, H., Sullivan, W. E., and Carr, C. E. (1988). Neurophysiological and anatomical substrates of sound localization in the owl. In *Auditory Function: Neurobiological Bases of Hearing* (G. W. Edelman, W. E. Gall, and W. M. Cowan, eds.). Wiley, New York.

Konishi, T., 125

Konishi, T., and Nielsen, D. W. (1978). The temporal relationship between basilar membrane motion and nerve impulse initiation in auditory nerve fibers of guinea pigs. *Jpn. J. Physiol.* **28**, 291–307.

Kramer, M.; *see* Wightman

Kuhn, G. F., 181

Kuhn, G. F. (1987). Physical acoustics and measurements pertaining to directional hearing. In *Directional Hearing* (W. A. Yost and G. Gourevitch, eds.). Springer-Verlag, New York. (*Fig. 12.3*)

Laming, D., 160

Laming, D. (1988). *Sensory Analysis.* Academic Press, London.

Lane, C. E.; *see* Wegel

Langner, G., 250

Langner, G. (1992). Periodicity coding in the auditory system. *Hear. Res.* **60**, 115–143.

Larsen, O. N.; *see* Ryals

Lawrence, M.; *see* Wever

Liberman, M. C., 126, 129, 131, 145; *see* Warren

Liberman, M. C. (1978). Auditory-nerve response from cats raised in a low-noise chamber. *J. Acoust. Soc. Amer.* **63**, 442–455. (*Fig. 9.1*)

Liberman, M. C. (1980). Morphological differences among radial afferent fibers in the cat cochlea: An electron-microscope study of serial sections. *Hear. Res.* **3**, 45–63.

Liberman, M. C. (1982). Single-neuron labeling in the cat auditory nerve. *Science* **216**, 1239–1241.

Liberman, M. C. (1991). Effects of chronic de-efferentation on auditory-nerve response, *Hear. Res.* **49**, 209–223.

Liberman, M. C., and Brown, M. C. (1986). Physiology and anatomy of single olivocochlear neurons in the cat. *Hear. Res.* **24**, 17–36.

Liberman, M. C., and Kiang, N. Y.-S. (1978). Acoustic trauma in cats. *Acta Otolaryngol.* **358**, 1–63. (*Fig. 9.5*)

Licklider, J. C. R., 202, 203

Licklider, J. C. R. (1954). Periodicity by "pitch" and place "pitch." *J. Acoust. Soc. Amer.* **26**, 945–951.

Lindsay, H., 230

Lindsay, H., and Norman, D. A. (1972). *Human Information Processing: An Introduction to Psychology.* Academic Press, New York. (*Fig. 15.2*)

Lippe, W., 251

Lippe, W., and Rubel, E. (1983). Development of the place principle: Tonotopic organization. *Science* **219**, 514–516.

Litovsky, R., 191

Litovsky, R., Colburn, S., Yost, W. A., and Guzman, S. (1999). The precedence effect. *J. Acoust. Soc. Amer.* **106**, 1633–1654.

Lonsbury-Martin, B. L., 125

Lonsbury-Martin, B. L., and Martin, G. K. (1990). The clinical utility of distortion-product otoacoustic emissions. *Ear Hear.* **11**, 144–154.

Lutfi, R. A., 177

Lutfi, R. A. (1988). Complex interactions between pairs of forward maskers. *Hear. Res.* **35**, 71–78.

Mair, A.; *see* Taylor

MacKenzie, A.; *see* Ryals

Martin, G. K., 115; *see* Lonsbury-Martin

Martin, G. K.., Probst, R., and Martin-Lonsbury, B. L. (1990). Otoacoustic emissions in human ears: Normative Findings. *Ear Hear.* **11**, 106–120. (*Fig. 8.8*)

Matthies, M. L.; *see* Bilger

McAdams, S.; *see* Hartmann

McFadden, D., 192

McFadden, D., and Passenan, E. G. (1975). Binaural beats at high frequencies. *Science* **190**, 394–397.

McGee, T.; *see* Wightman

Meddis, R., 146

Meddis, R. (1986). Simulation of mechanical to neural transduction in the auditory receptor. *J. Acoust. Soc. Amer.* **79**, 702–711.

Melnick, W., 258, 266, 267

Melnick, W. (1991). Human temporary threshold shifts (TTS) and damage risk, *J. Acoust. Soc. Amer.* **90**, 147–155. (*Fig. 16.5*)

Merzenich, M. M.; *see* Brugge

Miller, J. D., 267; *see* Carder

Miller, J. D. (1970). Audibility curve of the chinchilla. *J. Acoust. Soc. Amer.* **48**, 513–523.

Miller, J. D., Watson, C. S., and Cosell, W. (1963). Deafening effects of noise on the cat. *Acta Otolaryng. Suppl.* **176**.

Miller, J. M., 267

Miller, J. M., and Spelman, F. A., eds. (1990). *Cochlear Implants, Models of the Electrically Stimulated Ear*. Springer-Verlag, New York.

Mills, A. W., 182, 187, 191

Mills, A. W. (1972). Auditory localization. In *Foundations of Modern Auditory Theory*, Vol. 2 (J. V. Tobias, ed.). Academic Press, New York. (*Fig. 12.7*)

Mills, J. H.; *see* Henderson

Moller, A. R., 70

Moller, A. R. (1970). The middle ear. In *Foundations of Modern Auditory Theory*, Vol. 2 (J. V. Tobias, ed.). Academic Press, New York. (*Fig. 6.4*)

Moller, A. R. (2000). *Hearing: Its Physiology and Pathophysiology*, Academic Press, New York.

Moody, D. B., 262

Moody, D. B., Stebbins, W. C., and Hawkins, J. R. (1976). Noise-induced hearing loss in the monkey. In *Effects of Noise on Hearing* (D. Henderson, D. Hamernik, D. Dosanjh, and J. Mills, eds.). Raven, New York. (*Fig. 16.6*)

Moore, B. C. J., 19, 160, 171, 174, 176, 177, 178, 191, 202, 224, 225, 266; *see* Glasberg; *see* Patterson

Moore, B. C. J. (1978). Psychophysical tuning curves measured in simultaneous and forward masking. *J. Acoust. Soc. Amer.* **63**, 524–532. (*Fig. 11.13*)

Moore, B. C. J. (1986). *Frequency Selectivity in Hearing*. Academic Press, London.

Moore, B. C. J. (1993). Frequency processing. In *Human Psychoacoustics* (W. A. Yost, A. N. Popper, and R. R. Fay, eds.). Springer-Verlag, New York.

Moore, B. C. J. (1995). *Perceptual Consequences of Cochlear Damage*. Oxford University Press, Oxford.

Moore, B. C. J. (1997). *An Introduction to the Psychology of Hearing*, 3rd ed. Academic Press, London. (*Fig. 11.8*)

Moore, B. C. J., and Glasberg, B. R. (1982). Interpreting the role of suppression in psychophysical tuning curves. *J. Acoust. Soc. Amer.* **72**, 1374–1379.

Moore, B. C. J., and Patterson, R. D., eds. (1986). *Auditory Frequency Selectivity*. Plenum, London.

Moore, B. C. J., Peters, R. W., and Glasburg, B. R. (1985). Thresholds for the detection of inharmonicity in complex tones. *J. Acoust. Soc. Amer.* **77**, 1861–1867.

Moore, E. J., 251

Moore, E. J. (1983). *Bases of Auditory Brain-Stem Evoked Responses*. Grune & Stratton, New York.

Morest, D. K., 250

Morest, D. K., and Oliver, D. L. (1984). The neuronal architecture of the inferior colliculus in the cat: Defining the functional anatomy of the auditory midbrain. *J. Comm. Neurol.* **222**, 209–236.

Moushegian, G., 246

Moushegian, G., Rupert, A., and Whitcomb, M. (1972). Processing of auditory information by medial superior olivary neurons. In *Foundations of Modern Auditory Theory* (J. Tobias, ed.). Academic Press, New York. (*Fig. 15.15*)

Moxon, E. C.; *see* Kiang

Mulroy, J. J., 144

Mulroy, J. J., Altmann, D. W., Weiss, T. F., and Peake, W. T. (1974). Intracellular electrical response to sound in a vertebrate cochlea. *Nature* **249**, 482–485.

Munson, W. A.; *see* Fletcher

Naunton, R. F., 116

Naunton, R. F., and Fernandez, C., eds. (1978). *Evoked Electrical Activity in the Auditory Nervous System*. Academic Press, New York. (*Fig. 8.9*)

Nedzelnitsky, V., 74

Nedzelnitsky, V. (1980). Sound pressures in the basal turn of the cat cochlea. *J. Acoust. Soc. Amer.* **68**, 1676–1680. (*Fig. 6.8*)

Neff, D. L., 224

Neff, D. L., and Green, D. M. (1987). Masking produced by spectral uncertainty with multicomponent maskers, *Percep. Psychophysiol.* **41**, 409–415.

Neti, C., 250

Neti, C., Young, E. D., and Schneider, M. H. (1992). Neural network models of sound localization based on directional filtering by the pinna. *J. Acoust. Soc. Amer.* **92**, 3140–3157.

Newman, E. B.; *see* Stevens

Nielsen, D. W.; *see* Konishi

Norman, D. A.; *see* Lindsay

Northern, J.; *see* Jerger

Nuttal, A. L., 125

Nuttal, A. L., and Dolan, D. F. (1996). Steady-state sinusoidal velocity response of the basilar membrane in guinea pig. *J. Acoust. Soc. Amer.* **99**, 1556–1565.

Oliver, D., 232; *see* Morest

Oliver, D. L., and Huerta, M. F. (1992). Inferior and superior colliculi. In *The Auditory Pathway: Neuroanatomy* (D. Webster, R. R. Fay, and A. N. Popper, eds.). Springer-Verlag, New York. (*Fig. 15.4*)

Ortel, D.; *see* Wickesburg

Osborne M., 94; *see* Pickles

Osborne M., Comis, S. D., and Pickles, J. O. (1988). Further observations on the fine structure of tip links between stereocilia of the guinea pig cochlea. *Hear. Res.* 35, 99-108. (*Fig. 7.11*)

Passanen, E. G.; *see* McFadden

Patterson, J. H., 266, 267

Patterson, J. H. (1991). Effects of peak pressure and energy of impulses. *J. Acoust. Soc. Amer.* **90**, 205–209.

Patterson, R. D., 143, 146, 171, 177, 208; *see* Griffiths; *see* Moore

Patterson, R. D., and Moore, B. C. J. (1986). Auditory filters and excitation patterns as representations of frequency resolution. In *Frequency Selectivity in Hearing* (B. C. J. Moore, ed.). Academic Press, London. (*Figs. 11.7 and 11.8*)

Patterson, R. D., Allerhand, M., and Giguere, C. (1995). Time-domain modeling of peripheral auditory processing: A modular architecture and a software platform. *J. Acoust. Soc. Amer.* **98**, 1890–1895. (*Figs. 9.15 and 14.1*)

Peake, W. J.,; *see* Guinan; *see* Lynch; *see* Mulroy; *see* Weiss

Pearson, J.; *see* Taylor

Perkins, M. E.; *see* Wightman

Perrott, D. R.; *see* Saberi

Pestalozza, G., 108

Pestalozza, G., and Davis, H. (1956). Electric responses of the guinea pig ear to high audio frequencies. *Am. J. Physiol.* **185**, 595–600. (*Fig. 8.3*)

Peters, R. W.; *see* Moore

Pickles, J. O., 19, 78, 102, 124, 125, 144, 250, 266; *see* Osborne

Pickles, J. O. (1988). *An Introduction to the Physiology of Hearing*, 2nd ed., Academic Press, London.

Pickles, J. O., Comis, S. D., and Osborne, M. (1984). Cross-links between stereocilia in the guinea pig organ of Corti, and their possible relation to sensory transduction. *Hear. Res.* **15**, 103–112.

Plack, C.; *see* Viemeister

Popper, A. N.; *see* Fay; *see* Webster; *see* Yost; *see* Rubel

Postman, L. J., 258

Postman, L. J., and Egan, J. (1949). *Experimental Psychology: An Introduction.* Harper & Row, New York. (*Fig. 16.4*)

Probst, R.; *see* Martin

Pythagorous, 4

Rabiner, L., 225

Rabiner, L., and Juang, B.-H. (1993). *Fundamentals of Speech Recognition.* Prentice Hall, Englewood Cliffs, NJ.

Ranke, O. F., 97

Ranke, O. F. (1942). Das Massenverhaltnis zwischen Membran und Flussigkut im Innenohn. *Akust. Zeits.* **7**, 1–11. (*Fig. 7.13*)

Rasmussen, T., 126

Rasmussen, T. (1943). *Outlines of Neuro-Anatomy.* William C. Brown, Dubuque, IA.

Rauch, I.; *see* Rauch, S.

Rauch, S., 103

Rauch, S., and Rauch, I. (1974). Physicochemical properties of the inner ear, especially ironic transport. In *Handbook of Sensory Physiology*, Vol. 5 ((W. D. Keidel and W. D. Neff, eds.). Springer-Verlag New York.

Rayleigh, L., 4, 182

Reed, C. M., 168, 169

Reed, C. M., and Bilger, R. C. (1973). A comparative study of S/N and E/No. *J. Acoust Soc. Amer.* **53**, 1039–1045. (*Figs. 11.4 and 11.5*)

Reico, A.; *see* Ruggero

Rich, N. C.; *see* Robles; *see* Ruggero

Riesz, R. R., 162

Riesz, R. R. (1928). Differential intensity sensitivity of the ear for pure tones. *Phys. Rev.* **31**, 867–875.

Robles, L., 100; *see* Ruggero

Robles, L., Ruggero, M. A., and Rich, N. C. (1986). Basilar membrane mechanics at the base of the chinchilla cochlea, I: Input–output functions, tuning curves, and response phase. *J. Acous. Soc. Amer.* **80**, 1363-1374. (*Fig. 7.17*)

Rose, J. E., 131, 134, 243; *see* Anderson; *see* Hind

Rose, J. E., Galambos, R., and Hughes, J. R. (1959). Microelectrode studies of the cochlear nuclei of the cat. *Bull. Johns Hopkins Hos.* **104**, 211-251. (*Fig. 15.13*)

Rose, J. E., Brugge, J. F., Anderson, D. J., and Hind, J. E. (1967). Phase-locked response to low-frequency tones in single auditory nerve fibers of the squirrel monkey. *J. Neurophysiol.* **30**, 769–793. (*Fig. 9.8*)

Rose, J. E., Hind, J. E., Anderson, D. J., and Brugge, J. F. (1971). Some effects of stimulus intensity on response of auditory nerve fibers in the squirrel monkey. *J. Neurophysiol.* **34**, 685–699. (*Fig. 9.4*)

Rosen, S., 50, 61, 172, 177

Rosen, S., and Baker, R. J. (1994). Characterising auditory filter nonlinearity. *Hear. Res.* **73**, 231–244. (*Fig. 11.9*)

Rosen, S., and Howell, P. (1991). *Signals and Systems for Speech and Hearing.* Academic Press, London.

Ross, D. A.; *see* Wiener

Rossing, T., 36

Rossing, T. (1990). *The Science of Sound*, 2nd ed. Addison-Wesley, Reading, MA.

Rosowski, J. J., 78, 266

Rosowski, J. J. (1991). The effects of external- and middle-ear filtering on auditory threshold and noise-induced hearing loss. *J. Acoust. Soc. Amer.* **90**, 124–136.

Rubel, E., 251, 266; *see* Lippe; *see* Werner

Rubel, E., Popper, A. N., and Fay, R. R., eds. (1997). *Development of the Auditory System.* Springer-Verlag, New York.

Ruben, R. J., 251

Ruben, R. J., Van de Water, T. R., and Rubel, E. W., eds. (1986). *The Biology of Change in Otolaryngology.* Excerpta Medica, Amesterdam.

Ruggero, M. A., 101, 103 112; *see* Robles

Ruggero, M. A., and Santos-Sacchi, J. (1997). Cochlear mechanics and biophysics. In *Encyclopedia of Acoustics*, Vol. 3. Wiley, New York. (*Fig. 8.7*)

Ruggero, M. A., Robles, L. R., Rich, N. C., and Reico, A. (1992). Basilar membrane response to two-tone and broadband stimuli. *Phil. Trans. R. Soc. London* **336**, 307–315.

Ruggero, M. A., Rich, N. C., and Recio, A. (1996). The effect of intense acoustic stimulation on basilar-membrane vibrations. *Aud. Neurosci.*, **3**, 329–345. (*Fig. 7.18*)

Rupert, A.; *see* Moushegian

Russell, J. J., 144

Russell, J. J., and Sellick, M. (1977a). The tuning properties of cochlear hair cells. In *Psychophysics and Physiology of Hearing* (E. F. Evans and J. Wilson, eds.). Academic Press, London.

Russell, J. J., and Sellick, M. (1977b). Tuning properties of cochlear hair cells. *Nature* **267**, 858–860.

Ryals, B. M., 264

Ryals, B. M., Dooling, R. J., Westbrook, E., Dent, M. L., MacKenzie, A., and Larsen, O. N. (1999). Avain species differences in susceptibility to noise exposure, *Hear. Res.* **131**, 71–88. (*Fig. 16.8*)

Ryan, A. F., 298

Ryan, A. F., Axelsson, G. A., and Wolfe, N. K. (1992). Central auditory metabolic activity induced by intense noise exposure. *Hear. Res.* **61**, 24–30. (*Fig. F.4*)

Saberi, K., 191

Saberi, K., and Perrott, D. R. (1990). Minimum audible movement angles as a function of sound source trajectory. *J. Acoust. Sco. Amer.* **88**, 2639–2644.

Sachs, M.; *see* Winslow; *see* Young

Sachs, M. B., and Young, E. D. (1979). Encoding of steady-state vowels in the auditory nerve: Representation in terms of discharge rate. *J. Acoust. Soc. Amer.* **66**, 470.

Salvi, R.; *see* Hamernik

Santos-Sacchi, J.; *see* Dallos; *see* Ruggero

Saul, L., 124

Saul, L., and Davis, H. (1932). Action currents in the central nervous system, I: Action currents of the auditory tracts. *Arch. Neurol. Psychiat.* **28**, 1104–1116.

Schacht, J.; *see* Zajic

Schneider, M. H.; *see* Neti

Schreiner, C. E., 249

Schreiner, C. E., and Langner, G. (1988). Coding of temporal factors in the central auditory nervous system. In *Auditory Function: Neurobiological Bases of Hearing* (G. W. Edelman, W. E. Gall, and M. Cowan, eds.). Wiley, New York. (*Fig. 15.19*)

Schuknecht, H. F., 242

Schuknecht, H. F. (1974). *Pathology of the Ear.* Harvard University Press, Cambridge. (*Fig. 15.12*)

Schweitzer, L., 296

Schweitzer, L., and Cecil T. (1992). Morphology of HRP-labeled cochlear nerve axons in the dorsal cochlear nucleus in the developing hampster. *Hear. Res.* **60**, 34-44. (*Fig. F.2*)

Sellick, M.; *see* Russell

Sellick, M., and Russell, J. J. (1980). The response of inner hair cells to basilar membrane velocity during low-frequency auditory stimulation in the guinea pig cochlea. *Hear. Res.* **2**, 439–445.

Sensimeterics (1997). *Speech Production and Perception.* Sensimetrics Corporation, Cambridge, MA.

Shamma, S. A., 250

Shamma, S. A. (1985). Speech processing in the auditory system, II: Lateral inhibition and the central processing of speech evoked activity in the auditory nerve. *J. Acoust. Soc. Amer.* **78**, 1622.

Shannon, R. V., 175

Shannon, R. V. (1974). Two-tone unmasking and suppression in a forward-masking situation. *J. Acoust. Soc. Amer.* **59**,1460-1470. (*Fig. 11.14*)

Shaughnessy, J. J., 7

Shaughnessy, J. J., and Zechmeister, E. B. (2000). *Research Methods in Psychology*, 5th ed.. McGraw-Hill, New York.

Shaw, E. A. G., 71

Shaw, E. A. G. (1974). The external ear. In *Handbook of Sensory Physiology*, Vol. 5 (W. D. Keidel and W. D. Neff, eds.). Springer-Verlag, New York. (*Fig. 6.6*)

Sheft, S.; *see* Yost

Shower, E. G., 162

Shower, E. G., and Biddulph, R. (1931). Differential pitch sensitivity of the ear. *J. Acoust. Soc. Amer.* **3**, 275–277.

Sivan, L. D., 160

Sivan, L. D., and White, S. D. (1933). On minimum audible fields. *J. Acoust. Soc. Amer.* **4**, 288–321.

Sklar, D. L.; *see* Ward

Slepecky, N. B., 297

Slepecky, N. B., and Ulfendahl, M. (1992). Actin-binding and microtube-associated proteins in the organ of Corti. *Hear. Res.* **57**, 201–215. (*Fig. F.3*)

Sokolich, W. G.; *see* Zwislocki

Spelman, F. A.; *see* Miller, J. M.

Spoendlin, H., 103, 116, 118, 125, 126

Spoendlin, H. (1966). *The Organization of the Cochlear Receptor*. Karger, Basel.

Spoendlin, H. (1970). Structural basis of peripheral frequency analysis. In *Frequency Analysis and Periodicity Detection in Hearing* (R. Plomp and G. F. Smoorenburg, eds.). A. W. Sythoff, Leiden.

Spoendlin, H. (1973). The innervation of the cochlear receptor. In *Proceedings of a Symposium on Basic Mechanisms in Hearing*, pp. 185–234. Academic Press, New York.

Spoendlin, H. (1974). Neuroanatomy of the cochlea. In *Facts and Models in Hearing* (E. Zwicker and E. Terhardt, eds.). Springer Verlag, New York. (*Fig. 8.11*)

Spoendlin, H. (1978). The afferent innervation of the cochlea. In *Evoked Electrical Activity in the Auditory Nervous System* (R. F. Naunton and C. Fernandez, eds.). Academic Press, New York. (*Fig. 8.9*)

Spoendlin, H. (1979). Neural connections of the outer hair cell system. *Acta Otolaryng.* **87**, 381–387.

Stebbins, W.; *see* Moody

Steinberg, J. C., 194, 202

Steinberg, J. C., and Gardner, M. B. (1937). The dependence of hearing impairment on sound intensity. *J. Acoust. Soc. Amer.* **9**, 11–23. (*Fig. 13.2*)

Stevens, S. S., 7, 181, 182, 201, 202, 203, 281

Stevens, S. S. (1970). Neural events and the psychophysical law. *Science* **170**, 1043–1050.

Stevens, S. S., ed. (1975). *Psychophysics*. Wiley, New York.

Stevens, S. S., and Davis, H. (1983). *Hearing: Its Psychology and Physiology*. Acoustical Society of America, Melville, New York. (Originally published 1938.)

Stevens, S. S., and Newman, E. B. (1936). The localization of actual sources of sound. *Am. J. Psychol.* **48**, 297–306. (*Fig. 12.4*)

Suga, N., 250

Suga, N. (1988). Neuroethology and speech processing. In *Complex Sound Processing by Combination-Sensitive Neurons. Auditory Function: Neurobiological Bases of Hearing* (G. W. Edelman, W. E. Gall, and W. M. Cowan, eds.). Wiley, New York.

Sullivan, W. E.; *see* Konishi

Summerfield, Q., 225

Summerfield, Q., and Assmann, P. F. (1991). Perception of concurrent vowels: Effects of harmonic misalignment and pitch-period asynchrony. *J. Acoust. Soc. Amer.* **89**, 1364–1377.

Takahashi, T. T.; *see* Konishi

Tanaka, Y., 144

Tanaka, Y., Asanuma, A., and Yanagisawa, K. (1980). Potentials of outer hair cells and their membrane properties in cationic environments. *Hear. Res.* **2**, 431–438.

Tasaki, I., 107, 127

Tasaki, I. (1954). Nerve impulses in individual auditory nerve fibers of guinea pig. *J. Neurophysiol.* **17**, 97–122.

Tasaki, I., Davis, H., and Eldredge, D. H. (1954). Exploration of cochlear potentials with a microelectrode. *J. Acoust. Soc. Amer.* **26**, 765-773. (*Fig. 8.2*)

Teas, D. C.; *see* Klein

Thomas, E. C.; *see* Kiang

Tonndorf, J., 72, 78, 96, 98

Tonndorf, J. (1960). Dimensional analysis of cochlear models. *J. Acoust. Soc. Amer.* **32**, 493–497. (*Fig. 7.12*)

Tonndorf, J. (1962). Time/frequency analysis along the partition of cochlear models: A modified place code. *J. Acoust. Soc. Amer.* **34**, 1337–1350. (*Fig. 7.15*)

Tonndorf, J., and Khanna, S. M. (1972). Tympanic membrane vibrations in human cadaver ears studied by time-averaged holography. *J. Acoust. Soc. Amer.* **52**, 1221-1233. (*Fig. 6.7*)

Trahiotis. C.; *see* Elliott

Ulfendahl, M.; *see* Slepecky

Van de Water, T. R.; *see* Ruben

Van Opstal, A. J.; *see* Hofman

Van Riswick, J. G. A., 157, 159, 161–164; *see* Hofman

Viemeister, N. F.; *see* Bacon

Viemeister, N. F. (1974). Intensity discrimination in noise in the presence of band-reject noise. *J. Acoust. Soc. Amer.* **56**, 1594-1601. (*Fig. 10.7*)

Viemeister, N. F. (1979). Temporal modulation transfer functions based on modulation thresholds. *J. Acoust. Soc. Amer.* **66**, 1364-1381. (*Fig. 10.10*)

Viemeister N. F., and Plack, C. (1993). Temporal processing. In *Human Psychoacoustics* (W. A. Yost, A. N. Popper, and R. F. Fay, eds.). Springer-Verlag, New York.

von Bekesy, G., 4, 7, 72, 76, 78, 99, 102, 197

von Bekesy, G. (1936). Zur Psycik des Mittelohres und uber das Horen bei fehlerhaftem Trommelfell. *Akust. Zeits.* **1**, 13–23. (*Fig. 6.9*)

von Bekesy, G. (1941). Uber die Messung der Schwingungsamplitude der Gehorknochelchen mittels einer kapazitiven Sonde. *Akust. Zeits.* **6**, 1–16. (*Fig. 6.7*)

von Bekesy, G. (1947). The variation of phase along the basilar membrane with sinusoidal vibrations. *J. Acoust. Soc. Amer.* **19**, 452–460. (*Fig. 7.16*)

von Bekesy, G. (1989). *Experiments in Hearing*. Acoustical Society of America, Melville, New York. (Originally published 1960).

Wagner, H.; *see* Konishi

Ward, W. D., 256; *see* Tobias

Ward, W. D., Glorig, A., and Sklar, D. L. (1959). Temporary threshold shift from octave-band noise: Applications to damage-risk criteria. *J. Acoust. Soc. Amer.* **31**, 522–528. (*Fig. 16.3*)

Warr, W. B., 121, 126

Warr, W. B. (1992). Organization of olivoco-chlear efferent systems in mammals. In *The Auditory Pathway: Neuroanatomy* (D. Webster, R. R. Fay, and A. N. Popper, eds.). Springer-Verlag: New York. (*Fig. 8.14*)

Warr, W. B., Guinan, J. J., and White, J. S. (1986). Organization of the efferent fibers: The lateral and medial olivocochelar systems. In *Neurobiology of Hearing: The Cochlea* (R. D. Altschuler, D. Hoffman, R. Bobbin, and B. Clopton, eds.). Raven, New York.

Warren, M. L., 139

Warren, M. L., and Liberman, M. C. (1989). Effects of contralateral sound on auditory-nerve response, I: Contributions of cochlear efferents. *Hear. Res.* **37**, 89–104. (*Fig. 9.12*)

Warren, R. M., 78, 224

Warren, R.; *see* Warren, R. M.

Warren, R. M. (1999). *Auditory Perception: A New Analysis and Synthesis*, Cambridge University Press, Cambridge.

Warren, R. M., and Warren, R. (1968). *Helmhlotz on Perception: Its Physiology and Development*. Wiley, New York.

Watanabe, T.; *see* Kiang

Watson, C. S., 150, 153, 224; *see* Miller

Watson, C. S. (1976). Factors in the discrimination of work-length auditory patterns. In *Hearing and Davis: Essays Honoring Hallowell Davis* (S. K. Hirsh, D. H. Eldridge, and S. R. Silverman, eds.), pp. 175–188. Washington University Press, St. Louis.

Watson, C. S., and Gengel, R. W. (1969). Signal duration and signal frequency in relation to auditory sensitivity. *J. Acoust. Soc. Amer.* **46**, 989–997. (*Fig. 10.3*)

Watson, C. S. Franks, J. R., and Hood, D. C. (1972). Detection of tones in the absence of external masking noise, I: Effects of signal intensity and signal frequency. *J. Acoust. Soc. Amer.* **52**, 633–643. (*Fig. 10.1*)

Weber, D. L., 155, 177

Weber, D. L. (1983). Do off-frequency simultaneous maskers suppress the signal? *J. Acoust. Soc. Amer.* **73**, 87–894.

Webster, D., 124, 144

Webster, D., Fay, R. R., and Popper, A. N., eds. (1992). *The Auditory Pathway: Neuroanatomy*. Springer-Verlag, New York.

Webster, F. A., 190, 250

Webster, F. A. (1951). The influence of interaural phase on masked thresholds, I: The role of interaural time-deviation. *J. Acoust. Soc. Amer.* **23**, 452–461. (*Fig. 12.11*)

Wegel, R. L., 168, 169

Wegel, R. L., and Lane, C. E. (1924). The auditory masking of one sound by another and its probable relation to the dynamics of the inner ear. *Phys. Rev.* **23**, 266–285. (*Fig. 11.3*)

Weir, C. C., 156, 162; *see* Jesteadt

Weir, C. C., Jesteadt, W., and Green, D. M. (1977). Frequency discrimination as a function of frequency and sensation level. *J. Acoust. Soc. Amer.* **61**, 178–183. (*Fig. 10.5*)

Weiss, T. F., 144; *see* Mulroy

Weiss, T. F., Mulroy, M. J., and Altmann, D. W. (1974). Intercellular responses to acoustic clicks in the inner ear of the alligator lizard. *J. Acoust. Soc. Amer.* **61**, 178–183.

Werner, L. A., 251

Werner, L. A., and Rubel, E. W., eds. (1992). *Developmental Psychoacoustics*. American Psychiatric Association, Hyattsville, MD.

Westbrook, E.; *see* Ryals

Wever, E. G., 4, 79, 103

Wever, E. G. (1949). *Theory of Hearing*. Wiley, New York.

Wever, E. G., and Bray, C. (1930). Action currents in the auditory nerve in response to acoustic stimulation. *Proc. Nat. Acad. Sci.* **16**, 344–350.

Wever, E. G., and Lawrence, M. (1954). *Physiological Acoustics*. Princeton University Press, Princeton.

Whitcomb, M.; *see* Moushegian

White, S. D.; *see* Sivan

White, J. S.; *see* Warr

Whitfield, I. C., 228

Whitfield, I. C. (1967). *The Auditory Pathway.* Williams & Wilkins, Baltimore. (*Fig. 15.1*)

Wickesberg, R. E., 250

Wickesberg, R. E., and Oertel, D. (1990). Delayed, frequency-specific inhibition in the cochlear nuclei of mice: A mechanism for monaural echo suppression. *J. Neurosci.* **10**, 1762–1768.

Wiener, F. M., 78

Wiener, F. M. (1947). Sound diffraction by rigid spheres and circular cylinders. *J. Acoust. Soc. Amer.* **19**, 444–451.

Wiener, F. M., and Ross, D. A. (1946). The pressure distribution in the auditory canal in a progressive sound field. *J. Acoust. Soc. Amer.* **18**, 401–408.

Wightman, F. L., 70, 166, 181, 185, 191

Wightman, F. L., and Kistler, D. J. (1989a). Headphone simulation of free-field listening, I: Stimulus synthesi. *J. Acoust. Soc. Amer.* **85**, 858-867. (*Fig. 6.5*)

Wightman, F. L., and Kistler, D. J. (1989b). Headphone simulation of free-field listening, II: Psychophysical validation. *J. Acoust. Soc. Amer.* **85**, 868-87. (*Fig. 12.5 and Fig. 12.9*)

Wightman, F. L., and Kistler, D. J. (1993). Localization. In *Human Psychoacoustics* (W. A. Yost, A. N. Popper, and R. R. Fay, eds.). Springer-Verlag, New York.

Wightman, F., McGee, T., and Kramer, M. (1977). Factors influencing frequency selectivity in normal and hearing-impaired listeners. In *Psychophysics and Physiology of Hearing* (E. F. Evans and J. Wilson, eds). Academic Press, London. (*Fig. 11.1*)

Wightman, F. L., Kistler, D. J., and Perkins, M. E. (1987). A new approach to the study of human sound localization. In *Directional Hearing* (W. Yost and G. Gourevitch, eds.). Springer-Verlag, New York. (*Table 12.1*)

Williston, J. S.; *see* Jewett

Wilson, J., 115

Wilson, J. (1980). Evidence for a cochlear origin for acoustic re-emissions, threshold fine-structure and tonal tinnitus. *Hear. Res.* **2**, 233–252. (*Fig. 8.8*)

Winslow, R. L., 145

Winslow, R. L., and Sachs M. B. (1988). Single-tone intensity discrimination based on auditory-nerve rate responses in backgrounds of quiet, noise, and with stimulation of the crossed olivocochlear bundle. *Hear. Res.* **35**, 165–190.

Wolfe, N. K.; *see* Ryan

Wrightson, T., 85

Wrightson, T., and Keith, A. (1918). *An Enquiry into the Analytical Mechanism of the Internal Ear.* Macmillan, New York. (*Fig. 7.4*)

Yanagisawa, K.; *see* Tanaka

Yin, T. C. T., 246

Yin, T. C. T., and Chan, J. C. K. (1988). Neural mechanisms underlying interaural time sensitivity to tones and noise. In *Auditory Function: Neurobiological Bases of Hearing* (G. W. Edelman, W. E. Gall, and M. Cowan, eds.). Wiley, New York. (*Fig. 15.16*)

Yost, W. A., 7, 160, 161, 177, 186, 191, 192, 203, 219, 224; *see* Green; *see* Litovsky

Yost, W. A. (1992a). Auditory image perception and analysis. *Hear. Res.* **56**, 8–19.

Yost, W. A. (1992b). Auditory perception and sound source determination. *Curr. Dir. Psychol. Sci.* **1**, 12–15. (*Fig. 14.8*)

Yost, W. A. (1993). Overview: Psychoacoustics. In *Human Psychoacoustics.* (W. A. Yost, A. N. Popper, and R. R. Fay, eds.). Springer-Verlag, New York.

Yost, W. A. (1996). Pitch of iterated rippled noise. *J. Acoust. Soc. Amer.* **100**, 511–519.

Yost, W. A., and Dye, R. H. (1991). Properties of sound localization by humans. In *Neurobiology of Hearing: The Central Nervous System* (R. D. Altschuler, D. Hoffman, R. Bobbin, and B. Clopton, eds.). Raven, New York. (*Fig. 12.10*)

Yost, W. A., and Gourevitch, G., eds. (1987). *Directional Hearing*. Springer-Verlag, New York.

Yost, W. A., and Killion, M. C. (1997). Quiet absolute thresholds. In *Handbook of Acoustics* (M. Crocker, ed.). Wiley, New York.

Yost, W. A., and Sheft, S. (1993). Auditory processing. In *Human Psychoacoustics* (W. A. Yost, A. N. Popper, and R. R. Fay, eds.). Springer-Verlag, New York.

Yost, W. A., and Watson, C. S., eds. (1987). *Auditory Processing of Complex Sounds*. Erlbaum, Hillsdale, NJ.

Yost, W. A., Popper, A. N., and Fay, R. R., eds. (1993). *Human Psychoacoustics*. Springer-Verlag, New York.

Young, E. D., 141, 142, 244; *see* Neti

Young, E. D. (1984). Response characteristics of neurons in the cochlear nuclei. In *Hearing Science* (C. I. Berlin, ed.). College-Hill, San Diego. (*Fig. 15.14*)

Young, E. D., and Sachs, M. B. (1979). Representation of steady-state vowels in the temporal aspects of the discharge patterns of populations of auditory-nerve fibers. *J. Acoust. Soc. Amer.* **66**, 1381–1392. (*Figs. 9.13 and 9.14*)

Zajic, G., 112

Zajic, G., and Schacht, J. (1991). Shape changes in isolated outer hair cells: Measurements with attached microspheres. *Hear. Res.* **52**, 407–411. (*Fig. 8.7*)

Zechmeister, E.; *see* Shaughnessey

Zemlin, W. R., 82, 98, 111, 225

Zemlin, W. R. (1981). Speech and hearing science. In *Anatomy and Physiology*. Prentice-Hall, Englewood Cliffs, NJ. (*Figs. 7.1, 7.14, and 8.6*)

Zurek, M., 115

Zurek, M. (1985). Acoustic emissions from the ear: A summary of results from humans and animals. *J. Acoust. Soc. Amer.* **78**, 340-344. (*Fig. 8.8*)

Zwicker, E., 202, 203

Zwicker, E., and Fastl, H. (1991). *Psychoacoustics: Facts and Models*. Springer-Verlag, Berlin.

Zwislocki, J. J., 125, 150

Zwislocki, J. J., and Sokolich, W. G. (1973). Velocity and displacement responses in auditory nerve fibers. *Science* **182**, 64–66.

Glossary and Subject Index

The definitions of some of the terms are from the American National Standards Institute (ANSI). These definitions are reproduced with permission from American National Standard S.3.20-1974 (R 1978), copyright 1974 by the American National Standards Institute (copies of which may be purchased from the Acoustical Society of America, 335 East 45th Street, New York, NY 10017). Note that certain ANSI definitions below do not agree in detail with the uses of terms in this book. The first page listing for a term most likely contains the page where the term is defined within the context of this book.

masking, 165–178. (1) Masking is the process by which the threshold of audibility for one sound is raised by the presence of another (masking) sound. (2) Masking is the amount by which the threshold of audibility of a sound is raised by the presence of another (masking) sound. The unit customarily used is the decibel.

binaural, 189–192

critical band, 169–172

fringe data, 173, 174

masked threshold, 165, 166

noise, 169–171

patterns, 167

temporal, 172–176

tonal, 165–169

masking level difference (MLD), 189–192. Masking level difference is any decrease (improvement) in the masked threshold obtained when two ears are used instead of one. Masking level difference is expressed in decibels, as a function of the frequency of the masked tone.

mass, 19, 20

mass reactance, 30

mastoid, 161

matching procedure, 193, 286

meatus, 65

mechanical energy, 81

Meddis haircell, 146

medial, 287

medial geniculate body, 229

medial plane, 183. The medial plane is a vertical plane that divides the body into left and right. It is perpendicular to the frontal plane.

midsagittal, 183

medial olivary complex, 245

medial superior olive (MSO), 120–122, 245

mel, 196. The mel is a unit of pitch. One thousand mels is the pitch of a 1000-Hz pure tone for which the loudness level is 40 phons. The pitch of a sound that is judged by a subject to be n times that of a 1000-mel tone is n thousand mels; thus, the pitch of a sound subjectively judged to be n times that of a 1-mel tone is n mels.

mel scale, 196

Ménière's disease, 260

meningitis, 260

hydrops, 260

message, 3

method of adjustment, 282. The method of adjustment is a psychophysical method used primarily to determine thresholds; in this procedure, the subject varies some dimension of a stimulus until that stimulus appears equal to or just noticeably different from a reference stimulus.

method of constant stimuli, 282. The method of constant stimuli is a psychophysical method used primarily to determine thresholds; in this procedure, a number of stimuli, ranging from rarely to almost always perceivable (or rarely to almost always perceivably different from some reference stimulus), are presented one at a time. The subject responds to each presentation: "yes–no," "same–different," "greater than/equal to/less than," etc.

method of cross-modality matching, 285. The method of cross-modality matching is a psychophysical method used primarily to scale sensations; in this procedure, the subject adjusts a stimulus along some dimension until that stimulus appears equal to another stimulus received by a different send modality.

method of limits, 282. The method of limits is a psychophysical method used primarily to determine thresholds; in this procedure, some dimension of a stimulus, or of the difference between two stimuli, is varied incrementally until the subject changes his response.